PROPERTIES
OF MATERIALS

Mary Anne White

Department of Chemistry
Dalhousie University

New York • Oxford
OXFORD UNIVERSITY PRESS
1999

Oxford University Press

Oxford   New York
Athens   Auckland   Bangkok   Bogotá   Buenos Aires   Calcutta
Cape Town   Chennai   Dar es Salaam   Delhi   Florence   Hong Kong   Istanbul
Karachi   Kuala Lumpur   Madrid   Melbourne   Mexico City   Mumbai
Nairobi   Paris   São Paulo   Singapore   Taipei   Tokyo   Toronto   Warsaw

and associated companies in
Berlin   Ibadan

Published by Oxford University Press, Inc.
198 Madison Avenue, New York, New York 10016
http://www.oup-usa.org

Oxford is a registered trademark of Oxford University Press.

Library of Congress Cataloging-in-Publication Data
White, Mary Anne, 1953–
    Properties of materials / Mary Anne White.
        p.   cm.
    Includes bibliographical references and index.
    ISBN 0-19-511331-4 (pbk.)
    1. Materials.   I. Title.
    TA404.8.W458   1999
    610.1'12—dc21   98-38204
                                               CIP

9 8 7 6 5 4 3 2 1

Printed in the United States of America
on acid-free paper

Dedicated to my first teachers, my parents

CONTENTS

PREFACE

The idea for this book germinated at a public lecture in about 1989: the lecturer had mentioned in passing, "and you all know how a photocopier works." Most of the remainder of the lecture was lost on me, as I wondered what fraction of that educated audience knew how a photocopier worked. Then I began to realize that nowhere in our curriculum was such an important concept taught to our students. After that lecture I decided to take advantage of the revision of the physical chemistry curriculum going on in my Department to prepare a curriculum for a class based on properties of materials. The class was launched in 1991. Finding no appropriate textbook, I wrote out my lecture notes for the students to use. With subsequent revisions and additions these have become this book.

The purpose of this book is to introduce the principles of materials science through an atomic and molecular approach. In particular, the aim is to help the reader learn to think about properties of materials, in order to understand the principles behind new (or old) materials. A "perfect goal" would be to enable the reader to learn about a new material in the business pages of the daily newspaper, giving minimal scientific information, and from that decipher the scientific principles on which the use of the material is based.

Properties of materials have interested many people. When speaking of his youth, Linus Pauling[1] said[2]: "I mulled over the properties of materials: why are some substances colored and others not, why are some minerals or inorganic compounds hard and others soft." It is noteworthy that such deep considerations of the world around us eventually led Pauling to make seminal contributions in many areas. It is hoped that encouragement of wonder in the variety and nature of properties of materials will be of benefit to those who read these pages.

This textbook should be viewed as an introductory survey of principles in materials science. Although this book assumes basic knowledge of physical sciences, many of the concepts are presented with introductory mathematical and theoretical rigor. This is largely to refrain from use of tensors (avoided in all but a few places) in this presentation; the interested reader will find more detailed presentations and derivations in the Further Reading sections.

---

[1] Linus Pauling (1901–1994) was an American chemist and winner of the 1954 Nobel Prize in Chemistry for research into the nature of the chemical bond and its applications to the elucidation of the structure of complex substances. Pauling also won the 1962 Nobel Peace Prize.

[2] John Horgan (1993). Profile: Linus Pauling, *Scientific American,* March 1993, 36.

This book is divided into parts based on various properties of materials. After a general introduction to materials science issues (Part I), the origins of colors and other optical properties of materials are considered (Part II). Part III concerns thermal properties of materials, including thermal stability and phase diagrams. Part IV is about electrical and magnetic properties of materials, followed by Part V on mechanical stability.

This book has been used as the basis of a one-semester class in materials science, offered in a chemistry department, taken by students in chemistry, biochemistry, physics, engineering, and earth sciences. The prerequisite for the class is prior introduction to the laws of thermodynamics. A basic understanding of crystalline structure is assumed and presented in an appendix.

A feature of this book is the introduction of tutorials, in which the students, working in small groups, can apply the principles presented in the text to work out for themselves the physical principles behind applications of materials science.

An Instructors' Supplement containing a complete discussion of all the points raised in the tutorials and complete worked solutions to all the end-of-chapter problems is available through the publisher to instructors who have chosen the book for class use.

This book has aimed to present the principles behind various properties of materials, and, since it is a survey of a large subject, additional reading suggestions are given at the end of each chapter as sources of more detailed information. References to recent developments also are given to expose the readers to the excitement of current materials science research. Updates to these references will be made available through the Oxford University Press internet website.

The presentation for this book is by type of property—optical, thermal, electrical, magnetic, and mechanical. Types of materials—metals, semiconductors (intrinsic and extrinsic), insulators, glasses, orientationally disordered crystals, defective solids, liquid crystals, Fullerenes, Langmuir–Blodgett films, colloids, inclusion compounds, and more—are introduced through their various properties. As new types of materials are made or discovered, it is hoped that the approach of presenting principles that determine physical properties will have a lasting effect on future materials scientists and many others.

ACKNOWLEDGMENTS

This book has been helped by the comments of many people, especially students who have provided feedback concerning the class "Materials Science" given at Dalhousie University since 1991. I hope that readers will help me to continue to refine this text by bringing suggestions for improvement and inclusion to my attention.

I particularly want to thank N. Aucoin, P. Bessonette, R.J.C. Brown, R.J. Boyd, D.B. Clarke, A. Cox, J. Dahn, S. Dimitrijevic, A. Ellis, J.M. Honig, B. Kahr, B. London, M. Marder, T. Matsuo, K. Nassau, L. Schramm, J.-M. Sichel, E. Slamovich, and M. Tan for detailed constructive comments, C. Kittel for encouragement, M. LeBlanc for preparation of most of the diagrams, J.E. Burke, D. Eigler, F. Fyfe, Ch. Gerber, M. Gharghouri, M. Jericho, A. Koch, M. Marder, J. M. Philippe, A. Soldatov, R.L. White, and K. Worsnop for photographs or other assistance with graphic contributions, and G. Smith for photography. I also acknowledge the assistance of Randy Perry in researching certain aspects of this book. In that regard, I thank the Dalhousie University Faculty of Science for a teaching award (in 1993) that made it possible to hire Randy. I am especially grateful to Paul Bessonette for editorial assistance. In addition, I am grateful to J.S. Grossert, K.J. Lushington, A.J. Paine, D.G. Rancourt, and R.L. White for providing useful information.

Writing a book takes time. For their forbearance while I have been preoccupied, I sincerely thank my family—Rob, David, and Alice. This would not have been accomplished without their loving support.

Knowledge comes, but wisdom lingers.

—Alfred, Lord Tennyson

INTRODUCTION

Good material often stands idle for want of an artist.

—Lucius Annæus Seneca

1

MATERIALS
SCIENCE

1.1 HISTORY

In some sense, materials science began about two million years ago, when people began to make to make tools from stone, at the start of the Stone Age. At this time, emphasis was on applications of the materials, with no understanding of microscopic origins of material properties. Nevertheless, the possession of a stone axe or other implement certainly was an advantage to an individual.

The Stone Age ended about 5000 years ago in the Far East, with the introduction of the Bronze Age. Bronze is an **alloy** (a metal made up of more that one element), mostly composed of copper with up to 25% tin, and possibly other elements such as lead (which makes the alloy easier to cut), zinc, and phosphorus (which strengthens and hardens the alloy). Bronze is a much more workable material than stone, especially since it can be hammered, beaten, or cast into a wide variety of shapes. After a surface oxide film forms, bronze objects corrode only slowly.

Although bronze is still used today, about 3000 to 3500 years ago iron working began in Asia Minor. The Iron Age continues today. The main advantage of iron over bronze is its lower cost, bringing metallic implements into the range of the common person. The Iron Age ushered in the common use of coins, which greatly improved trade, travel, and communications. Even today these three activities are strongly tied to materials usage.

Throughout the Iron Age many new types of materials have been introduced. Today we may take for granted the properties of glass, ceramic, semiconductors, polymers, composites, etc.

One major change over the millennia of the "Materials Age" has been our understanding of the properties of materials, and our consequent ability to develop and prepare materials for particular applications. Materials science has been defined as the relationships among structure, properties, processing, and performance of materials (see Figure 1.1). Understanding the principles that give rise to various properties of materials is the aim of our journey.

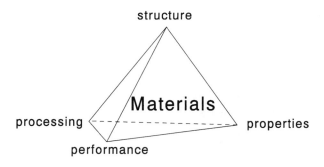

Figure 1.1. Materials science is the investigation of the relationships among structure, properties, processing, and performance of materials.

1.2 MORE RECENT TRENDS

A report written in 1953[1] listed objectives for "future" materials research: materials to behave well at extremely high and extremely low temperatures; more basic knowledge about the behavior of metals (e.g., to predict strength, fatigue); production of metals of higher purity; preparation of new alloys to replace stainless steel (due to the then shortage of nickel); preparation of better conductors and heat-resistant insulators for miniature electronic circuits; improvements in welding and soldering; development of adhesion methods to allow applications of fluorocarbons and other new alloys and plastics. Most of these goals have been achieved, but the list, which now seems quite dated, shows the emphasis on metals and metallurgy that was prevalent at that time.

About 20 years later, another report[2] again listed current topics of interest to materials scientists. This time the issues were quite different: strategic materials, fuel availability, biodegradable materials, and scrap recovery. This list shows the close connection between materials science and economic and political issues.

More recently it has been said that "[t]he field of materials science and engineering is entering a period of unprecedented intellectual challenge and productivity."[3] Although metals will always be important, emphasis in the subject has moved from metallurgy to ceramics, composites, polymers, and other molecular materials. Important considerations include semiconduction, magnetic properties, photonic properties, superconduction, and biomimetic properties. Furthermore, perceived barriers between types of materials are falling away: e.g., metals can be glasses and molecular materials can be conductors or magnets. Following the clues that nature provides we now understand that different end products depend on assembly and addition of other phases. This is an example of complexity depending on the different length scales—from nanostructures to macrostructures. With structure–composition–properties rela-

---

[1] Science and the Citizen, *Scientific American,* August 1953.

[2] *Materials and Man's Needs,* National Research Council's Committee on the Survey of Materials Science and Technology, 1975.

[3] *Materials Science and Engineering for the 1990s,* National Academy Press, Washington, DC, 1989.

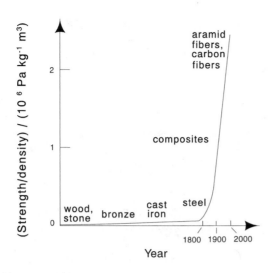

Figure 1.2. The dramatic progress in the strength-to-density ratio of materials has allowed a wide variety of new products, from dental materials to tennis racquets.

tions now more fully understood, some materials can be tailor-made. Nevertheless, there are still some elements of surprise in the subject; examples include the discoveries of high-temperature superconductors and the fullerenes.

Modern materials science is both important to our economies and intrinsically fascinating. The subject cuts across many traditional disciplines, including chemistry, physics, engineering, and earth sciences. For example, the dramatic recent improvement in the strength-to-density ratio in materials (see Figure 1.2) has required input from a wide variety of subjects and has led to considerable improvements in materials available to consumers.

1.3 IMPACT ON DAILY LIVING

Modern materials science has had an immeasurable impact on our daily lives. New materials and processes have led to diverse products and applications: fiber optics, better computers, durable outdoor plastics, cheaper metal alloys to replace gold in electrical connections, more efficient liquid crystal displays, and many microelectronic applications. Materials development is important to many industrial sectors: aerospace, automotive, biological, chemical, electronic, energy, metals, and telecommunications. Indeed, economic growth is no longer linked to consumption of basic materials, but rather to their use in goods and services.

The understanding of behavior and properties of materials underlies every major technology. For example, in the space industry, the following material properties must be considered: strength, thermal conductivity, outgassing, flammability, effect of radiation, and stability under thermal cycling. Every manufacturing industry depends on materials research and development.

Although fields like plastics started to have an impact on consumers a few decades ago, there are still areas for improvement. New materials are being developed for applications such as bulletproof vests; polyester clothing is being made from recycled pop bottles. So-called "smart" materials have led to products such as airplane wings that deice themselves and buildings that stabilize themselves in earthquakes. The drive to miniaturization, especially of electronic components, is opening new fields of science. For example, the recent development of blue light-emitting diodes has allowed more compact storage of digital information. New materials are being tested for applications as diverse as artificial body parts and new applications in semiconductor technology.

1.4 FUTURE OF MATERIALS SCIENCE

Future developments in materials science could include the predictable (cotton shirts that never require ironing!) and the unimaginable. We will learn some lessons from nature: spider silk mimics; composites that are based on the structure of rhinoceros horn, which is a natural composite similar to the Stealth aircraft, and shells, which are as strong as most advanced laboratory-produced ceramics; and biocompatible adhesives fashioned after mollusc excretions. We may also learn from nature how molecules assemble themselves into complex three-dimensional arrays. Electronic components will certainly be miniaturized further; but what will happen when the length scale is comparable to atomic sizes and no longer divisible? Quantum mechanics will govern the processes, and studies of these nanomaterials undoubtedly will lead to new materials.

Some questions are of immediate importance. For example, we need a fundamental understanding of high-temperature superconductors. With this, room-temperature superconductivity might be achieved. We also need to understand complex structures, such as quasicrystals, composites, and fullerenes. In the interests of our planet, we need to find plastics with increased strength that are biodegradable and/or photodegradable. In the development of new chemical sensors, issues are sensitivity, longevity, and selectivity. In the automotive area, electric power and light-weight aerodynamic bodies with a high degree of recyclability are some of the goals. Other questions are too preposterous to even pose at present!

With a firm understanding of the principles on which materials science is based, new materials and processes will take us confidently into the future.

COMMENT

UNITS AND UNIT PRESENTATION

For the most part, units in this book are given according to Système International (SI) conventions. Fundamental physical constants are given in Appendix 1 and unit conversions are presented in Appendix 2.

The presentation of units in tables and graphs in this book is according to **quantity calculus** (i.e., algebraic manipulation method), as recommended by IUPAC. Briefly, a **quantity** is treated as a product of its **value** and its **units**:

(*continued*)

COMMENT

UNITS AND UNIT PRESENTATION (continued)

$$quantity = value \times units \tag{1.1}$$

so, for example, the value of the strength/density ratio for composites (Figure 1.2) is 1×10^6 Pa kg^{-1} m^3, i.e.,

$$strength/density = 1 \times 10^6 \; Pa \; kg^{-1} \; m^3 \tag{1.2}$$

so the value (=1) plotted on the y-axis in Figure 1.2 is given by

$$1 = (strength/density)/(10^6 \; Pa \; kg^{-1} \; m^3). \tag{1.3}$$

Labels on graphs and tables are treated similarly throughout this book. [See M.A. White (1998). Quantity Calculus: Unambiguous Presentation of Data in Tables and Graphs. *Journal of Chemical Education* **75,** 607.]

FURTHER READING

See also the lists at the end of individual chapters.

General References

General sources include *Advanced Materials, Annual Review of Materials Science, Chemical & Engineering News, Chemistry of Materials, Journal of Chemical Education, Journal of Materials Education, Materials Research Society (MRS) Bulletin, Physics Today, Scientific American.*

The Encyclopedia of Advanced Materials (1994). Pergamon Press.
Scientific American, September 1967, Special Issue: Materials.
Scientific American, October 1986, Special Issue: Advanced Materials and the Economy.
P. Ball (1997) *Made to Measure.* Princeton University Press.
R.W. Cahn, P. Haasen and E.J. Kramer (1991–). *Materials Science and Technology,* A Comprehensive Treatment. VCH.
B.S. Chandrasekhar (1998). *Why Things Are the Way They Are.* Cambridge University Press.
R. Cotterill (1985). *The Cambridge Guide to the Material World.* Cambridge University Press.
A.B. Ellis, M.J. Geselbracht, B.J. Johnson, G.C. Lisensky, and W.R. Robinson (1993). *Teaching General Chemistry: A Materials Science Companion.* American Chemical Society.
M.L. Good, Ed. (1988). *Biotechnology and Materials Science.* American Chemical Society.
D.L. Goodstein (1985). *States of Matter.* Dover, New York.
L.V. Interrante, L.A. Casper, and A.B. Ellis, Eds. (1995). *Materials Chemistry.* American Chemical Society.
J.S. Langer (1992). Issues and Opportunities in Materials Research. *Physics Today,* October 1992, 24.
W.J. Moore (1967). *Seven Solid States.* W.A. Benjamin, Inc.
W.W. Mullins (1996). Remarks on the Evolution of Materials Science. *Materials Research Society Bulletin,* July 1996, 20.

P.A. Thrower (1992). *Materials in Today's World.* McGraw-Hill, Inc.

D.B. Williams, A.R. Pelton, and R. Gronsky (1991). *Images of Materials.* Oxford University Press.

A. Wold and K. Dwight (1993). *Solid State Chemistry: Synthesis, Structure, and Properties of Selected Oxides and Sulphides.* Chapman & Hall.

Biomaterials

Special Issue on Biomaterials. *Chemistry in Britain,* March 1992.

Reference Textbooks

M.F. Ashby and D.H.R. Jones (1986). *Engineering Materials 2: An Introduction to Microstructures, Processing and Design.* Pergamon Press.

D.J. Barber and R. Loudon (1989). *An Introduction to the Properties of Condensed Matter.* Cambridge University Press.

R.S. Berry, S.A. Rice, and J. Ross (1980). *Physical Chemistry.* John Wiley & Sons, Inc.

R.J. Borg and G.J. Dienes (1992). *The Physical Chemistry of Solids.* Academic Press.

D.W. Bruce and D. O'Hare, Eds. (1992). *Inorganic Materials.* John Wiley & Sons, Inc.

W.D. Callister, Jr. (1994). *Materials Science and Engineering,* 3rd Ed. John Wiley & Sons, Inc.

A.K. Cheetham and P. Day, Eds. (1992). *Solid State Chemistry: Compounds.* Clarendon Press.

J.H.W. de Wit, A. Demaid, and M. Onillon, Eds. (1992). *Case Studies in Manufacturing with Advanced Materials,* Vols. 1 and 2. North-Holland.

R.A. Dunlap (1988). *Experimental Physics: Modern Methods.* Oxford University Press.

H.F. Franzen (1994). *Physical Chemistry of Solids: Basic Principles of Symmetry and Stability of Crystalline Solids.* World Scientific.

A. Guinier and R. Jullien (1989). *The Solid State.* Oxford University Press.

G.G. Hall (1991). *Molecular Solid State Physics.* Springer-Verlag.

C. Kittel (1996). *Introduction to Solid State Physics,* 7th Ed. John Wiley & Sons, Inc.

D.V. Ragone (1995). *Thermodynamics of Materials,* Vols I and II. John Wiley & Sons, Inc.

C.N.R. Rao, Ed. (1993). *Chemistry of Advanced Materials.* Blackwell Scientific Publications.

C.N.R. Rao and J. Gopalakrishnan (1997). *New Directions in Solid State Chemistry.* Cambridge University Press.

H.M. Rosenberg (1988). *The Solid State,* 3rd Ed. Oxford University Press.

J.F. Shackelford (1988). *Introduction to Materials Science for Engineers,* 2nd Ed. Macmillan Publishing Company.

J.P. Sibilia (1988). *A Guide to Materials Characterization and Chemical Analysis.* VCH.

L. Smart and E. Moore (1992). *Solid State Chemistry: An Introduction.* Chapman & Hall.

W.F. Smith (1990). *Principles of Materials Science and Engineering,* 2nd Ed. McGraw-Hill.

D. Tabor (1991). *Gases, Liquids and Solids and Other States of Matter,* 3rd Ed. Cambridge University Press.

L.H. Van Vlack (1989). *Elements of Materials Science and Engineering,* 6th Ed. Addison-Wesley.

A.J. Walton (1983). *Three Phases of Matter,* 2nd Ed. Oxford University Press.

M.T. Weller (1994). *Inorganic Materials Chemistry.* Oxford University Press.

Interactive Media

C.J. McMahon, Jr. (1996). *Interactive Glossary for Materials Science and Engineering.* Merion Media, Inc.

J.C. Russ (1996). *Materials Science: A Multimedia Approach.* (CD-ROM). PWS Publishing Company.

II

COLOR AND OTHER OPTICAL PROPERTIES OF MATTER

It is the pure white diamond Dante brought
*   To Beatrice; the sapphire Laura wore*
When Petrarch cut it sparkling out of thought;
*   The ruby Shakespeare hewed from his heart's core;*
The dark, deep emerald that Rosetti wrought
*   For his own soul, to wear for evermore.*

—Eugene Lee-Hamilton
"What Is a Sonnet?"

2

ATOMIC AND MOLECULAR
ORIGINS OF COLOR

2.1 INTRODUCTION

For many, the colors in nature and in the laboratory are one of the inspirations for studying science. Perhaps this is partly because of our perception of color. Including hue, saturation, brightness, and intensity, we can distinguish about ten million colors with our eyes.

Color in materials is caused by interactions of light waves with atoms, and especially with their electrons. In fact, color is a manifestation of many subtle effects that are important in determining the structure of matter.

In practical terms, it is the relative contributions of light of various wavelengths that determines the color of a material. To pursue this further, it is useful to review the electromagnetic spectrum (Figure 2.1), keeping in mind that light is an electromagnetic wave. The energy of the radiation, $E,$ is related to its wavelength, λ, and frequency, ν, by

$$E = h\nu = \frac{hc}{\lambda} \tag{2.1}$$

where h is Planck's constant and c is the speed of light. (See Appendix 1 for values of physical constants.)

As shown by Equation 2.1, when the wavelength is longer, the energy per photon is lower. Very long wavelength radiation (e.g., radio waves) passes through us without damage, whereas radiation that does biological damage is of much shorter wavelength and higher energy. For example, high doses of x-rays can be very damaging, because their high energy can initiate unfavorable chemical reactions in the body. It is easy to see from the above it is safe to stand under an infrared lamp to dry oneself after a bath, but an ultraviolet light (e.g., used for tanning purposes) requires more safeguards.

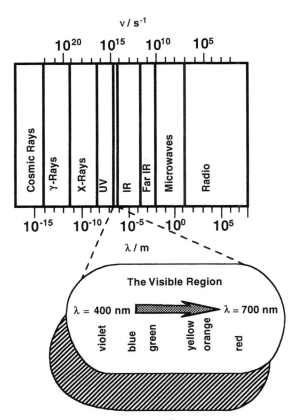

Figure 2.1. The electromagnetic spectrum. Note that the boundaries between regions are not sharp, and the spacing of the visible colors is not even.

Color is often the result of transitions between electronic states; this can involve energy absorption or emission. Most sources of color can be categorized as originating from absorption or emission of light, and it is useful to keep in mind which of these dominates the discussion in this chapter and in Chapters 3, 4, and 5. A useful general consideration is that emission can be seen in the dark, whereas absorption requires a source of light. (Color also can arise from either transmission or reflection of light. See Comment: Transmission or Reflection? Other sources of color are light scattering, dispersion, and interference; these are discussed in Chapter 4.) In considering these changes in states, it is useful to keep in mind that electronic states are much more widely separated than vibrational states, which are themselves more widely separated than rotational states. This is shown schematically in Figure 2.2. If an electronic transition involves localized electrons in a closed-shell electron configuration, the energy involved is so high that it will be in the ultraviolet (UV) or even in the x-ray region of the electromagnetic spectrum. Many colors that we see are caused by electronic transitions involving valence electrons, the electrons that form chemical bonds, as these transitions can be at low enough energy for us to see their effects with our eyes.

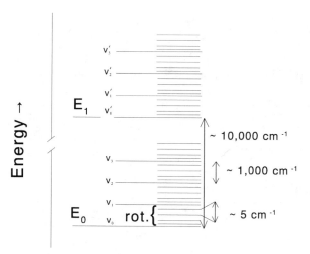

Figure 2.2. Schematic representations of the relative spacings of electronic energy levels (marked E_0 and E_1), vibrational energy levels (marked v_0, v_1, . . . and v_0', v_1', . . .) and rotational energy levels (marked rot.). The approximate energies are given in units of cm^{-1}, called **wavenumbers**, and abbreviated $\bar{v}$, where $\bar{v} = 1/\lambda$. 1 $cm^{-1} = 1.97 \times 10^{-5}$ eV $= 3.15 \times 10^{-24}$ J.

TRANSMISSION OR REFLECTION?

When an opaque material derives its color primarily from absorption of light, the color observed is due to the absorption of certain colors from the ambient white light. The light that reaches the eye and causes color perception is due to reflection and scattering, and it is depleted of the absorbed color(s). The color absorbed is the **complementary color** (see Figure 2.3) to the color that is observed.

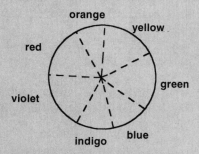

Figure 2.3. The "color wheel" of Newton. Colors opposite to each other are complementary. If one color is absorbed by a material, the material will bear the complementary color to the color absorbed.

If a material is transparent, we may perceive its color due to the wavelengths of light that are transmitted. Again, relative to ambient light, the transmitted light will be depleted in the colors of the absorbed light. Once again, the material will appear as the color that complements the absorbed color(s).

(continued)

TRANSMISSION OR REFLECTION? (continued)

If the color of a transparent object is due only to absorption, the transmitted and reflected colors would be the same. However, surface reflectivity can have some dependence on wavelength, and this can lead to different colors in reflection and transmission. Reflection and transmission are illustrated in Figure 2.4.

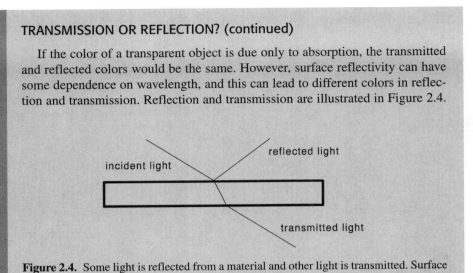

incident light

reflected light

transmitted light

Figure 2.4. Some light is reflected from a material and other light is transmitted. Surface scattering and internal scattering are omitted here.

In this chapter, we introduce atomic and molecular origins of color in insulating materials.

2.2 ATOMIC TRANSITIONS

Electronic transitions within a given atom can give rise to color through emission of light. A well-known example of atomic emission is the yellow color of sodium in the flame test for this element. The ground electronic configuration of Na is $1s^2 2s^2 2p^6 3s^1$, so the "outermost" electron is the $3s^1$ electron. The ground state of this valence electron has two closely spaced energy levels above it, one 2.105 eV above the $3s^1$ level and one 2.103 eV above the $3s^1$ level. (The electron volt, or eV, is a convenient method for measuring energies of electrons. 1 eV = 1.602×10^{-19} J; see Appendix 2 for more energy conversions.) The corresponding wavelengths of light of these energies are 589.1 and 589.6 nm, where, of course, the longer wavelength corresponds to the lower energy emission shown in Figure 2.5.

These wavelengths of light emitted from excited sodium atoms, at about 590 nm, correspond to the yellow part of visible light. Therefore, when a sodium-containing material is heated in a flame such that $3s^1$ electrons are promoted to the 3p excited levels, each electron can then return to the ground state ($3s^1$ configuration), concurrent with emission of yellow light. This light is often referred to as the sodium D line, in reference to it being the fourth prominent line (hence D line) recorded by Fraunhofer[1] as missing from the otherwise nearly continuous emission from sunlight. We now know that this wavelength of light is missing because it is absorbed by sodium atoms present in the sun's atmosphere. Close examination also shows that the D "line" is really two closely spaced absorptions (see Figure 2.5).

[1] Joseph Fraunhofer (1787–1826) was a German physicist who studied optical properties of matter in order to design and produce fine optical and mechanical instruments.

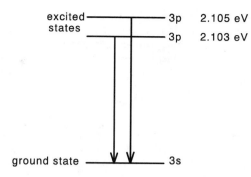

Figure 2.5. The electronic energy levels of Na. Excitation is caused by heat and emission of yellow light results.

The light emitted in this way from an atom is characteristic of the particular element, since each element has its own characteristic electronic energy levels. For example, the red glow of a neon sign is characteristic of the electronic energy scheme of Ne. The blue color of a mercury lamp is caused by two emissions, one green and one violet (green and violet mixed together produce blue). Other examples of colors arising from atomic emission include the colors of lasers making use of monatomic gases (such as the Ar-ion laser). In addition, lightning and arcs (e.g., around spark welding or a skate-sharpening device) derive some of their color from electronic excitations of the atoms in the surrounding gases. Northern lights (*aurora borealis*) arise from atomic emissions caused by interaction of atmospheric atoms or molecules (mostly oxygen, nitrogen, and hydrogen) with the solar wind, i.e., charged species from the sun entering our atmosphere. Emissions from various elements are used in fireworks, e.g., strontium for red and barium for green.

In general, gases have relatively sharp emission and absorption lines (see Figure 2.6). This is, at least in part, caused by the low density of gas molecules. At higher pressures there would be more collisions, which would increase the linewidth.

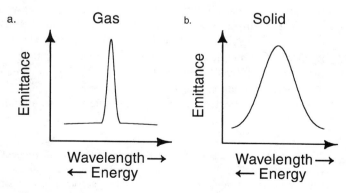

Figure 2.6. Schematic emissions from (a) a gas and (b) a solid.

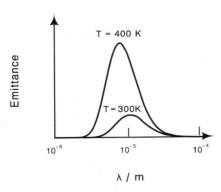

Figure 2.7. Typical ideal black-body radiation spectra at two temperatures. Note the increase in the emitted radiation (curve area) and the shift in the peak to lower wavelength (higher energy) when the temperature is increased.

2.3 BLACK-BODY RADIATION

As the name implies, black-body radiation also involves emission. Whereas gases have sharp emission lines, condensed matter (solids and liquids) usually gives a broad emission over a wide range of wavelengths. This is shown schematically in Figure 2.6.

A "black body" is an idealized material that absorbs light of all wavelengths (it does not reflect or transmit any light, hence the adjective "black"), and it also is a perfect emitter of light of all wavelengths. The study of the intensity distribution of radiation from black bodies by Max Planck[2] in 1900 led to the development of quantum theory. An ideal black body emits a "spectrum" of light, and this spectrum depends only on the temperature of the black body. Black-body radiation is often referred to as **incandescence.**

For example, at $T = 0$ K, all the atoms and subatomic particles in the black body are in their ground state, so there can be no light emitted at any wavelength.

At a temperature T, where $T > 0$ K, the occupancy of states from which the black body can emit light depends on the energy of the state. The body both absorbs and emits light, but when its temperature exceeds the temperature of the surroundings, emission will dominate. Therefore, there is a distribution of emission (spectrum) as shown in Figure 2.7.

As the temperature is increased, the distribution of emitted light moves, as shown in Figure 2.7, such that the maximum intensity is at shorter wavelength (higher energy) at the higher temperature. Of course, because of the higher occupancy of excited states at the higher temperature, the overall intensity of emitted light (i.e., the area under the entire curve) is greater at higher temperature than at lower temperature.

The exact spectrum is determined solely by the temperature of the black-body emitter,

---

[2] Max Karl Ludwig Planck (1858–1947) was a German theoretical physicist who made major contributions to the fields of thermodynamics, electrodynamics, radiation theory, and relativity. He also contributed to the philosophy of science and was an accomplished musician. He was awarded the Nobel Prize in Physics in 1918.

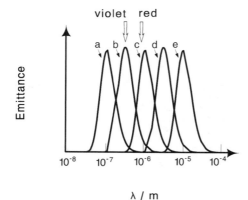

violet red

a b c d e

Emittance

10^{-8} 10^{-7} 10^{-6} 10^{-5} 10^{-4}

λ / m

Figure 2.8. Black-body radiation as a function of temperature, highlighting how the color changes as the temperature changes. The corresponding temperatures of the black bodies are a, $T = 30,000$ K; b, $T = 10,000$ K; c, $T = 3000$ K; d, $T = 1000$ K; e, $T = 300$ K. To have all the peaks of the same area, they have been scaled as follows: a, ×1; b, ×270; c, ×10^5; d, ×2.7×10^7; e, ×10^{10}. Note the shift in the peak to longer wavelength (lower energy) and the smaller peak area with lower temperature.

independent of the type of material. This can be of considerable importance, as we will see below.

At room temperature, black-body radiation is negligible to our eyes. At a temperature of about 700°C, the emission maximum is still in the infrared (IR) region, but some long wavelength (red) light is perceptible, and an object at this temperature glows red. We are familiar with this in glowing red embers in a fire. As the temperature of the black-body emitter increases, the "glow" moves from red to orange to yellow to white (the spectrum covers the whole visible range and when all colors are emitted it appears to be white[3]) to very pale blue. The progression of the emission over a range of temperatures is shown in Figure 2.8.

Examples of color caused by black-body radiation can be found in high-temperature bodies in our world. For example, carbon of a candle flame or log on a fire glows orange-yellow at a temperature of about 1700°C. The tungsten filament of an incandescent light bulb glows yellow-white at a temperature of about 2200°C. The filament of a flash bulb provides virtually "white" light because it is at a higher temperature still, about 4000°C. The light appears white because its emission peak covers the whole visible range and almost exactly matches the sensitivity of the eye to various colors.

As mentioned earlier, the ideal black-body radiation spectrum is characteristic of the temperature of an object and does not depend on the type of material. This knowledge, combined with a chance observation of unexpectedly high levels of radiation at a particular wavelength, led researchers from Bell Laboratories to determine that the average temperature of our universe is 3 K, a souvenir of the Big Bang that created the universe 12 billion years ago. The discovery of this black-body radiation has considerably advanced theories of the origins of our universe.[4]

---

[3] When white light is broken into its parts, e.g., with a prism, it can be seen to be composed of red, orange, yellow, green, blue, and violet constituents. In a similar, but reverse manner, when all of the colors of the rainbow are emitted, they produce white light.

[4] For further details, see Jeremy Bernstein (1984). *Three Degrees Above Zero: Bell Labs in the Information Age,* Charles Scribner's Sons.

2.4 VIBRATIONAL TRANSITIONS AS A SOURCE OF COLOR

In general, vibrational transitions are relatively rare as a source of color to our eyes. The energies of most vibrational excitations fall in the infrared region, so this does not noticeably affect the color of white light that falls upon a sample. However, there are some exceptions, and one rather prominent one in our lives is H_2O, both in the liquid and in the solid state.

Excitation of a complex many-molecule motion in the hydrogen-bonding network in water or in ice involves absorption of a little red-orange light. This is a rather high-energy excitation (higher energy than the usual IR range for molecular vibrations) because it is to a highly excited vibrational overtone, i.e., leading to a much higher vibrational level. The excitation is also high in energy because hydrogen bonding is an unusually strong intermolecular interaction. The absorption of red-orange light depletes white light of red-orange, and leaves the complementary green-blue color. (The visible colors and their complements were shown in Figure 2.3.)

The absorbance, A, is related to the incident light intensity, I_0, and the exiting light intensity, I, through the Beer[5]–Lambert[6] Law (also known as the Beer–Bouguer[7]–Lambert Law):

$$\log_{10}\left(\frac{I_0}{I}\right) = A = \varepsilon c l \tag{2.2}$$

where ε is the absorption coefficient, c is the concentration, and l is the length of the sample in the light path. Because ε is rather small for this excitation in H_2O, it takes a large path length, l, to produce a detectable color. We are familiar with this as a glass of water and an ice cube both appear colorless, whereas the blue color of water in a lake or ice in a glacier is readily apparent. For example, with a 3 m path length, 56% of incident red light is absorbed by water and the liquid appears blue-green in transmitted light. (The color seen in reflection is caused by both absorption of red light and light scattering; see Chapter 4 for a discussion of the latter effect.)

Although vibrational excitation is not a common source of visible color, it gives water its familiar color.

2.5 CRYSTAL FIELD COLORS

Another case of absorption as a source of color involves ions present in solids. These may be ions in a pure solid, or impurity ions.

When atoms are free as in the gas phase, their valence electrons may be thermally excited, giving rise to emission colors, as in the case of sodium in a flame. However, when valence electrons combine with one another to form chemical bonds, the ground state energy of the atom

[5] August Beer (1825–1863) was a German physicist and optical researcher who worked at Bonn University.
[6] Johann Heinrich Lambert (1728–1777) was a German mathematical physicist who discovered a method for measuring light intensities.
[7] Pierre Bouguer (1698–1758) was a French mathematician who invented the photometer.

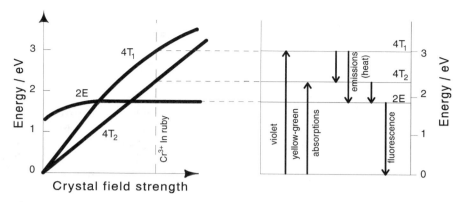

Figure 2.9. The effect of crystal field strength on the splitting of the energy levels of Cr^{3+} ions in a lattice. Styled after K. Nassau (1980), *Scientific American,* October 1980, 134.

is lowered, and much more energy is then required to promote them to excited states. In many compounds the outermost electrons are so stabilized by chemical bonding that their excitation takes energy in the UV region, so no visible color is associated with these materials, although they do absorb UV light.

One exception is compounds containing transition metals with incomplete d-shells. The excitation of these valence electrons can fall in the visible region of the spectrum, and these compounds can have beautiful colors. This is also true of compounds containing transition metal impurity ions.

All electronic energy levels are sensitive to their environment; electrostatic and other interactions can make the levels move relative to one another. An ion in the electric field in a crystal (i.e., in a **crystal field,** or if the field arises from the presence of ligands it is called a **ligand field**) can have different energy levels from the free ion. The interaction of an ion with an electric field can provide information concerning the electronic environment of the ion. In the cases considered here, the crystal field can give rise to observable colors.

As an example, consider the beautiful red color of a ruby. Chemically, a ruby is Al_2O_3 (known in its pure single crystal form as corundum), doped with about 1% Cr^{3+} in place of Al^{3+}. (It takes some of the glamor away from high-priced gems to know their chemical composition!) The Cr^{3+} ion is especially important because its presence gives rise to the color.

The Cr^{3+} ion in a ruby lies at the center of a distorted octahedron of O^{2-} ions. The Cr–O distance is about 1.9 Å, and the Cr–O bonds are quite ionic (estimated at more than 50% ionic character), so the Cr^{3+} ion exists in a strong electric potential (crystal field). The effect of the crystal field on the electronic energy levels of Cr^{3+} is shown in Figure 2.9.

At the crystal field strength of Cr^{3+} in a ruby, the higher-energy absorption is in the violet region of the spectrum and the lower-energy absorption is yellow-green. Therefore, white light is depleted of violet and yellow-green, and the ruby appears red. The main color of a ruby is due to absorption of violet and yellow-green light; it is only coincidental that the colour of **fluorescence**[8] in a ruby (Figure 2.9) is also red.

---

[8] See section 2.7 for a discussion of fluorescence.

Interestingly, emeralds, which are green, derive their color from Cr^{3+} ions, this time in a material with composition $Be_3Al_2Si_6O_{18}$ (known as beryl when pure). The Cr^{3+} in this structure is in a very similar environment to that in a ruby, but it experiences a different crystal field strength. The difference in color subtly illustrates the importance of the magnitude of the crystal field strength. Determination of whether the crystal field strength in an emerald is less than or greater than in a ruby is left as an exercise for the reader.

Electronic transitions of other ions in solids can give rise to crystal field colors. Some further examples are alexandrite (Cr^{3+} in $BeAl_2O_4$), aquamarine (Fe^{3+} in $Be_3Al_2Si_6O_{18}$), the jadeite form of jade [Fe^{3+} in $NaAl(SiO_3)_2$], citrine quartz (Fe^{3+} in SiO_2), blue and green azurite [$Cu_3(CO_3)_2(OH)_2$], malachite [$Cu_2CO_3(OH)_2$], and red garnets [$F_{e3}Al_2(SiO_4)_3$]. In the last three cases the color comes from the transition metal ions (Cu^{3+} or Fe^{3+}) of the pure salt rather than impurity ions. When the color of a mineral is inherent (said to be **ideochromatic;** the color will remain even if finely divided) it is often caused by its pure composition, which is usually independent of the origin of the mineral. A mineral of variable color (called **allochromatic**) generally derives its color from impurities (also called **dopants**) or other factors that depend on location and geological history.

COMMENT

DEFECTS IN CRYSTALS

In contrast with the perfect crystalline structures shown in Appendix 3, real crystals have imperfections. These can be **point defects** (i.e., occupational imperfections) or **line defects** (i.e., imperfections in translational symmetry). Some examples are shown in Figure 2.10.

a.

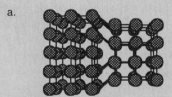

b.

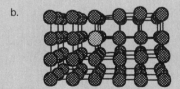

Figure 2.10. Defects in a solid: (a) line defect due to a packing fault; (b) point defect due to impurity (shown as a lighter-colored atom).

Defects play an important role in crystals, and can greatly modify optical, thermal, electrical, magnetic, and mechanical properties. Defects are also discussed in Chapter 14.

2.6 COLOR CENTERS (F-CENTERS)

Electronic excitation can give rise to visible colors, but the electrons being excited do not need to be bound to ions, they also could be "free." For example, the "electric blue" color of Na dissolved in liquid ammonia comes from the absorption of light in the yellow region by solvated electrons in solution (the solution contains NH_3, Na^+, and e^-).

Unattached electrons also can exist in solids. For example, if there is a structural defect, such as a missing anion, an electron may locate itself in the anion's usual position, as shown in Figure 2.11 for the salt M^+X^-.

An "extra" electron also could be associated with an impurity of mismatched valence. For example, if Mg^{2+} is an impurity in the M^+X^- lattice, there would be an extra electron located near Mg^{2+}, to balance electrical charge, as shown in Figure 2.12.

Similarly, the absence of an electron, which is termed a **hole** (see Figure 2.13), locates itself near an impurity ion or structural defect. Holes are as important as electrons in a discussion of the electrical properties of solids, because both electrons and holes are charge carriers. Whereas electrons carry negative charge, holes carry positive charge (see Comment: Holes).

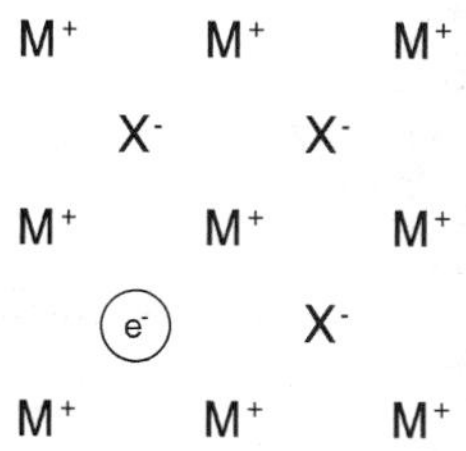

Figure 2.11. Presence of a trapped electron in an idealized two-dimensional lattice of salt MX.

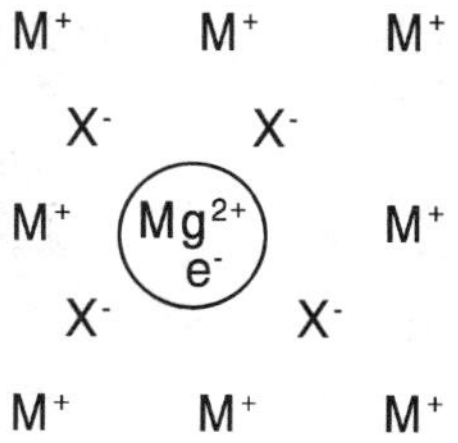

Figure 2.12. Presence of an extra electron associated with a **hypervalent impurity** (i.e., an impurity with an extra charge, shown here as Mg^{2+}) in the idealized two-dimensional salt MX.

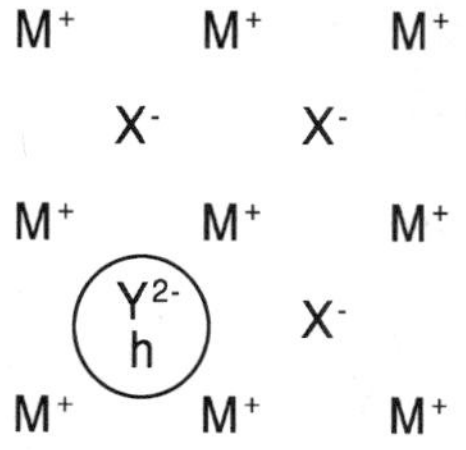

Figure 2.13. A hole, h, is present in a lattice of MX to compensate for the charge of the impurity Y^{2-}.

COMMENT

HOLES

In some ways it is difficult to imagine holes, since they are really not matter but more like the absence of matter. However, there are instances in which we are familiar with the motion of the absence of matter. One example is the bubbles that are seen to rise in a fish tank. Although most of the matter in view is water, the bubbles can be considered to be an "absence of water"; as a bubble rises from the bottom to the top of the tank, the net flow of water is in the opposite direction, i.e., toward the bottom. This is similar to the case of electrons and holes: while the negative charges (electrons) flow in one direction, the positive charges (holes) "flow" in the opposite direction.

Extra charge centers (electrons or holes) are called **color centers** or sometimes **F-centers** ("F" is an abbreviation of "Farbe," the German word for "color"). They give rise to color centers because the excitation of electrons (or holes) falls in the visible range of the spectrum.

As an example, consider the sometimes violet color of fluorite, CaF_2. This color arises when some F^- ions are missing from their usual lattice sites, giving rise to an extra electron in the lattice. This can happen in one of several ways: the crystal may have been grown in the presence of excess Ca, the crystal may have been exposed to high-energy radiation that displaced an ion from the usual lattice site, or the crystal may have been in an electric field that was strong enough to remove F^- electrochemically. In any case, electrons exist in the places where F^- would be in a perfect lattice. These electrons give rise to the color, as the electron absorbs yellow light, leaving behind a violet color. Figure 2.14 shows color centers induced in fluorite.

Because a color center could arise from radiation, the intensity of the color of a material such as fluorite can be used to monitor radiation dosage. This has laboratory applications (e.g., in radiation safety badges worn by people who work with x-rays), and it can be used in the field to determine the radiation dosage that a geological sample has received from its environment.

The color of amethyst also is caused by a color center, this time by the absorption of energy associated with the electronic energy levels of a hole. The hole exists because of the presence of Fe^{3+} in SiO_2. Fe^{3+} substitutes for Si^{4+} in the lattice, and this requires the presence of a hole for electrical neutrality. The energy levels of the hole allow absorption of yellow light, leaving amethyst with its familiar violet color. The intensity of the color is an indication of the concentration of iron impurity ions.

Color centers are often irreversibly lost on heating because increased temperature can mobilize imperfections and cause them to go to more regular sites. For example, amethyst turns

Figure 2.14. A fluorite crystal showing intense violet coloration at the upper left vertex due to the formation of color centers.

from violet to yellow or green on heating; the loss of violet is the result of the loss of the color centers, and the remaining color, which was there all along but was not as intense as the violet, is because the Fe^{3+} transition metal ion absorbs light due to crystal field effects. If it is yellow, this form is citrine quartz, described previously under crystal field effects.

2.7 CHARGE DELOCALIZATION, ESPECIALLY MOLECULAR ORBITALS

When electrons are paired in chemical bonds, their electronic excited states are usually so high that electronic excitations are in the UV range, and these materials are not coloured (i.e., there is no absorbance in the visible range). This is especially true if the pair of electrons is localized to a particular chemical bond.

In molecular systems with extensive conjugation, the electrons are delocalized throughout a number of chemical bonds. The electronic excitation is from the highest occupied molecular orbital (**HOMO**) to the lowest unoccupied molecular orbital (**LUMO**). For a conjugated system the HOMO–LUMO transition is of lower energy than for electrons localized to one bond; if the transition is low enough in energy, absorption of visible light can cause the promotion of an electron to this excited state. If visible light is absorbed, the material with conjugated bonds will be coloured. (If only more energetic UV light is absorbed, the material may have promise as a component in a sunscreen.)

Many organic molecules derive their colors from the excitation from electronic states that involve large degrees of conjugation. For example, the more than 8000 dyes in use today derive their color from this mechanism. The part of the molecule that favors color production is called a **chromophore** (meaning "color bearing").

The color of organic materials can be changed substantially by the presence of electron-withdrawing and/or electron-donating side groups, as this changes the electronic energy levels. These are called **auxochromes** ("color enhancers").

As an example of the dependence of the color of an organic compound on structure, we can consider the case of an organic acid–base indicator. These materials are often weak organic acids, which will be deprotonated in basic solution. The loss of the proton can be detected visually, as the color of the acid and its conjugate base can be substantially different, especially due to different degrees of delocalization. This is taken up further in a problem at the end of this chapter.

Retinal is the pigment material of the photoreceptors in our eyes. Its color changes in the presence of light, as retinal undergoes a photo-induced transformation from the *cis* to the *trans* form. In the absence of light, it converts from the *trans* form back to the *cis* form, assisted by an enzyme. The dominance of the *cis* or the *trans* form signals the light level.

COMMENT

CARBONLESS COPY PAPER

Carbonless copy paper, used to make countless forms, is a 10^9 kg business annually, and it is based on simple chemistry in which organic reactants change color on protonation. Copies are made when the pressure of a pen ruptures

(continued)

COMMENT

CARBONLESS COPY PAPER (continued)

microcapsules that are coated on the lower surface of the top page. These microcapsules contain organic dye precursor molecules that, after rupture of the microcapsules, can react with a proton-donating reagent layer that is coated on the upper surface of the lower paper. In the acidic conditions, the organic dye precursor becomes colored, making a copy of the top page's written message onto the bottom page.

COMMENT

DALTON'S COLOR BLINDNESS

In 1794, the great chemist John Dalton (1766–1844) gave the first formal account of color blindness—his own. He believed that he was unable to distinguish red from green because he had a blue tint in the vitreous humor of his eyes, which could selectively absorb the long wavelengths of light. He had ordered that his eyes be examined after his death, but no blue tint was found. More recent DNA tests have shown that Dalton's color blindness was caused by alteration of the red-responsive visual pigment. Although Dalton would not have been able to distinguish the color red, he possessed the red receptor and would have perceived light throughout the visible region, including red; this has been confirmed from the writings of his contemporaries.

Another example of charge delocalization and its associated color can be seen in the case of **charge transfer** from one ion to another, for example in a blue sapphire. This material is Al_2O_3 with Fe^{2+} and Ti^{4+}. An excited electronic state is formed when adjacent Fe^{2+} and Ti^{4+} ions exchange an electron to give Fe^{3+} and Ti^{3+} by a photochemically induced oxidation–reduction reaction. It takes energy of about 2 eV to reach this excited state; this corresponds to light of wavelength 620 nm. This light is yellow, and since the material is absorbing yellow from the incident white light, a sapphire appears blue. Only a few hundredths of 1% of Fe^{2+} and Ti^{4+} is required to achieve a deep blue color. The deep colors of many mixed valence transition metal oxides (e.g., Fe_3O_4, which can be represented as $FeO\cdot Fe_2O_3$) arise from homonuclear charge transfer mechanisms.

Some molecules with delocalized electrons also are capable of **luminescence** (i.e., light emission from a cool body[9]), which includes fluorescence, phosphorescence, and chemiluminescence. In **fluorescence** an excited state decays to an intermediate state that differs from the ground state by an energy that corresponds to rapid emission of light in the visible range of the spectrum, as shown in Figure 2.15. An example is compounds that are added to detergents as fabric brighteners. These absorb ultraviolet components of daylight (to get to an excited state) and then fluoresce, giving off blue light. This blue light makes the material appear bright by combining with the yellow of dirt (blue and yellow are complementary colours) to give white.

[9] The temperature of luminescence distinguishes it from incandescence associated with black-body radiation.

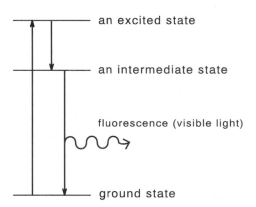

Figure 2.15. Schematic representation of fluorescence.

Phosphorescence is the slow emission of light, following excitation to a higher energy state. Phosphorescence is discussed further in Chapter 3.

Chemiluminescence is fluorescence initiated by chemical reactions that give products in an excited state. The material may decay to its ground state energy level by a fluorescent pathway, as described above. Examples of this are the reactions giving rise to the glow of fireflies, **bioluminescence** of some deep sea fish, "cold light," and glow sticks.

LASERS

Lasers derive their name from **l**ight **a**mplification by **s**timulated **e**mission of **r**adiation. The light produced can be very intense, and it has the special property of **coherence,** i.e., the waves produced are all in phase.

The action of a laser can be understood in terms of the energy levels of the system. A laser requires excitation (pumping) from the ground state energy to an excited state (absorption level); this excitation can be electrical, optical, or chemical; it is shown schematically in Figure 2.16. From this absorption level, the

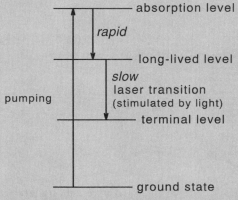

Figure 2.16. Typical energy level diagram for a laser.

(continued)

LASERS (continued)

energy rapidly decreases as the system goes to a long-lived excited state. From this state, the energy can decrease further as the system goes to the terminal energy level; this process can be stimulated by light, and more light is emitted in this process.

In the laser cavity, which is a tube with perfectly aligned end mirrors, the emitted light that is directed exactly along the length of the cavity will cause additional stimulations along the length of the cavity. Emission also will take place in other directions, but it will not be reflected by the mirrors, as the emission along the length of the cavity is. The emission in this particular direction will be coherent and very intense. One of the mirrors is slightly transparent and typically 1% of the coherent stimulated emission escapes from this end.

In a helium–neon laser, direct current or radio-frequency radiation is used to excite the helium atoms, which then collide with neon atoms exciting them into a long-lived excited state. The emission from this state gives rise to the familiar red light of a helium–neon laser.

Light of Our Lives: A Tutorial

The most common lights in our homes are incandescent lights. These have a wire inside them that glows when current is passed through it.

a. The early developers of light bulbs had to struggle to get bulbs that would last for a long time. Suggest some of their difficulties and how they have been overcome.

b. Tungsten is now the preferred material for the filaments of incandescent bulbs, whereas carbon was initially used. Why is tungsten superior?

c. Incandescent light bulbs get hot when they have been operating for some time. Why?

d. If a drop of water falls on an operating light bulb, the bulb will crack. Within a few minutes or less the bulb will dim, sputter, and then stop working. Describe the process of this failure.

Fluorescent lights are otherwise known as gas discharge tubes. Electrical discharge along the tube excites the Hg atoms in the vapor phase in the tube. These atoms emit some visible light and considerable light in the ultraviolet region, which we cannot see. The UV light causes excitations in the fluorescent coating (e.g., magnesium tungstate, zinc silicate, cadmium borate, cadmium phosphates) on the inside of the tube; this coating then emits additional light in the visible range.

e. Draw energy level diagrams to depict the fluorescent light processes.

f. Discuss factors that are important in choosing materials for the vapor phase.

g. Discuss factors that are important in choosing materials for the fluorescent coating.

h. How does the temperature of a fluorescent light compare with that of an incandescent light during use? Comment on relative energy efficiencies.

i. How can incandescent lights be made more energy efficient?

FURTHER READING

General References

Most introductory physical chemistry textbooks describe aspects of electromagnetic radiation and black-body radiation. Books on rocks and minerals often contain information concerning color and composition of geological materials.

F.W. Billmeyer, Jr. and M. Saltzman (1981). *Principles of Colour Technology,* 2nd Ed. John Wiley & Sons Ltd.

R.M. Evans (1948). *An Introduction to Color.* John Wiley & Sons Ltd.

P.L. Gourley (1998). Nanolasers. *Scientific American,* March 1998, 56.

M. Gouterman (1997). Oxygen Quenching of Luminescence of Pressure Sensitive Paint for Wind Tunnel Research. *Journal of Chemical Education,* **74,** 697.

A. Javan (1967). The Optical Properties of Materials. *Scientific American,* September 1967, 239.

C.G. Mueller, M. Rudolph, and the Editors of LIFE (1966). *Light and Vision.* Life Science Library, Time Inc.

K. Nassau (1980). The Causes of Colour. *Scientific American,* October 1980, 124.

K. Nassau (1983). *The Physics and Chemistry of Colour.* John Wiley & Sons Ltd.

H. Rossotti (1983). *Colour.* Princeton University Press.

D.R. Tyler (1997). Organometallic Photochemistry: Basic Principles and Applications to Materials Science. *Journal of Chemical Education,* **74,** 668.

V.F. Weisskopf (1968). How Light Interacts with Matter. *Scientific American,* September 1968, 60.

G. Wyszecki and W.S. Stiles (1967). *Color Science.* John Wiley & Sons Ltd.

Color of Water

C.L. Braun and S.N. Smirnov (1993). Why is Water Blue? *Journal of Chemical Education,* **70,** 612.

Carbonless Copy Paper

M.A. White (1998). The Chemistry Behind Carbonless Copy Paper. *Journal of Chemical Education* **75,** 1119.

PROBLEMS

1. If six paints (red, orange, yellow, green, blue, and violet) are mixed in equal portions, the resulting paint is black. However, if six lights (red, orange, yellow, green, blue, and violet) of equal intensity shine at the same spot, the resultant spot is white. Explain why two different "colors" (white vs. black) arise.

2. The jadeite form of jade is composed of $NaAl(SiO_3)_2$ and its green color results from the presence of a small amount of iron.
    a. It would appear that the color in jade could be caused by a color center or transition metal absorption. Describe the source and physical processes involved in each.
    b. Suggest an experiment that would distinguish between the above possibilities. Explain what you would do, what you would expect to observe in each case (and why), and how this would distinguish between the two possibilities.

3. The equation describing radiant emittance (M) for black-body radiation as a function of temperature (T) and wavelength (λ) is

$$M = \frac{2\pi hc^2}{\lambda^5}\left\{\frac{1}{e^{(hc/kT\lambda)}-1}\right\} \tag{2.3}$$

where h is Planck's constant, k is Boltzmann's constant, and c is the velocity of light.
    a. Calculate M for $T = 1000$ K at the following wavelengths (in m): 10^{-4}, 10^{-5}, 7×10^{-6}, 5×10^{-6}, 3×10^{-6}, 2×10^{-6}, 1.3×10^{-6}, 10^{-6}, 9×10^{-7}, 8×10^{-7}, 10^{-7}. Show a sample calculation (complete with working out of units, three significant figures in M) and put the results in a table that includes $\log_{10}(\lambda/m)$.
    b. Calculate M for $T = 2000$ K at the following wavelengths (in m): 10^{-4}, 10^{-5}, 3×10^{-6}, 2×10^{-6}, 1.5×10^{-6}, 10^{-6}, 9×10^{-7}, 8×10^{-7}, 7×10^{-7}, 6×10^{-7}, 5×10^{-7}, 3×10^{-7}. Put the results in a table that includes $\log_{10}(\lambda/m)$.
    c. Calculate M for $T = 4000$ K at the following wavelengths (in m): 10^{-4}, 10^{-5}, 3×10^{-6}, 2×10^{-6}, 1.5×10^{-6}, 10^{-6}, 9×10^{-7}, 8×10^{-7}, 7×10^{-7}, 6×10^{-7}, 5×10^{-7}, 3×10^{-7}, 2×10^{-7}. Put the results in a table that includes $\log_{10}(\lambda/m)$.
    d. On the same graph, plot the results of (a), (b), and (c). Plot the graph as M vs. $\log_{10}(\lambda/m)$. Use different symbols for the three sets of results and draw smooth curves. Use three sets of ordinate axes, one for (a), one for (b), and one for (c), so that each nearly fills the page.
    e. Mark the visible light region on your graph.
    f. Comment, based on your graph, on the relative areas of the three peaks. If they are different, explain why.
    g. Comment, based on your graph, on the color(s) of black-body radiation from an object at $T = 1000$ K, at $T = 2000$ K, and at $T = 4000$ K.
    A spreadsheet can be used to do this question.

4. Rubies are red because of the absorption of light by Cr^{3+} ions. The color of emeralds also is the result of absorption of light by Cr^{3+} ions; emerald is $Be_3Al_2Si_6O_{18}$ doped with Cr^{3+}. Based on the fact that emeralds are green, explain whether the crystal field experienced by Cr^{3+} is less in emeralds or in ruby. Refer to Figure 2.9.

5. Some types of old glass contain iron and manganese. After many years of intense exposure to sunlight, this type of glass turns violet through the development of color centers. Why is exposure to light needed?

6. Some materials develop color from exposure to radiation. This is because of the formation of color centers, and the number formed is proportional to the radiation dosage received. Suggest a method that could be used to quantify the radiation exposure, leaving the material in "new condition" to be used again.

7. When heated, some materials, such as diamond, give off a faint glow that decays with time. This effect is called **thermoluminescence.** Suggest a plausible reason for it. Explain how thermoluminescence could be used to determine the age and location of geological materials.

8. An old "magic" trick involves taking a seemingly normal piece of pale pink cloth and holding it over a glowing light bulb. While the audience watches, the cloth turns blue. The "magician" can return its color to pale pink by blowing on it. The cloth has been treated by dipping it in cobalt(II) chloride. $CoCl_2$ is a blue salt and $CoCl_2 \cdot 2H_2O$ and $CoCl_2 \cdot 6H_2O$ are red salts. Suggest an explanation for the "magic" trick, including the physical basis for the origins of the colors.

9. A large organic molecule with an acidic group can be used as an acid–base indicator. It has an equilibrium between its acid form (call it HA) and its conjugate base (call it A^-), as follows:

$$HA \rightleftarrows H^+ + A^-.$$

In the presence of acid, the HA form is favored. In the presence of base, the A^- form is favored (as H^+ is consumed, pulling the equilibrium to the right). The color depends on pH such that in one pH range the color is red and in another it is blue.
   a. In the red form, what color is absorbed from white light?
   b. In the blue form, what color is absorbed from white light?
   c. Are electrons in organic acids more delocalized in the acid form or in the base form? Why? The following may help:

$$\underset{R-C-OH}{\overset{O}{\underset{\|}{}}} \quad \rightleftarrows \quad H^+ + \underset{R-C-O^-}{\overset{O}{\underset{\|}{}}}.$$

   d. When the electrons are more delocalized, does this cause the absorption of visible light to be at lower or higher energy? Explain briefly.
   e. On the basis of your answers to (a), (b), (c), and (d), which is red and which is blue (of HA and A^-)? Explain.

10. It is possible to make a dill pickle "glow" by attaching both ends to leads to a 110 V power source. (Safety Note: This experiment requires safe handling procedures; see *Journal of Chemical Education* (1993), **70,** 250 and *Journal of Chemical Education* (1996), **73,** 456 for details of a safe demonstration.) The glow is yellow. What is the origin of the color?

11. Comment on the following which was taken from a syndicated newspaper article:

   Why are there no purple Christmas lights? . . . . [P]urple light has the shortest wavelength and is not visible to the human eye. When you see a red light, it is a result of the transparent material filtering out all wavelengths of light except red. When the transparent material is purple, it filters out everything except purple and since that is invisible to the human eye you would only see black.

12. What color would a flower that is white in daylight appear to be when observed in red light? Explain.

13. Most clothing stores have fluorescent lighting. In this light, two items can appear to be color-matched but, when viewed in sunlight, they are not well matched. Explain.

14. The window panes in some very old houses have a violet tint. Faraday noticed that this was because of the effect of sunlight on the glass, and it has been noticed that this effect requires a small amount of manganese. If you heat the glass, the color disappears and it does not reappear on cooling. What is the cause of the color?

15. Stars with different surface temperatures are different colors, e.g., bluish-white, orange, red, yellow, white. Arrange these star colors in order of increasing surface temperature.

16. By differentiating the expression for $M(\lambda)$ given in Equation 2.3, show that λ_{max}, the wavelength of the most intense black-body radiation, is inversely proportional to the temperature, i.e.,

$$\lambda_{max} = \frac{0.0029 \text{mK}}{T} \tag{2.4}$$

where T is in kelvin and λ_{max} is in meters. Equation 2.4 is known as Wien's law.

17. Some types of white plastic turn yellow over time. Explain the physical origin of the color, including an explanation of the yellow color, and why it has turned yellow rather than, for example, red.

18. A convenient new type of thermometer determines body temperature by placement of a probe in the ear. The device must be placed accurately in the ear canal, as the thermometer measures infrared energy from the eardrum and "translates" this into a temperature reading. Using the Stephan–Boltzmann law that the radiant emittance (M) for black-body radiation is given by

$$M = \sigma T^4 \tag{2.5}$$

where T is the temperature in kelvin and σ, the Stephan constant, has a value of 5.67×10^{-8} W m^{-2} K^{-4}, calculate the increase in energy flux with a 1 K increase in temperature from 37 to 38°C.

19. Some polymers, such as polycarbonate, change color when exposed to gamma radiation. Explain the physical principle that gives rise to the color change and suggest an application of this property.

20. A glass head representing a Pharaoh, made about 3400 years ago, is presently housed in an American Museum. The work of art was once bright blue, but now is faded. Discuss possible origins of the color and reasons why it has changed.

21. From the data given in Figure 2.17 for a particular object, what is the observed color of the object (i.e., observed by reflected light; assume that the surface is smooth so that scattering of light can be neglected)?

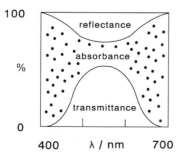

Figure 2.17. Absorption, transmission, and reflection as functions of wavelength.

22. Some brown or yellow topaz (aluminium fluorosilicate) gemstones have their color created artificially by irradiation, and this has been controversial because some of the resulting gemstones have been left highly radioactive. Suggest the origin of the color.

23. Would a mercury-based fluorescent light be expected to work at very low temperature? Explain.

24. On the basis of Equation 2.2 and the data in Section 2.4, what percent of red light would be absorbed by a 1 cm path length of water?

3

COLOR IN METALS
AND SEMICONDUCTORS

3.1 INTRODUCTION

In the last chapter we looked at atomic and molecular origins of color, emphasizing insulating materials. For the most part, the principles involved only a single atom (e.g., an electronic transition) or a small group of atoms (e.g., colors due to electronic transitions in molecules and crystal field effects). In this chapter we look at colors and luster in metals and origins of color in semiconductors. In both metals and semiconductors, very large numbers of atoms are usually responsible for the presence of color.

3.2 METALLIC LUSTER

Electrons have their maximum possible delocalization in metals, where the **free electron gas model** (i.e., electrons modeled as virtually floating in a sea of positive charges) has been used to describe many features.

In a metal, all the conduction electrons are essentially equivalent because they can freely exchange places with one another, but they do not all have the same energy due to the quantum mechanical limitation that no two electrons in the same system (meaning the metal, in this case) can have the same set of quantum numbers. Therefore, for a chunk of metal, there will be of the order of 10^{23} electrons, and nearly as many energy levels. (If all the levels were filled, there would be half as many levels as electrons since each level can accommodate only two electrons, one of each spin.) With this many energy levels in a metal, the energy ladder can be considered to be a virtual continuum.

Figure 3.1. Boltzmann's gravesite in Vienna. Photo by Robert L. White.

Fermi[1] and Dirac[2] determined that $P(E)$, the probability that a state with energy E will be occupied in a free electron gas, is

$$P(E) = \frac{1}{e^{(E-E_\mathrm{F})/kT} + 1}$$

(3.1)

where E_F, the **Fermi** energy, is the energy at which $P(E) = \frac{1}{2}$, k is the Boltzmann[3] constant, and T is temperature. Equation 3.1 is known as the **Fermi–Dirac distribution function.**

[1] Enrico Fermi (1901–1954) was an Italian-born physicist who made major contributions to the foundations of quantum mechanics and the first controlled self-sustaining nuclear reactor; he was the winner of the 1938 Nobel Prize in Physics.

[2] Paul A. M. Dirac (1902–1984) was a British theoretical physicist, noted for his contributions to quantum mechanics. Dirac shared the 1933 Nobel Prize in Physics with Schrödinger. From 1932 to 1969, Dirac was the Lucasian Professor of Mathematics at Cambridge University, a position previously held by Sir Isaac Newton.

[3] Ludwig Boltzmann (1844–1906) was an Austrian theoretical physicist. Boltzmann was an engaging lecturer who demanded discipline of his students and, in return, they were devoted to him. Although revered today for his work in the areas of thermodynamics, statistical mechanics, and kinetic theory of gases, Boltzmann's work was not well received by his colleagues in his lifetime. This was largely due to his espousal of the unfashionable corpuscular theory of matter. Boltzmann took his own life in 1906. Today his grave site, in the same Viennese cemetery where Beethoven lies, bears Boltzmann's famous equation: $S = k \log \omega$ (Figure 3.1).

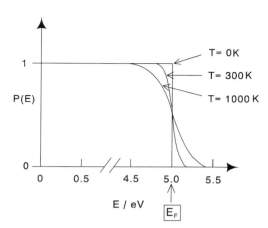

Figure 3.2. Probability distribution for electrons in a metal, according to the Fermi–Dirac distribution function (Equation 3.1), at various temperatures. E_F is the Fermi energy, where E_F separates the filled levels from the unoccupied levels at $T = 0$ K. Here, $E_F = 5.0$ eV.

This function looks "square" at $T = 0$ K, as one would expect since all the energy levels are filled up one by one at $T = 0$ K, up to the Fermi energy. This is shown in Figure 3.2.

COMMENT

BAND STRUCTURE

In going from localized molecular orbitals to delocalized molecular orbitals, the increased number of electrons makes the electronic energy levels become more numerous, until they become a virtual continuum, as shown schematically as bands in Figure 3.3.

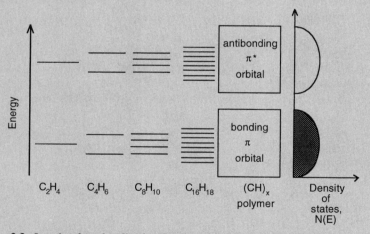

Figure 3.3. In going from localized molecular orbitals (smaller molecules, left side of diagram) to delocalized molecular orbitals (larger molecules, right side of diagram), electronic energy levels become more numerous until they become a virtual continuum, as shown schematically here. The **density of states,** $N(E)$, is defined as $N(E) \, dE$ being the number of allowed energy levels per unit volume in the range between E and $E + dE$. $N(E)$ is zero in the **band gap,** the forbidden region between the bonding and antibonding orbitals in this case. Metals have energy bands but no band gap (see Figure 3.2).

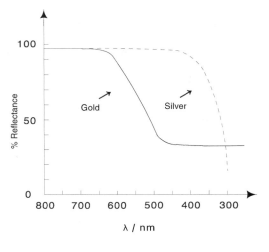

Figure 3.4. Reflectance spectra of gold and silver as functions of wavelength. Gold reflects red, orange, and yellow strongly but less so at blue and violet, and it appears yellow. In contrast, silver reflects incident light over most of the visible region of the spectrum. Adapted from A. Javan (1967), *Scientific American,* September 1967, 244.

At higher temperatures, some energies above the Fermi energy are occupied. The spillover (to states beyond E_F) depends on T, as shown in Figure 3.2.

Because a metal has a near continuum of excited energy levels (energies above the Fermi energy), a metal can absorb light of any wavelength, including visible wavelengths. If this were the only effect operable here, all metals would appear black because all visible light would be absorbed by them. However, when an electron in a metal absorbs a photon of light, it is promoted to an excited state that has many other energy levels available for deexcitation (due to the virtual continuum of levels). Light can be absorbed easily and since light is an electromagnetic wave, and a metal is a good conductor of electricity, the absorbed light induces alternating electrical currents on the metal surface. These currents rapidly emit light out of the metal. This rapid and efficient reradiation means that the surface of a metal is **reflective.**

Although all metals are shiny, and this is due to the effects described above, not all metals have the same color. For example, the colors of gold and silver are quite distinct from one another. This is due to subtle differences in the number of states above the Fermi level. The reflectances of gold and silver are compared in Figure 3.4. Silver reflects all colors of visible light with high efficiency; gold does not reflect very much high-energy visible light (blue and violet) because of the absence of levels in this energy region. Since gold reflects predominantly at the low-energy end of the visible range, it appears yellow.

COMMENT

POLISHING CHANGES COLOR

Almost every material has some degree of **luster,** as some incident light can be reflected back to the viewer. The degree of luster can depend on such matters as surface smoothness as well as electronic energy levels.

(continued)

POLISHING CHANGES COLOR (continued)

The change in the look of a piece of wood when it has been polished with wax or varnish is a familiar effect. Two processes are taking place here to increase the luster and color intensity of the polished wood—a reduction in diffuse reflection from the outer surface and introduction of multiple reflections under the polish.

The bare wood surface is not smooth, so much of the incident light is **diffusely scattered** (i.e., in all directions). However, the polished surface is much flatter, so it will reflect primarily at a particular angle (the same as the angle of incidence; this is called **specular reflection**) giving the wood its luster. The color also is more intense because the colored light from the rough wood surface will not be polluted with diffusely scattered light.

The polish layer also allows **multiple internal scattering** of the light (i.e., incident light reflected from wood and polished surfaces several times before escaping), as shown schematically in Figure 3.5. Multiple reflection narrows the wavelength range of the observed color, and this again intensifies the richness of a particular hue.

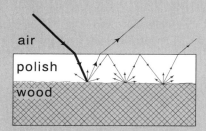

Figure 3.5. A schematic view showing some of the paths of a ray of light striking a polished wood surface. Specular reflection and multiple internal scattering are emphasized in this diagram.

It is possible for metals to transmit light, if the sample is sufficiently thin. The transmission will vary with the wavelength of the light, and this can give rise to color. A thin gold film (1000 Å or less) will transmit blue-violet light, again because blue-violet is not of the correct energy to be reflected.

On the other hand, if very small particles of gold, less than 10% of the wavelength of light across (typically 1 to 10 nm in size), are suspended in glass, the electrical currents required for high reflectivity in a bulk metal cannot develop and the resulting red color is due to the absorption of green light in a process called **Mie scattering.** This material is commonly called "ruby glass" due to its red color, although it is chemically distinct from the gem ruby.

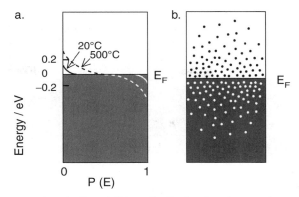

Figure 3.6. Other ways to view the Fermi–Dirac distribution for electrons in a metal: (a) occupational probability, $P(E)$, for energy levels in a typical metal at two different temperatures; (b) occupational probability schematically for $T > 0$ K, with ● representing excited electrons and ○ representing holes.

3.3 COLORS OF PURE SEMICONDUCTORS

When an electron in a metal is at an energy higher than the Fermi energy, it is said to be conducting as it contributes to the electrical and thermal conductivities of a metal (see Chapters 8 and 12). The probability diagram of Figure 3.2 can be turned on its side to reveal an energy diagram for a metal as shown in Figure 3.6.

Materials called **semiconductors** do not conduct electricity very well due to the gap between their **valence band** and their **conduction band,** as shown in Figure 3.7. The distance marked E_g is the **energy gap** (or **band gap**); there are no energy levels in this region. A similar gap appeared in the energy range between the bonding and antibonding orbitals (see Comment: Band Structure). The semiconductor energy gap changes from element to element, e.g., it increases on going through the series Sn, Ge, Si, C. This is the same order as increasing bond

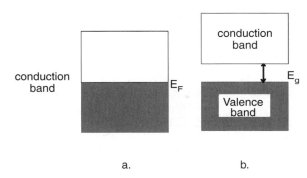

Figure 3.7. Energy bands in (a) a metal and (b) a semiconductor. E_g is the energy gap.

TABLE 3.1.
Periodic Trends in Diamond Structures in Group 14

Element	Lattice parameter /Å	Bond dissociation energy/kJ mol^{-1}	E_g/eV
C (diamond)	3.57	346	5.4
Si	5.43	222	1.1
Ge	5.66	188	0.66
α–Sn	6.49	146	0.1

interaction, as shorter, stronger bonds lead to a larger band gap energy. For a given element, application of intense pressure or dramatic lowering of temperature can shorten the interatomic distance and increase the band gap.

If there is sufficient thermal energy available, large numbers of electrons in a semiconductor could be promoted from the valence band to the conduction band, but this thermal energy might correspond to a temperature in excess of 1000 K, and the material may very well melt before achieving electrical conduction comparable to that in a metal.

However, visible light may provide sufficient energy to promote electrons from the valence band to the conduction band of a semiconductor, and this gives semiconductors their colors.

For a pure semiconductor, the color depends on the value of E_g. If E_g is less than the lowest energy of visible light (red light, $\lambda \sim 700$ nm, $E \sim 1.7$ eV), then any wavelength of visible light will be absorbed by this semiconductor. (Excess energy beyond E_g will be used to promote the electrons higher in the conduction band.) Because all visible light is absorbed, the semiconductor will appear either black or metallic. If reemission is rapid and efficient (this depends on the energy levels), then this material will have a metal-like luster; Si is an example. If reemission is not rapid then this semiconductor will appear black and lusterless (e.g., CdSe).

If E_g is greater than the highest energy of visible light (violet light, $\lambda \sim 400$ nm, $E \sim 3.0$ eV), then no visible light is absorbed and the material is colorless. Diamond, with E_g of 5.4 eV (corresponding to a wavelength of 230 nm), is an example.

When a semiconductor has E_g falling in the energy range of visible light, the material has a color that depends on the exact value of E_g. The value of E_g depends on two factors: the strength of the interaction that separates bonding and antibonding orbitals and the spread in energy of each band. Periodic trends in lattice parameters, bond dissociation energies, and band gap are summarized for one isostructural group of the periodic table in Table 3.1. The shorter, stronger bonds can be seen to give rise to larger values of E_g.

Let us now consider the colors of pure semiconductors. For example, HgS (called by the mineral name cinnabar, or the pigment name vermilion) has an E_g of 2.1 eV, which corresponds to $\lambda = 590$ nm. So in HgS all light with energies greater than 2.1 eV (wavelengths shorter than 590 nm) is absorbed. This means that in white light, only wavelengths greater than 590 nm are transmitted, and HgS is red.

As another example, consider CdS. Here the band gap energy is 2.4 eV, which corresponds to light of wavelength 520 nm (blue light). Since blue and higher-energy (i.e., violet) light are absorbed, this leaves behind a region of the visible spectrum that is centered on yellow, and CdS appears yellow-orange.

To generalize, the colors of pure semiconductors in order of increasing E_g, are black, red, orange, yellow, colorless. Some examples are given in Table 3.2.

TABLE 3.2.
Examples of Colors and Band Gaps in Pure Semiconductors

Material	Color	E_g/eV
C (diamond)	Colorless	5.4
ZnS	Colorless	3.6
ZnO	Colorless	3.2
CdS	Yellow-orange	2.4
HgS	Red	2.1
GaAs	Black	1.43
Si	Metallic grey	1.11

If the semiconductor has a large gap, it will be colorless if it is pure, but the presence of impurities can introduce color in the semiconductor, as we will see in the next section.

3.4 COLORS OF DOPED SEMICONDUCTORS

There are two categories of impurities in semiconductors, and these are defined here. These definitions are also important to discussions of electrical properties of semiconductors (see Chapter 12).

An impurity that donates electrons to the conduction band of a semiconductor creates what is known as an **n-type semiconductor.** The "n" stands for negative as this impurity gives rise to negative charge carriers (electrons in the conduction band). This is also known as a **donor impurity.**

Conversely, when an impurity produces electron vacancies that behave as positive charge centers (holes) in the valence band, this is known as a **p-type semiconductor.** The "p" stands for positive as the holes act as positive charge carriers. Because the impurity accepts an electron (giving rise to the hole), this also is known as an **acceptor impurity.**

The impurity in a semiconductor often can introduce a set of energy levels intermediate between the valence band and the conduction band, as shown in Figure 3.8.

Because the transition to the impurity level(s) takes less energy than the transition across the entire band gap, this transition can affect the color of a wide-band-gap semiconductor.

a.   b.   c.   d.   e.

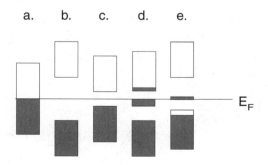

Figure 3.8. Energy bands in (a) a metal, (b) an insulator, (c) a pure semiconductor, (d) an n-type semiconductor, and (e) a p-type semiconductor. E_F is the Fermi energy.

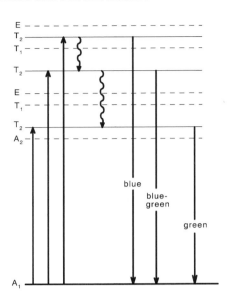

Figure 3.9. Schematic view of phosphorescence due to emission from excited states in the blue phosphor ZnS doped with Ag⁺. Adapted from Y. Uehara (1975). *Journal of Chemical Physics,* **62,** 2983.

For example, if a small amount of nitrogen is placed in the diamond lattice, the extra electrons can form a donor impurity level within the wide band gap of diamond, since nitrogen has one more electron than carbon. Excitation from this impurity band absorbs violet light, depleting white light of violet and making nitrogen-doped diamonds appear yellow at nitrogen levels as low as 1 atom in 10^5.

Conversely, boron has one electron less than carbon, and doping boron into diamond gives rise to acceptor (hole) levels in the band gap. These intermediate levels can be reached by absorption of red light (but not shorter wavelength light—do you know why?) so boron-doped diamonds appear blue. The Hope diamond is an example.

Doped semiconductors also can act as **phosphors,** which are materials that give off light with high efficiency when stimulated, for example, by an electrical pulse. (**Phosphorescence** results when light is emitted some time after the initial excitation; the delay is caused by the change in spin between the excited state and the ground state. In fluorescence, which is a faster process, the spins of the excited and ground states are the same.) On the surface of the picture tube of a color television there are three phosphors: one emitting red light, one emitting green light, and one emitting blue light; emission is stimulated by excitation with an electron beam. The energy level diagram for the blue phosphor ZnS doped with Ag⁺ is shown in Figure 3.9.

COMMENT

TRIBOLUMINESCENCE

When some materials are stimulated mechanically, they emit light; this phenomenon is known as **triboluminescence.** One of the best known examples is Wint-O-Green Lifesavers®, which can give rise to flashes of blue-green light when crushed. In this case, the sugar leads to triboluminescence, due to the production of large electric fields during the formation of a newly charged crystal surface,

(continued)

COMMENT

TRIBOLUMINESCENCE (continued)

the acceleration of electrons in this field, and the subsequent production of **cathodoluminescence** (fluorescence induced by cathode rays that energize a phosphor to produce luminescence) from the nitrogen atoms adsorbed on the surface. The blue luminescence is associated with N_2^+. Some of the light emitted during this process is in the UV range; the UV stimulates the wintergreen (methyl salicylate) molecules, which then fluoresce in the visible range. Therefore, the presence of the wintergreen flavoring enhances the observed triboluminescence.

When an n-type semiconductor is placed next to a p-type semiconductor, interesting features can result. Here we concentrate on color, but in Chapter 12 we will look at the importance of **p,n-junctions** to electronic properties of semiconductors. When an electrical potential is applied to a p,n-junction such that electrons are supplied through an external circuit to the n-side of the junction, the extra electrons in the conduction band of the n-type semiconductor fall into the holes in the valence band of the p-type semiconductor, as shown in Figure 3.10. As the electrons go to the lower energy state, they can emit energy in the form of light. This is the basis of the electronic device known as the **light-emitting diode** (LED), and it is $GaAs_{0.6}P_{0.4}$ that emits the red light that is commonly associated with these devices in applications such as digital displays. An LED shows **electroluminescence,** i.e., the production of light from electricity.

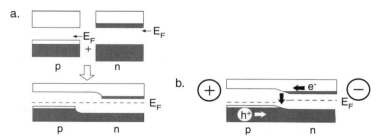

Figure 3.10. The electronic band structure of a p,n-type junction forms a light-emitting diode (LED). (a) When a p-semiconductor is placed beside an n-semiconductor, the Fermi levels become equalized. (For simplicity, the donor and acceptor levels are assumed to be negligible here.) (b) When a field is applied to a p,n-junction, as shown, electrons in the n-conduction band migrate to the positive potential on the p-side, and return to the valence band, recombining with the holes in the p-semiconductor. Similarly, the holes in the p-semiconductor migrate toward the negative n-side and recombine with electrons. The electrons dropping from the conduction band to the valence band emit light corresponding to the energy of the band gap. For example, the emission from $GaAs_{0.6}P_{0.4}$ is red.

NEW DIRECTIONS IN LIGHT-EMITTING DIODES

Researchers have recently devised a light-emitting diode that emits "white" light, "nearly as bright as a fluorescent lamp." The device uses layers of polymers, with blue, green, or red light, depending on the layer. The combination of blue, green, and red makes the emission appear white.

Some new LEDs are not limited to one color of emitted light. An LED composed of layers of cadmium selenide nanocrystals between layers of poly(p-phenylenevinylene) can have different colors, depending on the voltage and the preparation. At low voltage, CdSe is the light emitter, and the color can be changed from red to yellow by varying the size of the nanocrystals. (The latter is a consequence of quantum effects.) At higher voltages, the polymer is the emitter, and the emitted light is green. Materials such as these might one day be used for display devices in which the color of the **pixel** (dot of light on a screen) is determined by the applied voltage.

Semiconductors also are important in the photoelectric process that we know as photocopying.

The Photocopying Process: A Tutorial

The photocopying process, or xerography (which means "dry copying" literally from the Greek *"xeros graphy"*), relies on a few basic physical principles, and a fair bit of ingenuity. The process was developed by Chester Carlson in the mid-1930s.

One of the two main parts of a photocopy machine is a corona discharge wire, which has so strong an electric field, >30,000 V cm^{-1}, that it ionizes air to make O_2^+, and makes the air glow green due to photons emitted by excited electrons. The other main part is amorphous selenium, a semiconductor with a band gap of 1.8 eV.

a. What color is a bulk sample of amorphous selenium?

b. After you run a photocopying machine for quite a while, why does the surrounding air smell like ozone, O_3?

c. After the corona wire passes by the drum that is coated with amorphous selenium, the Se is charged positive. Why?

d. A very strong light then shines on the document to be copied, and this light is then reflected onto the Se drum. Does the light uniformly strike the drum, or is there an "image" from the document on the drum? Explain.

e. Where the light "image" hits the Se drum, electron/hole pairs are formed. Why? (This results in this area becoming electrically neutral since the drum is grounded and the holes move to the ground and the electrons move to the surface where they locally neutralize the positive charge.)

f. Where the dark "image" hits the drum, the surface stays positively charged. Why is it not neutral everywhere?

g. Devise a way to convert the **latent electrical image** (i.e., an electrical image that is present but not yet visible) on the Se drum to an image on paper.

h. Another form of Se has a band gap of 2.6 eV. Will this be useful as a direct replacement for amorphous selenium? What other factors could be important?

i. What areas of the xerographic process would you suggest for improvement or advancement?

The Photographic Process: A Tutorial

The basis of modern "wet" photography is the reactivity of silver halides, especially silver bromide and silver chloride, in the presence of light.

A photographic film contains silver halide particles (typically 50 to 2000 nm in size) embedded in an emulsion. When the silver halide particles are illuminated by a light of sufficient energy, conduction band electrons and valence band holes are created efficiently. The electrons move through the silver halide particle until, eventually, one becomes localized at a defect site. This trapped electron can combine with an interstitial silver ion to form a neutral silver atom. Because a single silver atom is unstable in the lattice, it is energetically favorable for additional electrons and interstitial Ag^+ ions to arrive at the trapped site, and their combination causes more silver atoms to form from Ag^+. Eventually a stable silver cluster of three to five silver atoms forms; this takes somewhere between 6 and 30 absorbed photons. At this site there is said to be a stable latent image in the film.

a. When a photographic film is exposed to a white object, is the corresponding area of the film more or less subject to Ag cluster formation than the area corresponding to exposure to a black object?

b. The band gap in silver halides is about 2.7 eV. What does this correspond to in terms of wavelength? Does this explain why silver halides are sensitive to visible light?

c. After a film is exposed to a subject, why can it not be handled in daylight? Why can it be handled in red light?

d. Once the latent image has been formed, the exposed film is placed in a developer, which contains a reducing agent such as hydroquinone. The regions of the latent image that have Ag clusters undergo further reduction of Ag^+ in their immediate region, "amplifying" the number of Ag clusters in highly exposed regions of the film by as much as a factor of 10^9. The unexposed grains of silver halide, unchanged by photographic development, are then removed by dissolution. Explain why the image formed on the film is a "negative" (dark subject appears light and vice versa).

e. How could a "positive" print be produced from the negative film? Consider the black and white process only, at this stage.

f. Chemical and spectral sensitizing agents, often containing sulfur and/or gold and/or organic dye molecules, are present with the silver halide in the film. These create sensitive sites on certain surfaces of silver halide microcrystals, thereby sensitizing the film to sub-band-gap light. Films of different "speeds" (i.e., meeting different light level requirements) exist; speculate on how films of different speeds can be made.

g. Formation of a color image is more complex than for a black-and-white image, although it still relies on silver halides. "Simple" color film contains three layers of emulsion, each containing silver halide and a "coupler" that is sensitive to particular colors. In the order in which light hits the film, the first layer is sensitive to blue and next there is a filter that stops blue and violet. The second emulsion is sensitive to green; the third emulsion is sensitive to red. (In a good quality color film there may be many more emulsions to get "truer" colors, but the principle is the same.) What color is the filter that stops the blue?

h. In the development of color film, the oxidized developer reacts with each coupler to produce a precursor to an organic dye molecule. The colors of the dye molecules produced in the development process are the complements of the light that exposed the emulsions [i.e. yellow color forms in the first layer since yellow is the complement to blue; magenta (red-violet), the complement to green, forms in the second layer; cyan (bluish-green), the complement to red, forms in the third layer]. Why do companies such as Kodak employ synthetic organic chemists?

i. Why do companies such as Kodak employ solid-state chemists and physicists who are interested in measurement and calculation of properties such as the effect of impurities and lattice defects on the mobility of ions and holes in silver halides, and topics such as the surface energies of different faces of silver halide crystals?

j. How do you expect the printing process for color film to work?

k. How does the Polaroid® (instant processing) process work?

l. The Polaroid® process depends rather critically on temperature. Why?

FURTHER READING

General References

Many introductory physical chemistry textbooks contain information concerning metals and semiconductors.

R. Cotterill (1985). *The Cambridge Guide to the Material World.* Cambridge University Press.
R. Dagani (1998). Light-Emitting Polymer Synthesis. *Chemical & Engineering News,* January 19, 1998, 9.
R.M. Evans (1948). *An Introduction to Color.* John Wiley & Sons Ltd.
A. Javan (1967). The Optical Properties of Materials. *Scientific American,* September 1967, 239.
T.A. Kavassalis (1995). The Application of Quantum Mechanics and Molecular Dynamics to the Design of New Materials in an Industrial Laboratory. *Physics in Canada,* March/April 1995, 92.

C. Kittel and H. Kroemer (1980). *Thermal Physics.* Freeman.

G.C. Lisensky, R. Penn, M.J. Geslbracht, and A.B. Ellis (1992). Periodic Trends in a Family of Common Semiconductors. *Journal of Chemical Education,* **69,** 151.

G.J. Meyer (1997). Efficient Light-to-Electrical Energy Conversion: Nanocrystalline TiO_2 Films Modified with Inorganic Sensitizers. *Journal of Chemical Education,* **74,** 652.

C.G. Mueller, M. Rudolph, and the Editors of LIFE (1966). *Light and Vision.* Life Science Library, Time Inc.

K. Nassau (1980). The Causes of Color. *Scientific American,* October 1980, 124.

K. Nassau (1983). *The Physics and Chemistry of Color.* John Wiley & Sons Ltd.

V.F. Weisskopf (1968). How Light Interacts with Matter. *Scientific American,* September 1968, 60.

E.A. Wood (1977). *Crystals and Light.* Dover Publications.

G. Wyszecki and W.S. Stiles (1967). *Color Science.* John Wiley & Sons Ltd.

Lasers

Physics Today, October 1988, Special Issue: Lasers.

M.G.D. Baumann, J.C. Wright, A.B. Ellis, T. Kuech, and G.C. Lisensky (1992). Diode Lasers. *Journal of Chemical Education,* **69,** 89.

LEDs

R.T. Collins, P.M. Fauchet, and M.A. Tischler (1997). Porous Silicon: From Luminescence to LEDs. *Physics Today,* January 1997, 24.

Light-Sensitive Glass

D.M. Trotter, Jr. (1991). Photochromic and Photosensitive Glass. *Scientific American,* April 1991, 124.

Photocopying Process

A.D. Moore (1972). Electrostatics. *Scientific American,* March 1972, 47.

J. Mort (1994). Xerography: A Study in Innovation and Economic Competitiveness. *Physics Today,* April 1994, 32.

D. Owen (1986). Copies in Seconds. *The Atlantic Monthly,* **257**(2), 64.

Photographic Materials and Processes

Silver Halides in Photography, *Materials Research Bulletin,* Volume XIV, Number 5, May 1989 (five research papers concerning this subject).

A.B. Bocarsly, C.C. Chang, and Y. Wu (1997). Inorganic Photolithography: Interfacial Multicomponent Pattern Generation. *Journal of Chemical Education,* **74,** 663.

C.A. Fleischer, C.L. Bauer, D.J. Massa, and J.F. Taylor (1996). Film as a Composite Material. *MRS Bulletin,* **14,** July 1996.

W.C. Guida and D.J. Raber (1975). The Chemistry of Color Photography. *Journal of Chemical Education,* **52,** 622.

J.F. Hamilton (1988). The Silver Halide Photographic Process. *Advances in Physics,* **37,** 359.

W.L. Jolly (1991). Solarization: The Photographic Sabatier Effect. *Journal of Chemical Education,* **68,** 3.

M.S. Simon (1994). New Developments in Instant Photography. *Journal of Chemical Education,* **71,** 132.

Triboluminescence

R. Angelos, J.I. Zink, and G.E. Hardy (1979). Triboluminescence Spectroscopy of Common Candies. *Journal of Chemical Education,* **56,** 413.
L.M. Sweeting, A.L. Rheingold, J.M. Gingerich, A.W. Rutter, R.A. Spence, C.D. Cox, and T.J. Kim (1997). Crystal Structure and Triboluminescence. *Chemistry of Materials,* **9,** 1103.

■ ──

PROBLEMS

1. A photographer's light meter is used to determine the intensity of light in a given area, to guide the exposure of the film. The light meter works on a **photoelectric principle:** light causes electrical response.
   a. What sort of material is likely used to produce the photoelectric response?
   b. Explain how the photoelectric effect might come about.

2. The color of the heating coil in a toaster dies off slowly when the toaster is unplugged. However, the color of the light-emitting diode in a clock dies off quickly when the clock is unplugged. Comment on the source(s) of the difference.

3. Two very different materials both absorb light of wavelength 580 nm, which corresponds to yellow light.
   a. One material is a pure semiconductor. What color is it?
   b. The absorption in the other material (which is an insulator) is due to a transition metal impurity. What color is this material?

4. Most camera film works on the silver halide process. Most silver halides are semiconductors with band gaps of around 2.7 eV ($\lambda = 460$ nm). There are films available that can take "infrared" pictures. Briefly explain how the silver halide process would have to be modified to give an infrared film.

5. Photographic films come in different speeds. High-speed film requires less light to obtain an image, whereas low-speed film requires more light. Some people worry about passing their camera through the X-ray process at security counters in airports, in case the X-rays damage the film. Is high-speed or low-speed film more susceptible to X-ray damage? Explain.

6. Our blood is red due to the presence of hemoglobin, a large molecular species with iron at the center. The blood of invertebrates in the phyla *Arthropoda* and *Mollusca* (e.g., crab, crayfish, lobster, snail, slug, octopus) contains hemocyanin as its oxygen carrier. Hemocyanin contains copper, which is bound directly to a protein. The oxygenated form, oxyhemocyanin, contains bound oxygen, and this species absorbs light. There are three regions of maximum absorbance, one at a wavelength of 280 nm (due to the protein), and two others due to the copper, one at 346 nm and one at 580 nm.
   a. Suggest the physical source of the color of blood. Give your reasoning.
   b. What color is the blood of species with hemocyanin? Give your reasoning.

7. Explain how the optical properties of a semiconducting chromophore can be influenced by different packing arrangements of the molecules in the solid state.

8. Lasers produce light by stimulated emission. Some lasers use semiconductors, with the light emission corresponding to the energy of the band gap. This makes the light for a given system fixed at one energy.
   a. Can you suggest how the addition of impurities to a semiconductor laser material can be used to change the wavelength of the emitted light?
   b. Can the wavelength be both shortened and lengthened using this approach? Explain.
   c. Stimulated emission leading to lasing requires a high density of lasing centers. How will this affect the intensity of lasing from a doped semiconductor?

9. The compound yttrium vanadate with 5% of the Y^{3+} ions replaced by Eu^{3+} gives the most efficient red phosphor known. This is the red phosphor that is commonly used as a red emitter in a color television screen (the other emitters are blue and green); its emission is perceptively pure red radiation at 612 nm from its $^5D_0$ electronic state on excitation. In a television screen, what excites the phosphor causing it to emit light?

10. Photochromic glass changes its transparency depending on the light level. The principle involved is similar to the effect of light on silver halides in photography. This glass contains copper-doped silver halide crystallites. When ultraviolet light hits the glass, copper gives up an electron, which is captured by the silver ion, leading to aggregates of silver atoms that darken the glass. This glass becomes transparent again when the UV source is removed (e.g., on moving indoors). Propose a mechanism that would allow for this reversion to the original state.

11. Two walls are painted with orange paint; one is "plain" orange and the other is fluorescent orange. Use energy level diagrams to show the difference in the two paints.

12. A science museum is building a new exhibit concerning the photocopying machine. A museum employee suggested that part of the exhibit be a working photocopier with the front panel made of glass so that visitors can see the interior workings of the copier when in use. Comment on this suggestion.

13. How would the color of a light-emitting diode be expected to change on cooling from room temperature (where it emits red light) to the temperature of the boiling point of liquid nitrogen ($T = 77$ K)? Consider that the lattice will have contracted, and this will affect E_g.

14. J.D. Bernal wrote, "All that gisters ["glistens"] may not be gold, but at least it contains free electrons." Give an example that refutes this statement.

15. From the data in Table 3.1, what color is Ge?

16. The pigment "zinc white" (ZnO) is white at room temperature (i.e., colorless with a white appearance due to scattering from the powder) and bright yellow when heated. Explain the color change.

17. The blue color of a diamond can be created by addition of boron as a dopant, or by irradiation and the creation of defects. How could measurement of the electrical conductivity distinguish between these sources of color in a blue diamond?

C H A P T E R

4

COLOR FROM INTERACTIONS
OF LIGHT WAVES
WITH BULK MATTER

4.1 INTRODUCTION

In Chapter 2 we saw that color can originate from single atoms or small numbers of atoms. Chapter 3 showed that very large numbers of atoms of metallic or semiconductor materials can give rise to energy bands that in turn can provide color when visible light is absorbed.

Light can interact in other ways with bulk matter (i.e., matter with of the order of the Avogadro[1] number of atoms), and this can also lead to color. In this chapter we explore some of these sources of color.

4.2 REFRACTION

Thus far the processes giving rise to color have involved absorption or emission of light. However, there are other physical interactions that give rise to color by change in the direction of the light, and one of these is considered here.

Refraction results from the change in speed of light when passing from one medium to another. When a uniform, nonabsorbing medium, such as air or glass or salt, transmits light, the photons are absorbed and immediately reemitted in turn by all the atoms in the path of the

---

[1] Amedeo Avogadro (1776–1856) was an Italian lawyer and scientist. In 1811, in a paper that should have settled the dispute concerning the relationship between atomic weights and properties of gases, Avogadro proposed that equal volumes of gases under the same conditions should contain equal numbers of molecules. This work was ignored for about 50 years, but its importance is recognized today in naming Avogadro's number.

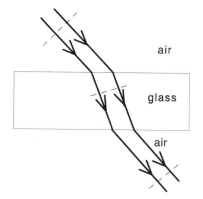

Figure 4.1. Monochromatic light is slowed as it passes from air to glass. When the light hits the glass at a glancing angle, one side of the light beam is slowed first, and this causes the light to be bent toward the perpendicular on entering the glass and away from the perpendicular on leaving the glass.

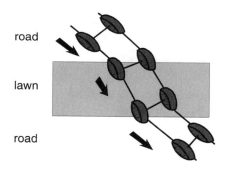

Figure 4.2. An analogy showing how light is bent as it changes speed on passing from one medium to another. When a pair of wheels hits a patch of soft lawn at a glancing angle, one side is slowed first, and this causes bending toward the perpendicular; it bends away from the perpendicular on leaving the lawn (slower medium).

ray, slowing down the light. Light travels fastest in a vacuum and more slowly in all other materials. Light also travels faster in air than in a solid.

Consider the case of **monochromatic** (i.e., single wavelength) light entering a piece of glass, as shown in Figure 4.1. One edge of the beam enters the prism and is slowed before the other edge hits the glass. This has the effect of bending the light toward the direction perpendicular to the edge of the glass. When exiting the prism, the side of the beam that exits first "pulls" the light in the direction that takes it away from the perpendicular direction. This bending of light as its velocity changes with a change in medium can be considered to be a consequence of **Fermat's[2] Principle of Least Time,** which states that light rays will always take the path that requires the least time. Bending of light on changing media is shown by analogy in Figure 4.2.

The **index of refraction** (also known as the **refractive index**), n, is defined as

$$n = \frac{v}{v'} \tag{4.1}$$

---

[2] Pierre de Fermat (1601–1665) was a French lawyer who studied mathematics as a hobby. He made several important contributions in the areas of number theory, analytic geometry (including the equation for a straight line), and calculus.

TABLE 4.1.
Refractive Indices for Some Common Materials[a]

Material	n
Air	1.00029
Ice	1.305
Water	1.33
Acetone	1.357
n-Heptane	1.385
Tetrahydrofuran	1.404
Fused quartz	1.456
Benzene	1.498
Aniline	1.584
Light flint glass (average)	1.59
Methylene iodide	1.738
Dense flint glass (average)	1.75
Diamond	2.42

[a] All materials are at 25°C except ice, which is at 0°C. All values were determined using the sodium-D line (yellow light).

where v is the speed of light in vacuum and v' is the speed of light in the medium under consideration; n depends on both the wavelength of the light used and the temperature. Values for a range of materials (for the Na yellow line) are given in Table 4.1.

Although we are perhaps not aware of it, many optical phenomena are associated with refractive index (see Figure 4.3). For example, if we drop a piece of glass in water, we see the edges of the glass due to bending of light because the refractive index of glass is different from that of water. If alcohol and water are mixed, their mixing can be traced by the refraction associated with the small differences in their refractive indices. Similarly, the change in refractive index of air as it is heated is apparent in viewing an air mass above hot asphalt.

Figure 4.3. A pencil entering a glass of water appears to be broken. This is because the light is refracted as it enters and as it leaves the water.

COMMENT

MIRAGES

The refractive index for air depends on its density, which in turn depends on temperature and pressure, such that lower density lowers the refractive index. Under conditions of density gradients, light bends toward the denser air. For example, in a strong vertical temperature gradient, such as warm air above air that is cold because it is in contact with cold water in the ocean, the refractive index can change so rapidly that it can induce curvature in the light rays exceeding the curvature of the earth. This makes objects at or below the horizon appear in the field of vision as a **mirage,** as shown in Figure 4.4.

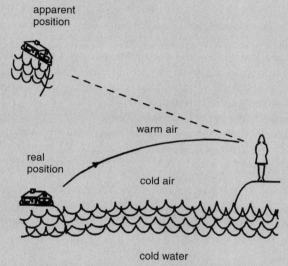

Figure 4.4. A mirage results when light bends toward the denser air in a strong vertical temperature gradient. Because light normally travels in a straight line, we perceive a mirage.

Refractive index measurement can be used in identifying materials, as in forensic work. A very small chip of glass from a crime scene can be identified and traced by its refractive index. The refractive index is determined by placing this chip of glass in a liquid of a particular refractive index. Another liquid of different refractive index is added to the first until the blended refractive index matches the glass chip—the chip is "invisible" at this point. When the refractive index of this solution has been determined with a **refractometer,** knowledge of the numerical value of the refractive index of the glass can be used to trace its origin, often to a specific make and model of car, for example.

Another rather peculiar example of a refractive index oddity results when using liquid helium. Most liquids have refractive indices very different from the gas above them, and this is why we can see their surfaces distinctly. However, in the case of helium (which boils at $T = 4.2$ K), the liquid and the gas have very similar refractive indices. Furthermore, the gas above liquid helium is just pure helium, as any other gas would condense to its solid state. When using liquid helium, which must be transferred carefully from one closed container to another

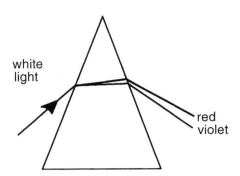

Figure 4.5. Refraction causes white light to be dispersed by a prism, since different colors (wavelengths) of light have different refractive indices. Shorter-wavelength light is refracted more than longer-wavelength light.

in an insulated siphon, it can sometimes be difficult to see how much liquid has been transferred, because the surface is almost "invisible."

Figure 4.1 showed refraction for monochromatic light, but refraction can lead to beautiful effects when dealing with white light. For example, it is responsible for the beautiful colors in a rainbow. In Cambridge, England in 1666, at the age of 23, Newton[3] described an experiment with a prism "to try therewith the phenomena of the colors." He wrote,[4] "In a very dark Chamber, at a round hole. . . I placed a Glass Prism, whereby the Beam of the Sun's Light, which came in at that hole, might be refracted upwards toward the opposite wall of the Chamber, and there form a colour'd Image of the Sun." The source of the spectrum of colors we now know is the refraction property of glass, as first described correctly by Newton, using the term **refrangibility** (known today as refraction).

When white light, which we now know to be composed of light of all visible colors, hits a prism, some colors are refracted more than others. In particular, higher-energy (shorter-wavelength) light is refracted more. This variable refraction causes white light to be **dispersed** by a prism; dispersion is illustrated in Figure 4.5.

The dispersion of white light by water droplets or ice crystals is the source of the colors of the rainbow. Similarly, the flashes of color characteristic of faceted gemstones (see Comment: Sparkling Like Diamonds) and high-quality "crystal" glassware are caused by dispersion effects associated with the high refractive index of the material.

COMMENT

SPARKLING LIKE DIAMONDS

When light leaves a material of refractive index n and goes to air with an incident light angle i, the refracted light angle, r, with both i and r measured with respect to the normal to the surface (see Figure 4.6a), can be determined from the general relation, known as Snell's[5] law:

(continued)

---

[3] Sir Isaac Newton (1642–1727) was an English natural philosopher and mathematician. Because of the plague, Newton spent 1665–1666 in the English countryside where he laid the foundations for his great contributions in optics, dynamics, and mathematics.

[4] I. Newton, *Optics*.

[5] Willebrod Snell van Royen (1581–1626) was a Dutch mathematician and scientist.

SPARKLING LIKE DIAMONDS (continued)

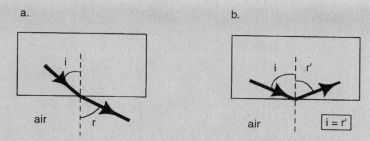

Figure 4.6. (a) When light leaves a medium to go to air, the angle of incidence, i, is related to the angle of refraction, r, by Snell's law (Equation 4.2). (b) If the angle of incidence is greater than the critical angle, there is no refraction but only internal reflection. Note that the angle of incidence equals the angle of reflection, $i = r'$.

$$\frac{\sin r}{\sin i} = n \tag{4.2}$$

where the refractive index of air has been assumed to be 1. From Equation 4.2, the maximum value of i is $\sin^{-1}(n^{-1})$, and this value of i is called the **critical angle.** For i greater than the critical angle, there is no refraction and total internal reflection results (see Figure 4.6b).

For diamond, $n = 2.42$ and this leads to a critical angle of 24.4°. Therefore a light ray that makes an angle of 65.6° (i.e., 90°–24.4°) or less with the outer surface of a diamond cannot pass through the surface to air, but is totally reflected, as shown for monochromatic light in a diamond in Figure 4.7a. A properly cut diamond will appear opaque when viewed from below, because no light from the upper surface will exit the bottom (Figure 4.7a). When white light hits the diamond's surface, it is dispersed into its colors. This is illustrated in Figure 4.7b.

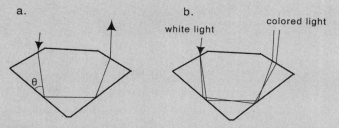

Figure 4.7. (a) Light within a diamond that makes an angle θ of 65.6° or less with the surface will be totally reflected. A properly cut diamond can appear opaque when viewed from the bottom due to total internal reflection of the impingent light, as shown. (b) White light enters a cut diamond and is dispersed. With an appropriate cut, colored light exits from the top face and no light exits from the bottom face.

This dispersion is not always a desired feature, however. One example of a situation where it would be preferable to avoid such dispersion is the **chromatic aberration** (i.e., unwanted color patterns) sometimes found in optical systems such as telescope lenses, camera lenses, and glass-mounted slides.

COMMENT

AN EXAMPLE OF REFRACTION: THE FRESNEL LENS

A **Fresnel[6] lens** is a flat, thin piece of plastic (typically a copolymer of polyvinyl acetate and polyvinyl butyrate) that can act as a lens (Figure 4.8). It does so using the principle of refraction. Into the lens are moulded a series of small concentric stepped prisms, each step only a few thousandths of a centimeter wide. The steps extend concentrically from the center of the lens to its outer edges and can easily be felt with a finger. Each prism causes refraction, and this bends the light. Taken together, they form a lens. A Fresnel lens is relatively inexpensive to produce and much thinner than conventional lenses. They have many applications, including use in overhead projectors.

Figure 4.8. Fresnel lenses. These lenses derive their optical properties from small concentric stepped prisms that refract light. The left sheet has 25 Fresnel lenses, and the right sheet is one large Fresnel lens.

4.3 INTERFERENCE

Newton also was the first to document an understanding of the sources of color due to interference.

---

[6] Augustin Jean Fresnel (1788–1827) was a French physicist. He invented his lens design to allow efficient and effective light signaling between lighthouses.

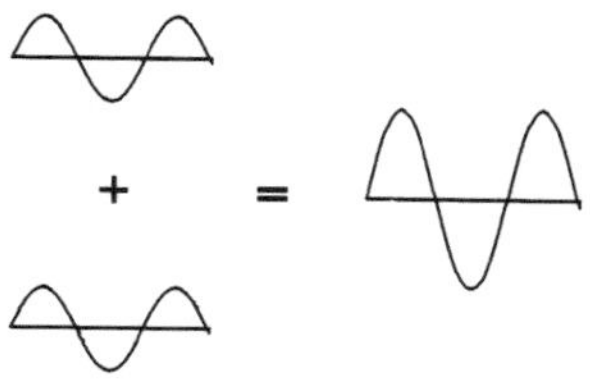

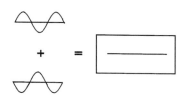

Figure 4.9. Constructive interference of two identical waves exactly in phase gives a resultant wave of the same wavelength and twice the amplitude.

Figure 4.10. Destructive interference: Two identical waves exactly out of phase give no resultant wave.

To understand interference patterns in light, it is instructive to first consider two extreme ways in which waves can interact. These interactions apply equally to light waves or any other waves such as water waves.

When two waves of exactly the same wavelength come together, the resulting wave depends very much on the **phases** of the waves. For example, if two identical waves exactly in phase (peaks and troughs at the same location, as shown in Figure 4.9) interact, the resulting wave will be of the same wavelength and its amplitude will be the sum of the amplitudes of the incident waves, as shown in Figure 4.9. These waves are said to be undergo **constructive interference.**

Although there is only one way in which waves can be in phase, there are many ways in which waves are "not in phase." The most extreme case is two identical waves that are exactly out of phase with one another: when one has a peak, the other has a trough. Two such waves of exactly the same wavelength and exactly the same amplitude will come together to result in total destruction: no wave results (see Figure 4.10). This is called **destructive interference.**

COMMENT

MOIRÉ PATTERNS

When two sets of equally spaced lines are overlaid with a slight displacement angle, as shown in Figure 4.11, an interference pattern, called a **moiré pattern,** from the French word "moiré" meaning water-spotted, is produced. The presence of moiré patterns has been used for more than a century to detect aberrations due to irregularities in optical components.

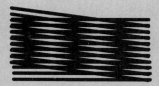

Figure 4.11. A moiré pattern.

COMMENT

SOUND INTERFERENCE

Interference effects apply to any kind of wave, including both light and sound. Destructive interference effects are used in some earphones that remove background noise: they work by generating sound waves that exactly destructively interfere with the sound waves of the background.

Interference effects are especially prominent when light interacts with thin films. In these cases, light may be reflected from the front and the back face of the film. Whether the front-reflected and back-reflected light cancels (destructive interference) or reinforces (constructive interference) depends on the nature and thickness of the film, the color of the light, as well as the viewing angle.

If the extra distance traveled by monochromatic light causes the front- and back-reflected light to be exactly out of phase, no light of this color will be seen. If the front- and back-reflected light are exactly in phase, constructive interference results and this color will be observed.

Therefore, in the case of interference due to front- and back-reflected white light from thin films, the color and intensity of light observed will depend on the viewing angle. We are familiar with this in such everyday experiences as the colors in a bubble, a butterfly wing, a compact disc, or oil on water. Even thin slices of inorganic materials can derive their colors from interference of front- and back-reflected light. For example, the colors of silicon oxide wafers range from blue-green at a thickness of 0.92×10^{-6} m to maroon-red at 0.85×10^{-6} m, back to blue-green again at 0.72×10^{-6} m. The colors repeat as the thickness changes, as additional layers add integer multiples of the wavelength to the extra path, allowing the same color to interfere constructively at different film thicknesses.

Interference colors in soap films bear further examination. Soap works as a cleansing agent because of its chemical structure: one end is polar and the other is nonpolar. This is shown schematically in Figure 4.12 for a "typical" soap.

The hydrocarbon "tail" of the soap molecule is hydrophobic and acts as a solvent for oils, and the polar end is attracted to water (hydrophilic). This property of being both oil loving and water-loving makes soap **amphiphilic** and allows soap to attract both water and oils, thereby allowing cleaning action.

A soap film is composed of both soap and water molecules. The soap molecules prefer to have their hydrophobic ends away from the water pointing out into the air, which can, at an appropriate concentration, give a bilayer structure, as shown schematically in Figure 4.13.

The main difference between a thick and a thin soap film is the number of water molecules between the soap layers.

The soap film can be modeled schematically as a front and back surface, each reflecting light, as shown in Figure 4.14. Views of constructive and destructive interference from front and back reflection are also shown.

 COO·Na⁺

Figure 4.12. Schematic structure of a "typical" soap molecule.

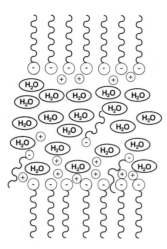

Figure 4.13. Schematic bilayer structure of a typical soap film. The alkyl chains of the soap molecules point into the air, and the polar head groups of the soap molecules point into the water. Much more water would be present in a real soap film.

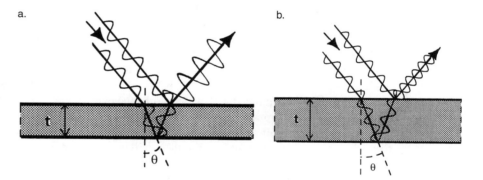

Figure 4.14. Schematic view of front- and back-surface reflection of light from a thin film, such as a soap film: (a) constructive interference; (b) destructive interference. Note that back reflection leads to a change in phase but front reflection does not.

For light of wavelength λ and incident intensity I_i, hitting the thin layer (soap film) of refractive index n, consideration of wave scattering gives the intensity of the reflected light, I_r, as[7]

$$I_r = 4I_iR \cos^2\left(\frac{2\pi \, nt}{\lambda} \cos \theta + \frac{\pi}{2}\right) \tag{4.3}$$

where t is the thickness of the soap film, θ is the angle defined in Figure 4.14, and R is the fraction of light reflected. Equation 4.3 shows that the intensity of reflected light observed depends

---

[7] See C. Isenberg (1992). *The Science of Soap and Soap Films.* Dover Publications, Inc., p. 36 for a derivation.

Figure 4.15. Cross section of a vertical soap film, showing a nonuniform thickness and constructive (solid line) interference and destructive (dashed line) interference from different regions of the film.

on many factors, including the thickness of the soap film, the wavelength of the light, and the viewing angle. This is immediately familiar to us as the beautiful range of colors seen in a soap bubble.

Equation 4.3 also shows that the intensity of reflected light will depend on the thickness of the film. Due to gravity, a vertical soap film will not be of uniform thickness (see Figure 4.15). In this case t varies with height. In monochromatic light, since the intensity of light will depend on t, there will be regions in which the thickness is such that maximum constructive interference will be observed (the extra path length traveled by the back-reflected light combined with its phase change keeps it exactly in phase with the front-reflected light) and the reflected light will be very intense. There will be other regions in the soap film in which maximum destructive interference will be observed (the extra path length traveled by the back-reflected light combined with its phase change puts it exactly out of phase with the front-reflected light) and there will be no net reflected light. There will also be intermediates between these two extremes, again depending on the thickness of the film. Therefore, for a wedge-shaped film, observed in monochromatic light, there will be an interference pattern of fringes of light and dark areas, depending on the thickness of the film.

For white light, colored fringes are observed, since the emerging beam will be the sum of all the contributions from each wavelength, and each wavelength has I_r given by Equation 4.3. This is the familiar pattern of color that we know from soap films and from other interference effects, such as the interference colors of oil on water.

When a soap film is allowed to sit in very stable conditions and drain over a period of time, much of the water between the bilayers drains away. If, over time, the thickness of the film becomes such that $t \ll \lambda$, Equation 4.3 leads to an interesting result. Since, for $t \ll \lambda$,

$$\cos^2\left(\frac{2\pi\,nt}{\lambda}\cos\theta + \frac{\pi}{2}\right) = \cos^2(0 + \pi/2) = \cos^2(\pi/2) = 0 \tag{4.4}$$

this implies that for very thin films ($t \ll \lambda$) there will be no observed reflected intensity. If this is the case, all light that interacts with the film is transmitted, and the film appears black in reflected light. This phenomenon is in fact observed, although it really does require very still surroundings to achieve such thinning of the film without it breaking! Black films have typical thicknesses in the range 50 to 300 Å (5 to 30 nm).

It should be apparent from Equation 4.3 that an analysis of color fringes in soap film can be used to determine the thickness of the film.

Although soap films were used as the basis for the discussion above, the approach applies equally to a discussion of any thin transparent film. The colors of a bevelled edge of a thin section of a mineral, gasoline on a puddle, and a soap bubble have common physical origins in interference effects. Colors of a fish scale or a butterfly wing have a similar basis, and are due to multilayer interference. Interference coloration is usually **iridescent,** i.e., displaying all the colors of the rainbow. Furthermore, the color observed changes with the viewing angle and on immersion in a fluid.

AN EXAMPLE OF INTERFERENCE EFFECTS: THE HOLOGRAM

The hologram was invented in 1947 by Gabor,[8] who was awarded the 1971 Nobel Prize in Physics for this invention.

The hologram is an example of the use of interference effects to produce not only color, but also a three-dimensional image.

The production of a hologram requires film, a light source (usually a laser), and a small three-dimensional model for the object to be captured on the hologram. To make a master hologram, the laser beam is split into two beams: the illuminating beam and the reference beam, as shown in Figure 4.16. The illuminating beam is aimed at the object, and then its reflection strikes a light-sensitive

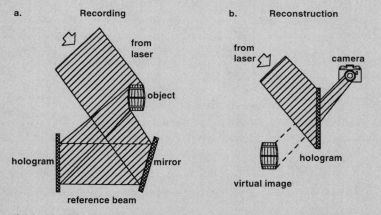

Figure 4.16. (a) Making a hologram: A hologram is made by splitting a laser beam into a reference beam and an illuminating beam. The illuminating beam "scans" the object to be captured. The reflected light hits a film where it again meets the reference beam, and they form an interference pattern. (b) Viewing a hologram: A hologram is viewed when ambient light from the same direction as the reference beam hits the interference image on the developed film, reconstructing a three-dimensional image of the original object. The hologram can be seen by eye or, as shown here, photographed by a camera. Adapted from R. Cotterill (1985). *The Cambridge Guide to the Material World,* Cambridge University Press.

(continued)

[8] Dennis Gabor (1900–1979) was a Hungarian-born electrical engineer who developed the holographic principles while carrying out research aimed to improve the resolution of electron microscope images.

COMMENT

AN EXAMPLE OF INTERFERENCE EFFECTS: THE HOLOGRAM (continued)

film. The reference beam meets the reflected beam at the film, and the interference between the reflected and reference beam leaves a pattern on the film, representative of the object scanned.

The master holographic image is converted to a physical image of microscopic pits in a fine layer of nickel. A special plate is then created to emboss these microscopic pits into a clear plastic film. Then the film is coated with a thin reflective layer of aluminum. The master hologram can be used to produce many such holograms.

Viewing the hologram requires ambient light from the same direction as the reference beam. This light interacts with the pattern on the film to reconstruct the three-dimensional image of the original object.

X-ray holograms with atomic resolution have been developed recently, and these could be used as data storage devices.

4.4 SCATTERING OF LIGHT

Light is bent or scattered when it hits the edge of an opaque object. This is a minor effect when the object is large, but becomes a major effect when the size of the object approaches the wavelength of light involved.

Scattering of light was put on a firm basis by the studies of Rayleigh[9] and Tyndall[10], and their names are often associated with light-scattering processes.

For a small object (about 10 to 300 nm in diameter) that scatters light of incident intensity I_i and wavelength λ, the intensity at a distance d from the scatterer and angle θ between incident and detected intensity, I_θ, is given by[11]

$$\frac{I_\theta}{I_i} = \frac{8\pi^4\alpha^2(1 + \cos^2\theta)}{d^2\lambda^4} \tag{4.5}$$

where α is the **polarizability** of the sample (a measure of how easy it is to displace electrons in the sample; this is important for interactions with light because light is electromagnetic radiation). Equation 4.5 describes **Tyndall scattering.**

Since I_θ is proportional to λ^{-4} in Tyndall scattering, the most intensely scattered light is

---

[9] Lord Rayleigh (born John William Strutt; 1842–1919) was a British mathematical physicist and winner of the 1904 Nobel Prize in Physics for accurate measurements of the density of the atmosphere and its component gases. This work led to the discovery of argon and other members of the previously unknown family of noble gases.

[10] John Tyndall (1820–1893) was an Irish-born physicist who carried out precise experiments on heat transfer and also on scattering of light by fine particles in air, known today as the Tyndall effect.

[11] See, for example, R.J. Hunter (1993), *Introduction to Modern Colloid Science,* Oxford, p. 42.

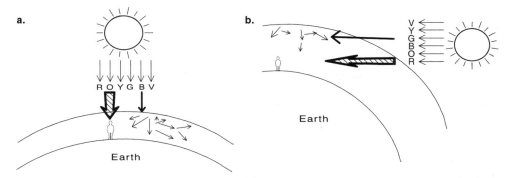

Figure 4.17. (a) The sky appears blue at noon because the shortest wavelengths of sunlight (corresponding to blue-violet light) are most highly scattered. The sun appears yellow because red to yellow light is much less scattered, and yellow light from the sun is more intense than red. (b) At sunset the light from the sun travels a longer path through the light-scattering atmosphere to reach us. This scatters away even more of the short-wavelength light than at noon, and the sun appears red.

that at the shortest wavelengths. For the visible range, this means that blue light is much more scattered than red light. (Violet is most scattered, but the intensity of violet light from the sun is much less than blue light.)

Tyndall scattering is responsible for the blue color of the daytime sky. The light from the sun contains all colors; we know that because we can see it dispersed into the whole spectrum when there is a rainbow. Of the visible wavelengths of light emitted by the sun, violet light will be most scattered by the particles in our atmosphere; blue will be less scattered and green even less so. Because violet, blue, and green together appear blue, when we stand beneath the noon sun and look at the sky, most of the colour that we see in the sky, not looking directly at the sun, is therefore blue (Figure 4.17a).

On the other hand, we are all familiar with the beautiful red sky at sunset. In the direction of the sun, much of the shorter wavelength light has been scattered by the atmosphere. This is particularly so at sunset because the sun's light has a longer path to travel through the atmosphere before getting to us. This leaves behind predominantly red color in the direction of the sun at sunset (Figure 4.17b).

The white color of clouds also is due to light scattering, but now the particle size is much larger (>1000 nm). The clouds are made of water droplets that can scatter light. The scattering is due to surface roughening and it is nearly equal at all wavelengths (as for a rough surface of paper or ground glass; see the Tutorial in Chapter 5) and clouds appear white.

Indeed, without an atmosphere, the sky would appear black even in daytime, as it does in space or on the moon.

COMMENT

COLOR-CHANGING SPECIES

The color-matching of the chameleon to its environment is largely a result of light scattering. The skin of a chameleon contains many layers—an outer yellow carotenoid layer, followed by a red chromophore layer, then a granular guanine
(continued)

COMMENT

COLOR-CHANGING SPECIES (continued)

layer, and finally an inner granular melanin layer. The light level alters the stimulation of the nerves in the skin, and this, in turn, changes the degree of aggregation of the guanine layer. In low light levels, granules aggregate, and the layer appears dark. In higher light levels, the granules are more dispersed, and the incoming light is scattered (Tyndall effect) by the silvery-blue guanine layer, making the skin appear more brightly colored. The degree to which the red and yellow layers are activated is in direct response to the stimulation of light from the surroundings.

The Hercules beetle changes color in response to humidity. The upper layer of its skin is spongy and, when dry, it scatters yellow light. In moist conditions, this layer becomes translucent, and the black underlayer dominates for a black appearance. This allows the beetle to look yellow in low humidity daylight conditions and black in the humid night air or amid rotting fruit. Once again, light scattering can provide camouflage.

4.5 DIFFRACTION GRATING

Interference and scattering can act together to produce colors in the case of a **diffraction grating,** which has a very small slit (or number of slits), with each opening of the order of the wavelength of light. When the light passes through the slit, it is scattered in all directions. Alternately, a diffraction grating can involve reflection from internal layers. In both cases, the optical effects arise from constructive and destructive interference.

For a single slit, the scattered light intensity varies with distance from the slit and angle. For many slits, the light coming from adjacent slits can interfere, either constructively or destructively. For a given wavelength of light there will be angles where there is complete constructive interference, and other angles where there is complete destructive interference (Figure 4.18; the Young[12] experiment). The exact angles will depend both on λ and on the spacing between the slits.

When white light is passed through a diffraction grating (or is reflected from it) each wavelength is enhanced in different sets of directions, so as the diffraction grating is rotated in different directions, different wavelengths come into view. This is the principle by which a grating **monochromator** (which means "single-color producer") works. The diffraction elements can be made, for example, by engraving fine rulings on a glass plate.

One example of a natural diffraction grating is the gemstone opal (Figure 4.19). Opal is composed of spheres of SiO_2 with a small proportion of water packed in a three-dimensional lattice with a repeat distance of about 250 Å. When white light enters this three-dimensional

[12] Thomas Young (1773–1829) was an English natural philosopher who made many contributions to the physical sciences including the principle of interference and principles of elasticity (memorialized in the term "Young's modulus"; see Chapter 14).

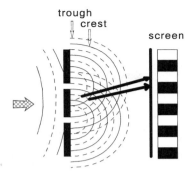

Figure 4.18. Light passing through two very small slits can interact in such a way that both constructive and destructive interference are possible. This gives "fringe patterns" as shown. This experiment was designed and first carried out by Thomas Young, and was the first proof of the wave nature of light.

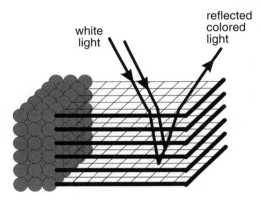

Figure 4.19. Spectral colors can arise from the interaction of light with a regular lattice array with a spacing close to or shorter than the wavelength of light. The reflected light interferes constructively, enhancing some colors, and also interferes destructively, depleting other colors. The color observed depends on the viewing angle. This is the source of color in opals.

grating, pure colors appear, but the colors change as the eye or the stone is moved (as this changes the viewing angle and the observed constructive interference). The source of color in opals was not understood until the structure of opal was elucidated in 1964 with the use of an electron microscope.

An Example of Diffraction Grating Colors: Liquid Crystals

Liquid crystals can provide another example of diffraction grating colors. Liquid crystals were first reported by the botanist Reinitzer[13] in 1888. He discovered that cholesteryl benzoate had two melting points. On heating, at 145°C it goes from a solid to a cloudy liquid; at 179°C, it transforms from a cloudy to a clear liquid. It is now known that the intermediate state (cloudy liquid) is liquid crystalline.

Liquid crystals derive their appropriate name from the fact that they are pourable ("liquid") yet structured ("crystal"). They owe these seemingly conflicting properties to their molecular

[13]Friedrich Reinitzer (1857–1927) was a Prague-born botanist who spent much of his working life in Austria. His insightful observation of a "double melting point" led to the discovery of liquid crystals.

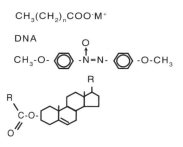

Figure 4.20. Some molecules that can form calamitic liquid crystals. Note that all have rod-like shapes.

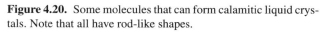

Figure 4.21. Liquid crystals are both pourable ("liquid") and ordered ("crystalline") because of their molecular shape. The molecular size is exaggerated here.

shapes: materials that form liquid crystals are long and thin (**calamitic liquid crystals**) or disc-shaped (**discotic liquid crystals**). Some typical molecular structures are shown in Figure 4.20.

Liquid crystals are **mesophase** (i.e., intermediate phase) materials. The liquid crystalline phase can be produced only in a certain temperature range (this is a **thermotropic** liquid crystal) or by addition of a nonliquid crystalline solvent (this is a **lyotropic** liquid crystal, as for concentrated aqueous soap solution).

The molecular shape of a pure liquid-crystal-forming molecule allows liquid crystals to be pourable, yet show some order (Figure 4.21).

There are many types of liquid crystals (and much nomenclature in this subject), and the categories are defined by the molecular arrangements.

Nematic (Greek for "threadlike") liquid crystals have their molecules aligned in the direction of the long axis, but they are interspersed, not in planes. This is shown schematically in Figure 4.22.

Smectic (Greek for "soap") liquid crystals have layers of molecules aligned in planes, as shown in Figure 4.22. A soap bubble is an example of a smectic liquid crystal, with the number of layers being reduced as the bubble expands.

Liquid crystals can be used to produce colors from the principle of a diffraction grating. This is due to the regularity of their stacking, which can act as a diffraction grating much as in the case of opal. As the temperature changes, the spacing in the liquid crystal changes, because at higher temperatures the molecules require a little more room to jostle around, and this change in spacing causes a change in color. The general phenomenon of change in color as a result of change in temperature, which is not limited to liquid crystals, is called **thermochromism.** In general, on cooling of a diffraction grating liquid crystal, the reflectance maximum moves from violet to blue to green to yellow to orange to red to colorless. So-called "mood" rings make use of thermochromism in liquid crystals; their color, which indicates hand temperature, and possibly correlates with mood, is due to the thin film of liquid crystalline material behind a bead of colorless transparent plastic. In some thermochromic liquid crystal devices, temperature changes as small as 0.001 K can be detected.

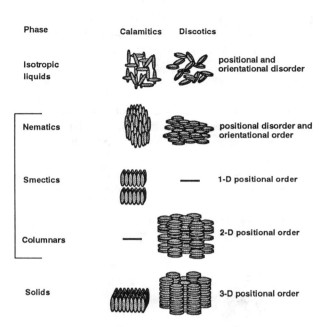

Phase   Calamitics   Discotics

Mesophases

Isotropic liquids — positional and orientational disorder

Nematics — positional disorder and orientational order

Smectics — 1-D positional order

Columnars — 2-D positional order

Solids — 3-D positional order

Figure 4.22. Liquid crystal structures, showing the wide variety of phases exhibited by liquid crystals. The highest-temperature phase, the isotropic liquid, is at the top, with the lowest-temperature phase, the crystalline solid, at the bottom; the mesophases are at intermediate temperatures.

COMMENT

COLORS OF LIQUID CRYSTALS

One of the earliest applications of liquid crystals was in the temperature dependence of the color of thin films of cholesteric liquid crystals. (See also Chapter 5.) These have been used to sensitively detect temperature changes, in applications as varied as detection of irregularities in blood circulation, metabolic changes due to breast cancer, and measurements of stress and strain in building components. An example is shown in Figure 4.23.

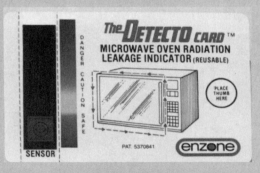

Figure 4.23. This microwave tester uses a liquid crystal thermometer. During operation of the microwave, the tester is passed around the microwave seal. If microwave leakage is negligible, the "happy face" is showing. If there is substantial microwave leakage, this will cause an increase in the temperature of the tester and one of the other liquid-crystal-covered patches will be observable.

Both the color of a liquid crystal and the temperature of its color change can be modified chemically. For example, 20% cholesteryl chloride with 80% cholesteryl nanoate is green, but added solvent can change the color: benzene changes the color to blue and chloroform changes the color to red.

One important factor concerning liquid crystalline phases is that they are intermediate in the temperature range between low-temperature crystalline forms and high-temperature "normal" liquid phases (see Figure 4.22). The latter is often referred to as an **isotropic** liquid phase, meaning that there is no preferential direction for molecular orientation in this phase. For materials that form liquid crystals, the isotropic liquid phase exists when the temperature is sufficiently high to allow the molecules to tumble freely in all directions. As the temperature is lowered, the molecules prefer to be aligned (e.g., in the smectic or nematic structure), but they are still relatively free to translate. As the temperature is lowered further, there is not enough thermal energy for the molecules to translate and the structure changes to a crystalline form. For this reason liquid crystal displays (LCDs; see Chapter 5 for a discussion of their operating principles) have a limited temperature range of operation, only that of the liquid crystal stability.

COMMENT

COLORED LIQUID CRYSTAL DISPLAYS

Colored liquid crystal displays can be produced using **pleochromic** dyes in a nematic phase. A pleochromic dye has different light absorptions, and therefore different colors, along different axes, due to its **anisotropic** nature (i.e., it has different properties in different directions). The dye molecules will tend to align themselves along the axis of alignment of the liquid crystals, and the color observed will depend on the orientation of the liquid crystal molecules with respect to the viewer. When an electric field is applied across this phase, the liquid crystal molecules can reorient, moving the dye molecules to another orientation, changing the observed color of the film. Some examples of pleochromic dyes are shown in Figure 4.24.

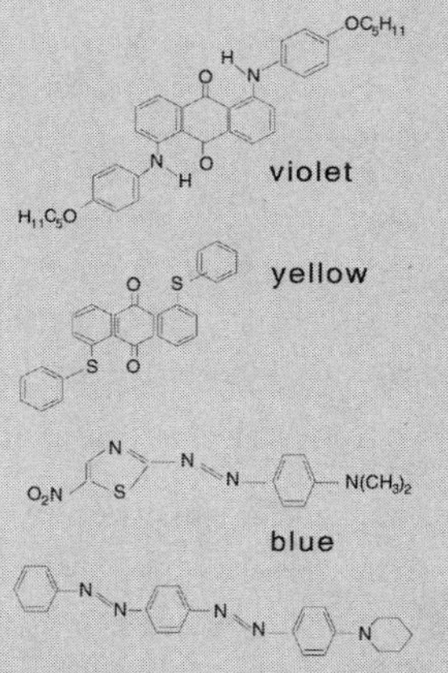

Figure 4.24. Some examples of pleochromic dyes, labeled with their dominant observed colors.

COMMENT

"SMART" GLASS

Although we are most familiar with transparent glass, we would probably find it convenient at times to be able to make the glass opaque. Two recently developed devices do just that—the glass changes transparency when a small electric field is applied.

The **electrochromic**[14] device, which is 2×10^{-6} m thick, has outer layers of a mixture of indium and tin oxides acting as electrodes, with a middle layer of tungsten oxide. When voltage is applied, tungsten oxide is reduced to metallic tungsten, which is bronze colored and reduces the amount of transmitted light. This process can be reversed with a change in the direction of the electric field.

Another "smart" glass contains a thin film of liquid crystals between transparent electrodes. When an electric field is applied, the crystals are aligned and the material is transparent. In the absence of an electric field the liquid crystalline molecules are randomly oriented, and light is scattered, giving the glass a frosted appearance.

Both these devices could have many uses, such as for windows on buildings that can be adjusted to achieve optimum lighting and privacy. However, cost is a major factor, as these smart glasses can cost $2,000 per m^2.

Fiber Optics: A Tutorial

One of the most important devices using optical properties of materials is the **optical fiber.** These fibers can transmit light signals for long distances with high efficiency. Internal reflection within the fiber can be used to propagate the light. A typical radius is of the order of 100 μm.

a. To decrease the scattering of light along the fiber, will it be better to use shorter or longer wavelengths of light? Explain.

b. The core of the optical fiber is made of a drawn fiber of SiO_2. What sorts of chemical impurities can be particularly harmful to the efficiency of transmission?

c. One advantage of **fiber optics** is that the fibers can transmit light of many different energies (wavelengths).
   i. How is this an advantage over transmission along metal wires?
   ii. Are there any limitations on possible energies for transmission? Explain.

d. Efficient transmission is possible as long as sufficient care is taken concerning the coating on the fiber. In particular, the coating (also called **cladding**) should allow for essentially complete internal reflection of the light that hits it. For the optical fiber shown in Figure 4.25, with a refractive index of n_1, and a coating of refractive index n_2, there is a critical angle, θ_c, for which any incident angle, $\theta_i > \theta_c$ leads to internal reflection. This is shown in Figure 4.25.

[14] An electrochromic material changes colour due to changes in electric field.

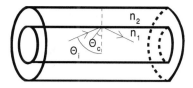

Figure 4.25. The indices of refraction of the optical fiber and its cladding are n_1 and n_2, respectively. The critical angle is θ_c, where $\theta_i > \theta_c$ leads to total internal reflection.

Extension of Equation 4.2 leads to the following relationship among θ_c, n_1 and n_2 is:

$$\sin \theta_c = \frac{n_2}{n_1}. \tag{4.6}$$

What should the relative values of n_1 and n_2 be to have maximum internal reflection?

e. What other properties of the optical fiber are important for its application?

FURTHER READING

General References

Many introductory physics and physical chemistry textbooks contain information concerning physical interactions of light with bulk matter.

B.S. Chandrasekhar (1998). *Why Things Are the Way They Are.* Cambridge University Press.
R.M. Evans (1948). *An Introduction to Color.* John Wiley & Sons Ltd.
A. Javan (1967). The Optical Properties of Materials. *Scientific American,* September 1967, 239.
M.G.J. Minnaert (1993). *Light and Colour in the Outdoors.* Springer-Verlag.
C.G. Mueller, M. Rudolph, and the Editors of LIFE (1966). *Light and Vision.* Life Science Library, Time Inc.
K. Nassau (1980). The Causes of Colour. *Scientific American,* October 1980, 124.
K. Nassau (1983). *The Physics and Chemistry of Colour.* John Wiley & Sons Ltd.
H. Rossotti (1983). *Colour.* Princeton University Press.
F. Vögtle, Ed. (1991). *Supramolecular Chemistry: An Introduction,* Chapter 8 "Liquid Crystals." John Wiley & Sons Ltd.
V.F. Weisskopf (1968). How Light Interacts with Matter. *Scientific American,* September 1968, 60.
E.A. Wood (1977). *Crystals and Light: An Introduction to Optical Crystallography.* Dover Publications.
G. Wyszecki and W.S. Stiles (1967). *Color Science.* John Wiley & Sons Ltd.

Fiber Optics

A.M. Glass (1993). Fiber Optics. *Physics Today,* October 1993, 34.
D. Hewak (1997). Travelling Light. *Chemistry in Britain,* May 1997, 26.

Films (including soap films)

F.J. Almgren and J.E. Taylor (1976). The Geometry of Soap Films and Soap Bubbles. *Scientific American,* July 1976, 82.

C.V. Boys (1959). *Soap Bubbles.* Dover Publications.

C. Isenberg (1992). *The Science of Soap Films and Soap Bubbles.* Dover Publications.

Interference Effects

R. Greenler (1994). Sunlight, Ice Crystals, and Sky Archeology. In *The Candle Revisited,* P. Day and R. Catlow, Eds. Oxford University Press.

A.J. Kinneging (1993). Demonstrating the Optic Principles of Bragg's Law with Moiré Patterns. *Journal of Chemical Education,* **70,** 451.

G. Oster and Y. Nishijima (1963). Moiré Patterns. *Scientific American,* May 1963, 54.

D. Psaltis and F. Mok (1995). Holographic Memories. *Scientific American,* November 1995, 70.

Liquid Crystals

Materials for Flat Panel Displays. Special Issue of *Materials Research Society Bulletin,* March 1996.

J.D. Brock, R.J. Birgeneau, J.D. Litser, and A. Aharony (1989). Liquids, Crystals and Liquid Crystals. *Physics Today,* July 1989, 52.

G.H. Brown (1972). Liquid Crystals and Their Roles in Inanimate and Animate Systems. *American Scientist,* **60,** 64.

P.G. Collings (1990). *Liquid Crystals: Nature's Delicate Phase of Matter.* Princeton University Press.

P.G. de Gennes and J. Prost (1993). *The Physics of Liquid Crystals.* Oxford University Press.

E. Demirbas and R. Devonshire (1996). A Computer Experiment in Physical Chemistry: Linear Dichroism in Nematic Liquid Crystals. *Journal of Chemical Education,* **73,** 586.

J.L. Fergason (1964). Liquid Crystals. *Scientific American,* August 1964, 76.

M. Freemantle (1996). Polishing LCDs. *Chemical & Engineering News,* December 16, 1996, 33.

M. Freemantle (1998). Photoluminescent Films Brighten Liquid-Crystal Displays. *Chemical & Engineering News,* February 9, 1998, 8.

G.H. Heilmeier (1970). Liquid-Crystalline Display Devices. *Scientific American,* April 1970, 100.

A.J. Leadbetter (1990). Solid Liquids and Liquid Crystals. *Proceedings of the Royal Institute of Great Britain,* **62,** 61.

S.G. Steinberg (1996). Liquid Crystal Displays. *Wired,* January 1996, 68.

R. Templer and G. Attard (1991). The World of Liquid Crystals. *The New Scientist,* May 4, 1991, 25.

Opals

P.J. Darragh, A.J. Gaskin, and J.V. Saunders (1976). Opals. *Scientific American,* April 1976, 84.

Thermochromism

D. Lavabre, J.C. Micheau, and G. Levy (1988). Comparison of Thermochromic Equilibria of Co(II) and Ni(II) Complexes. *Journal of Chemical Education, 65,* 274.

PROBLEMS

1. a. The depth of the water is not responsible for the difference in hue (color) between lake or ocean water close to the shore and water off-shore. What difference does the depth make to the optical properties of water?

    b. The difference in color of water near the shore and off-shore is related to the abundance of suspended particles in the water near the shore, due to the wave action on the shore. Explain briefly how the presence of suspended particles might affect the color of the water.

2. There is a commercial product available for fever detection. It is a strip of material that is placed on the forehead, and it changes color if the person is feverish. Comment on the principle(s) on which such a device could be based.

3. If you add a few drops of milk to a glass of water, darken the room, shine a flashlight directly at the glass, and observe the glass at right angles to the flashlight beam, what color will the milky suspension be? What color will be observed looking through the glass into the flashlight beam? Explain.

4. In a consumer column in a newspaper published in Ottawa, Canada, a reader reported trouble with the liquid crystal display of a car's dashboard features in the winter. The display was initially inoperable, but the problem was rectified after the car ran for some time. What do you think the problem was?

5. Toys that change color on exposure to sunlight are available commercially. Suggest an explanation for this phenomenon.

6. Toys that change color on exposure to water are available commercially. Suggest an explanation for this phenomenon.

7. Toys that change color on cooling are available commercially. Suggest an explanation for this phenomenon.

8. *The Invisible Man* is a famous science fiction novel. According to our understanding of scientific principles, should the invisible man be able to see? Explain.

9. A "new" gemstone is the ammolite, part of the crushed, fossilized mother-of-pearl shell of the prehistoric ammonite. These rare gemstones are found only near the banks of the St. Mary and Bow Rivers in southern Alberta, Canada. The main difference between ammolite and mother-of-pearl is that ammolite can display dazzling primary colors, especially red. Ammolite was formed from the mother-of-pearl when some elements were added or taken away. An important factor in the production is that the ammonite shell was crushed by ancient aquatic reptiles, leaving each fragment tilted at a slightly different angle. What physical process gives rise to the colors of ammolite?

10. If you were required to design a material for use in eye shadow, you might base your selection on light absorbance (for color) and light interference effects (for sparkle). One such material used in eye shadow is titanium-dioxide-coated mica. Would you expect this to meet the requirements?

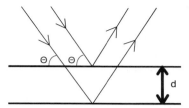

Figure 4.26. Light scattering from a thin film.

11. Hypercolour™ T-shirts change color (e.g., from pink to violet) when heated. It has been said that the color change cannot be based on thermochromic liquid crystals. Why not? Explain. Suggest another explanation for this effect.

12. Explain how a diffraction grating could be used to produce a monochromator. Use a diagram.

13. Consider the interference of light that is both front-reflected and back-reflected from a film. The thickness of the film is d. Consider the refractive index of the film to be the same as in the air surrounding the film. (Although a change in refractive index is required to get reflection, this assumption will simplify the mathematical model.) The angle of incidence of the light to the film surface is θ, as shown in Figure 4.26.
    a. Derive a general expression for the condition of constructive interference of the front- and back-reflected beams for this system, in terms of $m\lambda$, where m is the extra number of wavelengths for the back-reflected beam and λ is the wavelength of light under consideration. Keep in mind that the back-reflected beam will have a change in phase whereas the front-reflected beam will not.
    b. For light of $\lambda = 500$ nm, $\theta = 30°$, find d (in nm) for $m = 1, 2, 3$, and 10.
    c. For light of $\lambda = 500$ nm, and $d = 250$ nm, calculate the values of θ (in degrees) between 10° and 90° that give constructive interference.
    d. If change in the index of refraction on entering the film was considered, would the extra distance traveled by back-reflected light be shorter or longer than the case described above? Explain with a diagram.

14. Why do stars twinkle? (The light source is continuous, not pulsating at the frequency of the twinkling.)

15. During a lunar eclipse (earth positioned between the sun and the moon), the moon appears red. Why?

16. An announcement by Japanese car companies revealed "insect-inspired technologies," with colors of cars based on a South African butterfly. These insects appear to be brilliantly colored, but their fragile wings, which are covered with microscopic indentations, carry no pigment. Nissan produces colors by producing the car's surface with irregularities spaced only hundreds of nanometers apart.
    a. What is the principle on which the color is produced?
    b. Discuss advantages and disadvantages of this type of color compared with the more usual car color due to pigments in paint.

17. After the eruption of a volcano, the sunset can have even more spectacular color than usual. Why?

18. A particular polymer is thermochromic. When heated from room temperature to 240°C, the optical

absorption peak blue shifts, i.e., goes to shorter wavelength. This has been associated with a change in the degree of conjugation in the system.

    a. Is the HOMO–LUMO gap larger or smaller at higher temperature? Explain your reasoning.

    b. Is the system more or less conjugated at higher temperature? Explain your reasoning.

    c. Propose a physical explanation for the change in conjugation.

19. Liquid crystals can be used to orient molecules, e.g., to examine direction-dependent properties, such as IR, UV-vis, and nuclear magnetic resonance (nmr) parameters. How are liquid crystals useful for these studies? Suggest advantages and limitations of this method.

20. Why are there no white gases?

21. A coffee mug has been advertised as having a "disappearing STAR TREK® crew." The crew appears on the mug when it is empty, but, when the mug is filled with a hot beverage, the crew is no longer visible ("probably beaming down to some planet in distress"). Explain the principle of the disappearing crew.

22. The Paua shell, native to New Zealand waters, has beautiful colous, mostly in the blue-violet range. Part of its attraction is that the observed hue changes with changes in the angle of viewing. Suggest a plausible physical origin for the source of the color. Explain your reasoning.

23. Sketch the path of a beam of light being focused by a Fresnel lens. Show details of the light path through two stepped prisms in the lens, as the light enters, passes through, and exits. To simplify matters, consider the light to be monochromatic.

24. Ice crystals can cause a rainbow halo (circle of light with the full rainbow of colors) to appear around the sun. Would red or violet of the rainbow halo appear closer to the sun? Explain.

25. Titanium, niobium, and tantalum are all used by metalwork artists as materials that can have iridescent colors introduced by electrochemical oxidation of the surface. Explain why the color observed depends on the extent of the surface oxidation.

26. What color is the sky when viewed from outer space? Explain any difference from the color of the sky as viewed from earth.

27. A newly discovered mineral composed of Al, Ca, Mg, Fe, and O has a structure based on fibers somewhat like asbestos. This mineral has the remarkable quality of a striking blue color that changes to violet and cream on rotation. Draw a diagram that illustrates the interaction of light with this material and the resulting observed colors.

28. Air above hot pavement or desert sand has a steep temperature gradient, with cooler air at the higher elevation. Sketch the resultant mirage effect under these conditions, comparable with the diagram in Figure 4.4.

29. Titania (TiO_2) is sometimes added to laundry detergents to replace the TiO_2 that has washed away from synthetic fibers. It was added to the synthetic fibers in the manufacturing process to make them appear white instead of transparent or translucent.

    a. Describe the interaction of light with TiO_2. It has a very high refractive index (about 2.6).

    b. TiO_2 is somewhat photoactive in that it can absorb a high-energy photon (e.g., UV light) and release an electron. Would this limit textile applications of TiO_2? Explain.

30. Shampoos come in many colors but their foams are always white. Why?

31. At first glance, ordinary glass and a diamond look quite a bit alike. If both were cut to the classic diamond cut (see Figure 4.7), how would the dispersion of light compare for cut glass and the diamond? Note the indices of refraction: n(glass) = 1.4, n(diamond) = 2.4.

32. Is the refractive index for glass ($n \approx 1.4$) greater or less for light of wavelength 400 nm compared with light of wavelength 700 nm?

33. The blue coloration of most bird feathers is due to Tyndall scattering of light from barbs, which have a colorless transparent outer layer about 10^{-5} m thick over a layer of melanin-containing cells with irregularly shaped air cavities, <30 nm to about 300 nm across. Explain whether the blue color would be expected to remain if the feather is crushed, destroying the air cavities.

34. Fog lights are manufactured to produce yellow light, rather than white light, so that less scattering of the light occurs from fine droplets of water. Explain why the color makes a difference in the light scattering.

35. It is well known in the television business that plaids and small checked patterns should not be worn on camera, as they do not reproduce well on television. Explain why.

36. Merck KGaA in Germany has announced production of a new type of pigment capable of changing color depending on viewing angle. The material uses uniform-thickness silicon dioxide flakes. Explain the principle that gives rise to the color and the angular dependence.

C H A P T E R 5

OTHER
OPTICAL EFFECTS

5.1 INTRODUCTION

Although color is an aesthetically pleasing optical effect, and may also have many uses, there are other interesting optical properties of materials that do not necessarily involve color. Some of these are described in this chapter.

5.2 OPTICAL ACTIVITY AND RELATED EFFECTS

Normally, light is **unpolarized,** i.e., it contains waves with electromagnetic fields oscillating in all directions perpendicular to the direction of the light's motion, as shown schematically in Figure 5.1.

When this light passes through a filter called a **polarizer,** only one oscillation direction of the electromagnetic field is allowed through. On emerging from the polarizer, the light is said to be **plane** (or **linearly**) **polarized** (Figure 5.1).

Some materials have the property of rotating the oscillation direction of linearly polarized light; this property is called **optical activity.** Molecules or crystals that are **chiral** (i.e., not superimposable on their mirror image) have this property, but the degree of optical activity depends on the particular material. It also depends on the state of matter; a crystal may derive its optical activity from its chiral packing arrangement or from chirality of the constituent molecules. In the latter case, there will still be optical activity even if the crystal is melted or dissolved in solution. A 1 mm path length of quartz (which is an optically active crystal) can rotate the plane of polarization by about 20°.

Cholesteric liquid crystals (also called **twisted nematic liquid crystals**) have layers of chiral molecules, arranged so that each layer has a different orientation, as shown schematically in Figure 5.2. The optical activity of a cholesteric liquid crystal can be among the high-

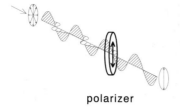

Figure 5.1. Normal unpolarized light contains waves with oscillating electromagnetic fields in all directions perpendicular to the direction of the light's motion as indicated by the wheel at the start of the light wave (for clarity, only two wave oscillation directions are shown here). Plane polarized light contains light of only one oscillation direction, and this can be selected from unpolarized (ordinary) light by a polarizer, giving linearly (or plane) polarized light.

polarizer

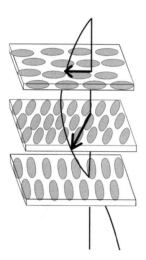

Figure 5.2. A schematic structure of a cholesteric liquid crystal. The arrows indicate the direction of the polarization vector, and this is rotated as it passes through the material.

est known, as much as 50 full rotations (i.e., $50 \times 360° = 18,000°$) for a 1 mm path length! This optical activity can be used to advantage in a liquid crystal display device, as detailed below.

In the **field-effect liquid crystal display** (LCD) each digit of the display device is composed of seven pairs of electrode sandwiches (see Figure 5.3). The two outer layers, like the bread of a sandwich, are composed of a transparent, electrically conducting material, typically

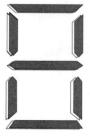

Figure 5.3. Seven pairs of electrodes are required for each digit in a liquid crystal display.

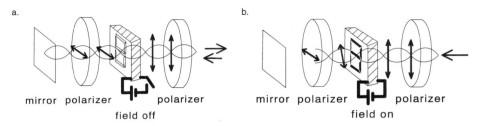

a.

b.

mirror polarizer   polarizer

field off

mirror polarizer   polarizer

field on

Figure 5.4. A schematic view of a field effect liquid crystal display (LCD) device. (a) When the field across the electrode containing the cholesteric liquid crystal is off, the light enters, is linearly polarized (the direction is indicated by the double-headed arrow), has its plane of polarization rotated by the liquid crystal, passes through a matched polarizer, is reflected by a mirror, and reverses the entire process on the way back to the outside world. Since no light is absorbed, the LCD digit has no color. (b) When a field is applied across the electrode containing the cholesteric liquid crystal, its optical activity is changed. This means that the entering polarized light is rotated by an amount different from in the field-off case, and the polarization no longer matches the second polarizer. Therefore the light cannot pass through this polarizer and cannot reach the mirror to be reflected out. Since light is absorbed in this process, the LCD digit appears black in the presence of the field.

a thin layer of indium tin oxide; between these layers is a layer of cholesteric liquid crystal. Activation of each of the seven portions of the digit is caused by the application of an electric field; independent activation or deactivation of the seven electrodes can allow display of the digits from 0 to 9.

Another important factor in this display involves plane polarized light. Outside light passing into a liquid crystal display device first passes through a polarizer (Figure 5.4). It then travels into the cholesteric liquid crystal cell, where the plane of polarization is rotated by the optically active liquid crystal. The display device is constructed so that this new direction of the plane of polarization exactly matches the direction of polarization of another polarizer in the light path. Because the polarization directions match, the light passes through this second polarizer unchanged. When the light hits a mirror at the back of the display device, it is reflected back toward the outside world. This light is still polarized in the direction of the innermost polarizer and it continues outward. When this polarized light encounters the liquid crystal cell, the polarization is changed by an amount equal but opposite to the change on passing through the cell in the opposite direction. For example, as shown in Figure 5.4a, the change in polarization was +90° on entering the cell and −90° on leaving the cell. This allows the polarization of the light to match that of the outermost polarizer, so the light can exit the device. Since light entered the device and it also can exit the device, the cell appears transparent.

In the situation just described there was no field applied across the seven electrode sandwiches in the cell. If a small electric field is applied, the color of the electrodes is modified for the following reasons. Light still enters from the outside world and is polarized by the first polarizer. However, the electric field has the effect of slightly changing the optical activity of the liquid crystal by slightly modifying the structure of the material. The change in optical activity means that the amount by which the liquid crystal rotates polarized light is not the same as it was without the field, and the new polarization on emerging from the liquid crystal no longer matches the innermost polarizer. Since this polarizer passes only light of the matched polarization, the light that hits this polarizer cannot go further. Therefore light is only absorbed with no reemission, and the electrode elements appear black. Each of the seven elements actually has its own electric field control, so each can be on or off, allowing the full range of digits from 0 to 9 to be selectively produced.

This simple liquid crystal display device was made possible technically by the development of thin coatings of transparent electrodes, usually indium tin oxide. Without this development, light could not pass into the liquid crystal cell. Currently work is underway in liquid crystal display devices to increase the number of pixels (color elements) by **multiplexing** (i.e., using each element to receive more than one signal); for example, liquid crystals are used in conjunction with colored semiconductors.

Liquid crystal display devices are often compared with light-emitting diode (LED) devices. As we have seen in Chapter 3, LEDs derive their color from emission of energy of excited electrons in semiconductors. LCDs have a major advantage over LEDs in their power consumption; the main power needed in an LCD is related to the small electric fields, and these consume very little energy. By contrast, LEDs require power to promote electrons, and this requires much more energy. For this reason, LCD display calculators can be solar powered, but this is impractical for LED display calculators. On the other hand, LCD displays are not very useful in the dark—they require external light to show the difference between activation and deactivation. So an LED display is a better bet for a bedside clock!

COMMENT

MORE NEW DIRECTIONS IN LIGHT-EMITTING DIODES

Whereas most light-emitting diodes emit light of all directions of polarization, researchers in Sweden have discovered a way to make polymer-based LEDs that emit polarized light. The polymer chains are normally randomly oriented, and the light they emit is not polarized. However, when the polymers are stretched they become oriented, and the light emitted has its electric field gradient polarized in a single direction. These devices could be important in liquid crystal displays that require a polarized light source.

5.3 BIREFRINGENCE

For materials that are **anisotropic** (not the same in all directions; this is the opposite of **isotropic**), an unusual optical property called **birefringence** (also known as **double refraction**) can occur. (Birefringence was first described in 1669 by Bartholin,[1] and explained by

---

[1] Erasmus Bartholin (1625–1698) was a Danish scientist who made contributions in mathematics and astronomy. His greatest contribution was his study of double refraction in Icelandic spar (collected by an expedition to Iceland in 1668), but his explanation was faulty: he assumed there was a double set of "pores" that led to two refraction rays.

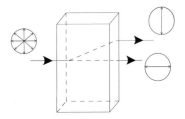

Figure 5.5. The effect of double refraction, also known as birefringence, in an hexagonal material. Because the material is anisotropic, the refractive index is not the same in all directions. This causes unpolarized incident light to be split into two polarized components with different velocities, giving rise to two emerging beams of light.

Figure 5.6. A sample of Iceland spar (also known as calcite, $CaCO_3$) showing birefringence, which makes a double image appear when looking through the crystal.

Huygens[2] in 1678 using the wave theory of light.) Examples of anisotropic materials are liquid crystals (e.g., the smectic and nematic structures have the molecules aligned along the long axis) and noncubic solids. In a liquid, the atomic arrangements are, on average, the same in the x-, y-, and z-directions and therefore they are isotropic, but this is not so for an anisotropic solid (see Appendix 3 for crystal types).

Birefringence in an anisotropic material arises because unpolarized incident light is split into polarized components (two components for tetragonal and hexagonal crystals; three components for less symmetric crystals); this occurs because the velocity of light is different in different directions (i.e., the refractive indices are different in different directions) in an anisotropic material (see Figure 5.5). Thus, parallel but displaced beams of polarized light emerge from the anisotropic material (Figure 5.5). Birefringence in a colorless transparent crystal of Iceland spar is shown in Figure 5.6.

The presence of birefringence is a simple way to detect anisotropy in a transparent material. In addition, birefringence can lead to **dichroism**, i.e., two different colors of material when viewed in different directions.

---

[2] Christiaan Huygens (1629–1695) was a Dutch scientist who contributed to many areas, including mathematics, statics, dynamics, astronomy, and optics. In addition, Huygens discovered Titan (a satellite of Saturn), was the first to realize that Saturn's odd shape was due to its rings, and also invented the pendulum clock. His wave theory of light was used to explain optical phenomena, including double refraction.

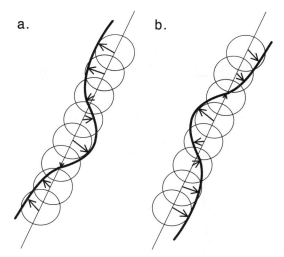

Figure 5.7. (a) Right- and (b) left-circularly polarized light. The polarization vector (shown as an arrow) rotates in opposite directions for the two forms of circularly polarized light.

5.4 CIRCULAR DICHROISM AND OPTICAL ROTATORY DISPERSION

We have seen in Figure 5.1 that light can be polarized in a particular direction; this is referred to as linearly (or plane) polarized light. Light also can be polarized such that the direction of polarization rotates in the plane perpendicular to the direction of motion of the light wave. This rotation can be clockwise or counterclockwise, giving rise to right- and left-**circularly polarized light,** respectively, as shown in Figure 5.7.

The refractive index, $n,$ and also the absorption coefficient, $\varepsilon,$ can depend on both the wavelength of light and whether the light is right- or left-circularly polarized. This is illustrated schematically in Figure 5.8.

The difference in absorption coefficient for right- and left-circularly polarized light is referred to as **circular dichroism** (CD). The term originates from the fact that the different absorption coefficients for left- and right-circularly polarized light lead to different colors (i.e., dichroism) depending on the circular polarization for the light used to make the observation.

The difference in refractive index for right- and left-circularly polarized light is called **optical rotatory dispersion** (ORD). (Note that ORD is seen only near absorption maxima; see Figure 5.8.) The presence of both CD and ORD is referred to as the **Cotton effect.**

Investigation of CD and ORD can provide information concerning structure, configuration, and conformation of complicated optically active molecules.

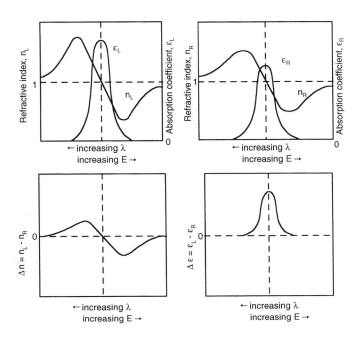

Figure 5.8. The variation of refractive index, *n,* and molar absorption coefficient, ε, as functions of wavelength, as measured with right- and left-circularly polarized light (referred to with subscripts R and L, respectively).

5.5 NONLINEAR OPTICAL EFFECTS

The term **nonlinear effect** refers to the nonlinear proportionality of the response to the input. It is a general concept and most physical processes, when examined closely, are nonlinear in detail, although they may be approximately linear.[3] Let us consider the case of an electron bound to an atom, and its potential energy, as this is important in considering the interaction of light with matter, especially when more than one photon is involved.

When an electron is pulled from an atom, to a first approximation the restoring force *F* is proportional to *x,* the displacement of the electron from its equilibium position:

$$F = -kx \tag{5.1}$$

where *k* is a constant, and the linear proportionality of Equation 5.1 makes this a **linear effect.** Because any force *F* is related to the potential energy, *V;*

$$F = -\nabla V \tag{5.2}$$

where ∇ denotes the gradient (i.e., $\nabla = \partial/\partial x$ for the one-dimensional case), it follows that the potential energy for a force given by Equation 5.1 is

$$V = \frac{1}{2} kx^2 \tag{5.3}$$

---

[3] Any effect is nonlinear if the response to an input (i.e., the output) changes the process itself. In the case of nonlinear optical effects, the interaction of light with a material causes the material's properties to change, such that the next photon that arrives will have a different interaction with the material.

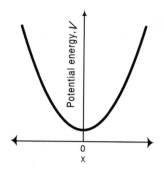

Figure 5.9. Potential energy as a function of displacement of an electron from an atom in the harmonic approximation.

where the potential energy has been chosen to be zero at $x = 0$. Equation 5.3 corresponds to a **harmonic potential,** as shown in Figure 5.9.

When an electric field E (e.g., light) interacts with electrons in a material, the electron density can be provoked to be displaced. When the optical effect is linear, this leads to an **induced dipole** (or **polarization**) of magnitude μ_i:

$$\mu_i = \alpha E \tag{5.4}$$

where α is the **linear polarizability.** This is illustrated in Figure 5.10.

a.

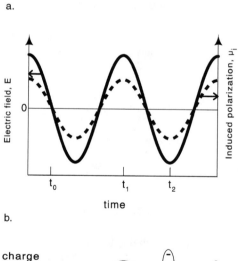

Figure 5.10. Linear optical effects. (a) The electric field (solid line) and the induced polarization wave (dotted line) as functions of time for an optically linear material. (b) Schematic of the induced polarization of an optically linear material interacting with light as a function of time. (c) The induced (linear) polarization as a function of applied field for an optically linear material.

b.

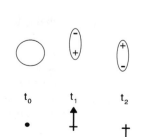

c.

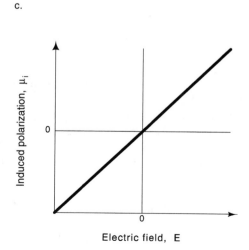

Two important optical properties arise directly from **linear optical effects**—the dielectric constant (ε) and the refractive index (n). The **dielectric constant** is defined as the ratio of the total electric field within the material (D, also called the displaced field) to the applied electric field of the light (E):

$$\varepsilon = \frac{D}{E} \tag{5.5}$$

and, in the absence of absorption of light, the refractive index is related to the dielectric constant ε by the following equation:

$$\varepsilon(v) = n^2(v) \tag{5.6}$$

where the dependence on the frequency of the light, v, is given explicitly. Both Equations 5.5 and 5.6 are written in their isotropic forms, but we know that refractive index may be anisotropic (as observed by birefringence, for example) and Equations 5.5 and 5.6 could be written in tensor form to account for this explicitly.

The harmonic restoring potential in a linear optical material gives rise to a motion of charge (electric field oscillation) that is of the same frequency as the light (Figure 5.11a). From a quantum mechanical perspective, the oscillating applied electric field induces a time-dependent induced polarization of the molecule that can be expressed as a mixing of the ground state and polarized excited states. (These are called **virtual transitions** because there is ex-cited state character but not long-lived populations of the excited states.) The motion of charge (or, equivalently, mixing of excited states) leads to reemission of the radiation at the same frequency as the frequency of the polarizing light.

Many times in physical science the harmonic model is a good first approximation, but closer examination indicates the importance of other terms in the potential. If the restoring force of Equation 5.1 can more accurately be written as

$$F = -kx - \frac{1}{2}k'x^2 \tag{5.7}$$

this is now nonlinear. Equation 5.7 leads to the following expression for potential energy:

$$V = \frac{1}{2}kx^2 + \frac{1}{6}k'x^3 \tag{5.8}$$

i.e., there is an additional anharmonic term compared to Equation 5.3. This potential is illus-trated in Figure 5.11.

Since the anharmonic potential depends on the direction of the displacement of the elec-tron (whereas Equation 5.3 gives the same value of V for x and $-x$, Equation 5.8 gives values of V that depend on the sign of x), so will the magnitude of the polarization. An example of such a material is one in which there is an anisotropic charge distribution in the absence of an electric field (e.g., a noncentrosymmetric polar molecule in the gas phase or a noncentrosym-metric polar crystal). In this case, the electric field of the light can displace the charge more in one direction than in the other and the resulting polarization can be a **nonlinear optical (NLO) effect** (Figure 5.12).

In a nonlinear optical material the polarization μ ($= \mu^0 + \mu_i$ where μ^0 is the **static dipole,**

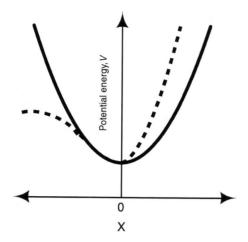

Figure 5.11. Potential energy as a function of displacement coordinate for a material with a harmonic potential (solid line, Equation 5.3) and for a material with an additional anharmonic term (broken line, Equation 5.8).

i.e., in the absence of the field, and μ_i is the induced dipole) can be written as a Taylor series expansion:

$$\mu = \mu^0 + \alpha_{ij}E_j + \frac{1}{2}\beta_{ijk}E_jE_k + \frac{1}{6}\gamma_{ijkl}E_jE_kE_l + \cdots \qquad (5.9)$$

where this is written in the tensor form. The terms beyond αE (which is the **linear polarizability;** see Equation 5.4) are not linear in E; they are referred to as nonlinear polarization terms and they give rise to nonlinear optical effects. β and γ are referred to as the first and second **hyperpolarizabilities,** respectively. For β to be nonzero, there must be no inversion center in the material. However, γ is nonzero for all materials, although it induces weaker effects. Since $\alpha E \gg \beta E^2$ and γE^3, observation of nonlinear optical effects requires large electric fields, and lasers have been very important in this developing subject.

In a linear optical material, $\beta = 0$, the induced electric field was at the same frequency as the light, and the light emitted was purely of the same frequency as the exciting light. In a nonlinear optical material, the induced polarization is not a pure sine wave at the frequency of the light (see Figure 5.12a), but it can be considered to be composed of that frequency (the **fundamental**) and its **harmonics,** as well as an **offset** (DC component), as shown in Figure 5.13. The static polarization gives rise to the DC component, the linear polarizability gives rise to the fundamental (as in linear optical materials), and the hyperpolarizabilities give rise to the harmonics.

This means that for a nonlinear optical material, if light of frequency ν is incident on this material, light of frequency ν and 2ν (and to a lesser extent 3ν, 4ν, etc.) will be observed. The production of light of frequency 2ν in this manner is known as **frequency doubling** by **second harmonic generation** (SHG). This is used in practice in nonlinear optical devices to achieve frequencies twice that of the incident light, for example, using $Ba_2NaNb_5O_{15}$, KH_2PO_4, $LiNbO_3$, or HIO_3. (The associated DC component of the electric field is referred to as **optical rectification.**) SHG is considered to be a **three-wave mixing** process since two photons with frequency ν are required to combine to give a single photon with frequency 2ν (energy is conserved). By analogy, third-order harmonics, at frequency 3ν from the γE^3 term of Equation 5.9, require **four-wave mixing.**

a.

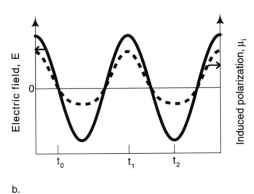

Figure 5.12. Nonlinear optical effects. (a) The electric field (solid line) and the induced polarization wave (dotted line) as functions of time for an optically nonlinear material. (b) Schematic of the dipole and induced dipole (which is the difference between the dipole at that time and at time t_0) of an optically nonlinear material interacting with light as a function of time. Note that the magnitude of the induced dipole depends on the direction of the polarization. (c) The induced (nonlinear) polarization as a function of applied field for an optically nonlinear material.

b.

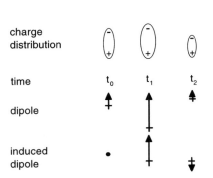

c.

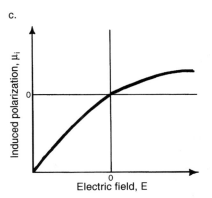

Figure 5.13. Fourier analysis of the asymmetric polarization wave of Figure 5.12a, showing its components: the fundamental (frequency ν), the second harmonic (frequency 2ν), and the offset (DC component).

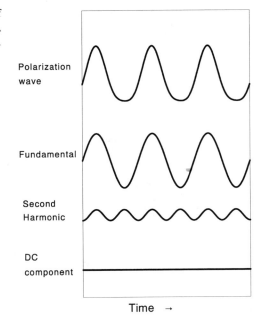

In the case of two beams of light, of frequencies v_1 and v_2, interacting with a nonlinear optical material, it also is possible to have emitted light at the **sum frequency** ($v_1 + v_2$) and at the **difference frequency** ($v_1 - v_2$) in processes referred to generally as **sum frequency generation** (SFG).

Other nonlinear optical effects include the dependence of the refractive index of a nonlinear optical material on the applied voltage [**linear electrooptic** (LEO) or **Pockels effect**] where the applied voltage distorts the electron density within the material.

Nonlinear optical properties present exciting challenges to the materials scientist. Advances in this area have been greatly helped by the use of high-powered lasers. Current research in this area includes design of materials to have enhanced nonlinear optical properties, design of materials to examine the role of ground state and excited state structure in nonlinear optical properties, and further research in structure/property relations. One promising area involves polymeric nonlinear optical devices that could be used to carry more information than present optoelectronic devices.

COMMENT

PHOTOREFRACTIVE EFFECT

Another nonlinear optical property is the **photorefractive effect.** This is the dependence of the index of refraction of the material on the local electric field. This property allows the imprinting of an interference pattern on a material, and it is expected that new materials with better photorefractive properties will lead to more nearly perfect resolution and retrieval of digitized images.

Transparency: A Tutorial

For application of an optical material, one of the most important considerations is transparency. Of course this is related in part to the color of the material.

a. Is it possible that a material can absorb visible light but be transparent to light outside the visible range? Consider this for narrow band-gap semiconductors and organic chromophores.

b. If a material absorbs negligibly in the visible range, does this guarantee that it is transparent? Consider that SiO_2 can be very transparent (as in quartz tubing) or quite opaque (as in sand at the beach).

c. If the material in question does not absorb visible light when pure, it can be made to have interactions that lead to opacity by the introduction of second-phase particles or pores. The basis of the opacity is refraction at the pores. Is the size of the pore important? If so, what size range (approximately) would be most important? What other factors concerning the pores are important?

d. Window glass can be frosted or dimpled, rendering it translucent. What principles are involved here?

e. Some materials are designed to be transparent to only certain polarizations of light. Explain the principles involved here.

f. Although some materials are quite transparent, a significant portion of the light impingent on them can be reflected. This reflection may be unwanted as, in use, it may lead to glare. One way to reduce glare is to coat the material with another material. What factors will be important in selecting and applying the coating?

g. Early crystallographers knew that repeat units in crystal structures were smaller than 400 nm, because otherwise they would appear frosted on the surface. Explain.

h. What does optical transparency of ordinary glass imply about the band gap in this medium?

i. Suggest a way in which a ceramic material could be made either transparent or opaque. See Figure 5.14 for an example.

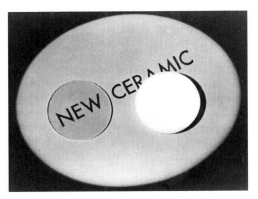

Figure 5.14. Aluminum oxide ceramic can be produced in both opaque and transparent forms. The latter is used in the Lucalox lamp, a high-efficiency, high-intensity sodium vapor lamp. Photo courtesy of Joseph E. Burke.

FURTHER READING

General References

Many introductory physics and physical chemistry textbooks contain information concerning polarized light and its interaction with matter.

R. Cotterill (1985). *The Cambridge Guide to the Material World.* Cambridge University Press.

M. Freemantle (1994). Novel Electro-optic Materials Discovered. *Chemical & Engineering News,* September 12, 1994, 28.

A. Javan (1967). The Optical Properties of Materials. *Scientific American,* September 1967, 239.

B. Kahr and J.M. McBride (1992). Optically Anomalous Crystals. *Angewandte Chemie International Edition in English,* **31,** 1.

J.M. Marentette and G.R. Brown (1993). Polymer Spherulites. I. Birefringence and Morphology. *Journal of Chemical Education,* **70,** 435.

K. Nassau (1983). *The Physics and Chemistry of Color.* John Wiley & Sons Ltd.

J.M. Rowell (1986). Photonic Materials. *Scientific American,* October 1986, 146.

S. Sengupta, J. Kumar, A. Jain, and S. Tripathy (1993). Functionalized Polymeric Materials for Electronics and Optics. Chapter 12 in *Chemistry of Advanced Materials,* C.N.R. Rao, Ed. Blackwell Scientific Publications.

V.F. Weisskopf (1968). How Light Interacts with Matter. *Scientific American,* September 1968, 60.

E.A. Wood (1977). *Crystals and Light: An Introduction to Optical Crystallography,* 2nd Ed. Dover Publications, Inc.

Devices

Physics Today, December 1992, Special Issue: The Physics of Digital Color.

R. Dagani (1996). Devices Based on Electro-optic Polymers Begin to Enter Marketplace. *Chemical & Engineering News,* March 4, 1996, 22.

S.W. Depp and W.E. Howard (1993). Flat-Panel Displays. *Scientific American,* March 1993, 90.

E. Wilson (1985). Nonlinear Optical Polymer Device May Speed up Information Superhighway. *Chemical & Engineering News,* August 14, 1995, 27.

Liquid Crystals

G.H. Brown (1983). Liquid Crystals—The Chameleon Chemicals. *Journal of Chemical Education,* **60,** 900.

J. Flnfschilling (1991). Liquid Crystals and Liquid Crystal Displays. *Condensed Matter News* **1**(1), 12.

Nonlinear Effects

Emergence of Chalcopyrites as Nonlinear Optical Materials. Special Iassue of *MRS Bulletin,* July 1998.

Physics Today, May 1994, Special issue on nonlinear optics.

R.T. Bailley, F.R. Cruickshank, P. Pavlides, D. Pugh, and J.N. Sherwood (1991). Organic Materials for Non-linear Optics: Inter-relationships between Molecular Properties, Crystal Structure and Optical Properties. *Journal of Physics D: Applied Physics,* **24,** 135.

N. Bloembergen (1998). Nonlinear Optical Instrumentation. *Journal of Chemical Education,* **75,** 555.

R. Dagani (1995). Photorefractive Polymers Poised to Play Key Role in Optical Technologies. *Chemical & Engineering News,* February 20, 1995, 28.

J. Feinberg (1988). Photorefractive Nonlinear Optics. *Physics Today,* October 1988, 46.

M. Jacoby (1997). Photorefractive Polymer Stacks. *Chemical & Engineering News,* July 28, 1997, 8.

B.G. Levi (1995). New Compound Brightens Outlook for Photorefractive Polymers. *Physics Today,* January 1995, 17.

S.R. Marder, J.E. Sohn, and G.D. Stucky, Eds. (1991). Materials for Nonlinear Optics: Chemical Perspectives. *American Chemical Society Symposium Series 455,* American Chemical Society.

S. R. Marder (1992). Metal Containing Materials for Nonlinear Optics. Chapter 3 in *Inorganic Materials,* D.W. Bruce and D. O'Hare, Eds. John Wiley & Sons Ltd.

S. Mukamel (1995). *Principles of Nonlinear Optical Spectroscopy.* Oxford University Press.

Transparency

J.E. Burke (1996). Lucalox Alumina: The Ceramic That Revolutionized Outdoor Lighting. *Materials Research Society Bulletin,* June 1996, 61.

PROBLEMS

1. When linearly polarized light is passed through a normally isotropic transparent object that is under stress, and the transmitted light is viewed through another polarizer, colored fringe patterns can be observed. Suggest an explanation for this phenomenon.

2. How could a field-effect liquid crystal display device be used to produce an LCD television screen? Do you foresee any limitations to its use? Can you suggest ways to get around these limitations using liquid crystals?

3. The absence of a center of symmetry in a crystal is essential for it to exhibit (second-order) nonlinear optical properties. However, this condition is not sufficient. Why not?

4. a. Give an example of a material (it can be hypothetical) that is (i) isotropic and centrosymmetric, (ii) anisotropic and centrosymmetric, and (iii) anisotropic and noncentrosymmetric.
    b. Is it possible to give an example of a material that is isotropic and noncentrosymmetric? Explain.

5. Leonard Cohen's song "Anthem" contains the following line in the refrain: "There is a crack, a crack in everything. That's how the light gets in." Comment.

6. Dry sand is light brown in color, and water is virtually colorless. When water is added to dry sand, the resulting wet sand is dark brown. Why does the color of the sand depend on how dry it is?

7. Some polymer films are transparent but become translucent when stretched. Suggest an explanation. Consider the ordering of the polymer molecules in the original form, compared with the stretched form.

8. Stained glass windows look rather dull from outside a building, but quite beautifully colored when viewed from inside the building. Why is there an apparent difference in color?

9. The difference in color of a material in reflected light, compared with transmitted light, is an important design consideration for inks to be used on colored viewgraphs for overhead transparencies. For example, when a dye is bound to a transparent sheet and projected light is passed through it, an image is produced on the screen that is colored due to transmitted light; this image may have very different color properties from what is observed looking at the transparency in reflected light. Suggest some of the factors that must be considered in development of color transparencies.

10. Iridescent plastic films are composed of thin layers of two or more types of polymer. These materials have different colors when viewed in reflection and transmission. Furthermore, the colors viewed in reflection depend on the viewing angle. Sketch diagrams that show these effects when white light is incident on such a film. For simplicity, just consider two polymer layers.

11. Why is solid candle wax translucent whereas molten candle wax is transparent?

12. When chiral molecules (i.e., those that are not superimposable on their mirror image) are incorporated into a cholesteric liquid crystal, the different **enantiomers** (i.e., nonsuperimposable mirror image pairs) perturb the **pitch**[4] of the liquid crystals (and hence the color of the system) in different ways. It is often difficult to distinguish enantiomers by usual chemical means, as they can have the same melting points, boiling points, etc. Can you suggest an application of the different colors in distinguishing enantiomers? (This has been done successfully: T. Nishi, A. Ikeda, T. Matsuda, and S. Shinkai (1991). *Journal of the Chemical Society: Chemical Communications 339.*)

13. Rubies can exhibit dichroism: purple-red vs. orange-red. The $^4T_1$ and $^4T_2$ levels (see Figure 2.9) can each be split in two due to distortion of the octahedral symmetry at the Cr^{3+} site; if a ruby crystal is viewed in polarized light, it can be rotated to see the effect of absorption to one set of levels in one direction and another set in another direction. In the direction in which the absorption is between the ground state and the higher excited state, is the color purple-red or orange-red?

[4] Pitch is the repeat distance along the direction perpendicular to the aligned layers of cholesteric liquid crystal molecules

THERMAL PROPERTIES

OF MATERIALS

Passed through the fiery furnace
As gold without alloy,
Triumphant over trials,
Accounting it all joy.

—Philip Carrington
from a Hymn, 1938

6

HEAT CAPACITY, HEAT CONTENT, AND ENERGY STORAGE

6.1 INTRODUCTION

The thermal properties of any material are among the most fundamental, whether directly or indirectly involved in the material's application. For example, it is almost always important to know how properties of materials change if the temperature is changed. In this section of the book we explore the fundamentals of thermal properties, from basic thermodynamics to applications. This chapter emphasizes heat capacity, heat content, and energy storage. Heat capacity, which is the most fundamental of all thermal properties, is related to the strength of intermolecular interactions, phase stability, thermal conductivity, and energy storage capacity.

6.2 EQUIPARTITION OF ENERGY

To understand the fundamentals of energy storage, it is first useful to examine the basic **theory of equipartition of energy.**

First, let us consider the information needed to know where a molecule is. This will lead to considerations of how the energy of that molecule is partitioned.

For a single monatomic species, three coordinate positions must be specified to know the atom's position. For example, this can be expressed as three Cartesian coordinates, (x,y,z) in orthogonal axes. We express the freedom of position by saying that a single atom has three degrees of freedom of motion.

For a polyatomic molecule with N atoms per molecule, $3N$ coordinates (three for each of the N atoms) must be known to specify the total molecular position. An N-atom polyatomic molecule has $3N$ degrees of freedom.

The molecule's motion can be expressed in terms of three types of movement: translational (motion of whole molecule), rotational (rotation of whole molecule), and vibrational (internal vibrations within the molecule).

For a monatomic species, there is only one meaningful type of motion, and this is translation. The translational motion can be considered to take place in three orthogonal directions (e.g., x, y, z), and any motion of the monatomic molecule can be described as a vector sum of the motions in these three directions. This is another way of expressing the three degrees of freedom of a monatomic species.

For a diatomic molecule, there are also three degrees of freedom of translational motion. In addition, there are two meaningful rotational motions for a diatomic molecule. (A third rotational motion, along the molecular axis, does not move the atoms and therefore does not contribute to the freedom of the molecule.) A diatomic molecule has one further vibrational degree of freedom due to an internal vibration. The sum of these degrees of freedom of motion (3 translational + 2 rotational + 1 vibrational = 6 total) is the same as the total number of degrees of freedom (for $N = 2$, $3N = 6$).

For a linear polyatomic, with N atoms, there are three translational degrees of freedom and two rotational degrees of freedom, leaving $(3N - 5)$ vibrational degrees of freedom.

For a nonlinear polyatomic, there can be three meaningful rotational motions. This gives, for N atoms per molecule, three translational degrees of freedom, three rotational degrees of freedom, and $(3N - 6)$ vibrational degrees of freedom.

From quantum mechanical considerations, virtually any translational energy value is allowed. Of course, translational energies are quantized, but their spacing is so close together that it results in a virtual continuum of energy levels.

It takes more energy to excite rotations than it does to excite translations, which is reflected in the wider spacing of rotational energy levels relative to the translational ladder.

Vibrational energy levels are even further apart than rotational levels, and it takes considerable energy to excite them. (Recall from the discussion of color of matter that electronic energy levels are even more widely spaced than vibrational levels, as shown in Figure 2.2, so we do not consider electronic excitation here.)

The **principle of equipartition** states that for temperatures high enough that translational, rotational, and vibrational degrees of freedom are all "fully" excited,[1] each type of energy contributes $\frac{1}{2}kT$ (where k is the Boltzmann constant and T is the temperature in kelvin) to the internal energy, U, per degree of freedom.

This means that each translational degree of freedom contributes $\frac{1}{2}kT$ to U, and each rotational degree of freedom contributes $\frac{1}{2}kT$ to U, **BUT** each vibrational degree of freedom contributes $2 \times \frac{1}{2}kT$ to U. The double contribution for vibration results from the fact that translation and rotation have only kinetic energies associated with them, whereas vibrational motion has both kinetic and potential energy.

This simple theory of equipartition can be useful to work out the contributions of degrees of freedom to U, and hence to the heat capacity. Two examples are given below.

Heat Capacity of a Monatomic Gas

For a monatomic gaseous species, there are only three degrees of freedom in total, and all of these are translational degrees of freedom. Therefore $U = 3 \times \frac{1}{2}kT$. This same result also is found from the kinetic theory of gases.

---

[1] Strictly speaking, all the degrees of freedom are fully excited only at infinite temperature.

TABLE 6.1.
Degrees of Freedom and Their Contributions to U, C_V, and $C_{V,m}$ for a Nonlinear Triatomic Molecule

Motion	Degrees of freedom	U	C_V	$C_{V,m}$
Translational	3	$3 \times \frac{1}{2}kT$	$3 \times \frac{1}{2}k$	$3 \times \frac{1}{2}R$
Rotational	3	$3 \times \frac{1}{2}kT$	$3 \times \frac{1}{2}k$	$3 \times \frac{1}{2}R$
Vibrational	3	$2 \times 3 \times \frac{1}{2}kT$	$3k$	$3R$
				Total $C_{V,m} = 6R$

Considering that the **heat capacity at constant volume,** C_V, is defined by

$$C_V = \left(\frac{\partial U}{\partial T} \right)_V \tag{6.1}$$

then $C_V = 1.5k$ per molecule for a monatomic gas.

To scale up to a mole of gas, we need to multiply by the Avogadro constant, N_A; this gives the molar heat capacity at constant volume, $C_{V,m}$:

$$C_{V,m} = C_V N_A \tag{6.2}$$

so, for a gaseous monatomic, $C_{V,m} = 1.5 \, N_A k = 1.5 \, R$ at any temperature, where $R \, (= N_A k)$ is the **gas constant.**

Heat Capacity of a Nonlinear Triatomic Gas

For a nonlinear triatomic molecule, such as H_2O, the high-temperature limit of the heat capacity can be calculated from equipartition theory. Since $N = 3$, there are $3N = 9$ degrees of freedom in total, partitioned as shown in Table 6.1.

Since equipartition theory considers only the case in which all the degrees of freedom are fully excited, this is the high-temperature limit of the heat capacity of gaseous H_2O. In practice, at moderate temperatures, $C_{V,m}$ is much less. For example, $C_{V,m}(H_2O$ gas, $T = 298.15$ K$) = 3.038R$. Since translational and rotational energies are most easily excited, and $3R$ is their summed contribution, the remaining $0.038R$ reflects the fact that the vibrations of gaseous water molecules are only slightly active at room temperature because the vibrational energy spacings are greater than the available thermal energy, kT.

We will see later (Sections 6.4 and 6.5) how equipartition theory can be used to assess heat capacity contributions in solids and liquids.

6.3 REAL HEAT CAPACITIES AND HEAT CONTENT OF REAL GASES

The heat content, or **enthalpy** (H), of a gas can be assessed through the relationships between H, T, and C_p (**heat capacity at constant pressure**),[2] defined as follows:

---

[2] Just as U and volume go together, H and pressure often appear together in thermodynamic equations.

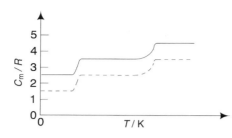

Figure 6.1. The temperature dependence of the molar heat capacity of an ideal diatomic gas at constant pressure ($C_{P,m}$; ——) and at constant volume ($C_{V,m}$: – – –). This is a stylized diagram in which the temperature should be considered to be a logarithmic scale.

$$C_p = \left(\frac{\partial H}{\partial T}\right)_p \tag{6.3}$$

and

$$\Delta H = \int_{T_1}^{T_2} dH = \int_{T_1}^{T_2} C_p \, dT. \tag{6.4}$$

For polyatomic gases, we can get high-temperature values of C_V from equipartition considerations. We will show later (Section 6.8) that for ideal gases $C_{V,m}$ and $C_{p,m}$ are related:

$$C_{p,m} - C_{V,m} = R. \tag{6.5}$$

At very low temperatures, not all degrees of freedom of a diatomic gas will be excited. The first modes to be excited will be those with the most closely spaced energy levels, the translational degrees of freedom, and in the limit of $T \to 0$ K, a gas will have $C_{V,m} = 1.5R$ due to translation only. As the temperature is increased, the contribution of the two rotational modes to $C_{V,m}$ will gradually turn on and $C_{V,m} = 2.5R$. As the temperature is increased further, the vibrational mode will contribute to the heat capacity and $C_{V,m} = 3.5R$. This is shown in Figure 6.1, where $C_{p,m}$ is shown as R higher than $C_{V,m}$ (i.e., the ideal gas approximation).

Returning to the theme of heat storage, and taking into account that the heat content over a temperature range is the integrated heat capacity (Equation 6.4), the higher the heat capacity, the greater the heat storage ability. This is true for all phases of matter, and we return to this later when considering heat storage applications.

First, let us consider what happens when a gas is cooled to very low temperatures. All gases liquefy when the temperature is sufficiently low. This may require a very low temperature; for example, helium liquefies at $T = 4.2$ K.

When a gas liquefies, this releases heat. Since enthalpy is a state function, the complete cycle from gas → liquid → gas would have $\Delta H = 0$. Therefore, the enthalpy change on liquefaction ($\Delta_{gas \to liquid} H$) is equal but of opposite sign to the enthalpy change on vaporization ($\Delta_{vap} H$):

$$\Delta_{gas \to liquid} H = -\Delta_{vap} H. \tag{6.6}$$

The enthalpy change on vaporization, $\Delta_{vap} H$, is a very useful quantity, and some typical enthalpies of vaporization are given in Table 6.2, but $\Delta_{vap} H$ is not known for all materials.

TABLE 6.2.
Enthalpy Changes on Vaporization, $\Delta_{vap}H$, and Entropy Changes on Vaporization, $\Delta_{vap}S$, for a Variety of Materials at Their Normal Boiling Points, T_b

Material	$\Delta_{vap}H$/J mol^{-1}	$\Delta_{vap}S$ /J K^{-1} mol^{-1}	T_b/K
He	8.37×10^2	19.7	4.2
N_2	5.577×10^3	72.13	77.3
CH_4	8.180×10^3	73.26	111.7
NH_3	2.335×10^4	97.40	239.7
CH_3OH	3.527×10^4	104.4	337.9
CCl_4	3.000×10^4	85.8	349.9
C_6H_6	3.076×10^4	87.07	353.3
H_2O	4.0657×10^4	108.95	373.2
Hg	5.812×10^4	92.30	629.7
Zn	1.148×10^5	97.24	1180.

However, since the **Gibbs**[3] **energy** change, ΔG, can be written at constant temperature as

$$\Delta G = \Delta H - T\Delta S \tag{6.7}$$

where ΔS is the change in **entropy,** and the Gibbs energies of any two phases in equilibrium are identical (so $\Delta_{trs}G$, the change in Gibbs energy due to transition, equals zero), then

$$0 = \Delta_{trs}H - T\Delta_{trs}S \tag{6.8}$$

i.e.,

$$\Delta_{trs}H = T\Delta_{trs}S \tag{6.9}$$

or $\Delta_{vap}H = T \Delta_{vap}S$ for the case of the vaporization transition. It is an experimental finding that many gases have values of $\Delta_{vap}S$ of about 90 J K^{-1} mol^{-1} (see Table 6.2). This finding was generalized by Trouton[4] and is usually called **Trouton's Rule.** It holds because the increase in disorder on going from the liquid phase (where molecules are translating and undergoing hindered rotations and perhaps muted vibrations) to the gas phase (where the molecules translate and rotate more freely than in the liquid) is approximately independent of the type of material involved.

There are exceptions to Trouton's Rule, and one is H_2O, with $\Delta_{vap}S$ of 108.95 J K^{-1} mol^{-1}.

[3] Josiah Willard Gibbs (1839–1903) was an American theoretical physicist. Gibbs was a recipient of one of the first American Ph.D. degrees, from Yale in 1863 (his thesis was titled "On the Form of the Teeth of Wheels in Spur Gearing"). After acquiring his Ph.D. he spent 3 years in Europe, then returned to New Haven and rarely traveled anywhere. This rather reclusive scientist made many great contributions, mostly in the area of thermodynamics. Gibbs' first scientific paper, published in 1873, set the record straight on the (then-confused) concept of entropy. Within 5 years, he published a 300-page memoir providing the basis of much of thermodynamics as it is taught to this day. Since most of the thermodynamic research at that time was carried out in Europe, especially Germany, it took some time (and Ostwald's 1892 translation of Gibbs' work to German) for Gibbs' work to make a lasting impression.

[4] Frederick Thomas Trouton (1863–1922) was an Irish physicist who, in 1902, took up a professorship at University College London. He discovered Trouton's Rule when he was an undergraduate student at Trinity College, Dublin.

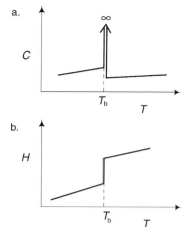

a.

b.

Figure 6.2. Boiling of a liquid at T_b at a fixed pressure. (a) Schematic of the heat capacity as a function of temperature of a material in the vicinity of the boiling point. (b) Schematic of the corresponding enthalpy as a function of temperature in the vicinity of the boiling point.

The main reason for the exceptionally large increase in disorder on vaporization of water is that its liquid phase has a relatively low entropy due to the order associated with hydrogen-bond networks, which hinder molecular motion. This is the same network that gives rise to the color of water through vibrational excitation, as seen in Chapter 2.

While we are looking at heat capacity and heat content, we should consider what happens to the heat capacity of a material at its boiling point. When a liquid is heated toward its boiling point, the material absorbs heat and the temperature rises until the boiling point is achieved. At that point, heat is absorbed without further increase in temperature until all the material has been converted to vapor. From either C_V or C_p considerations, with q_V as heat taken up at constant volume and q_p as heat taken up at constant pressure,

$$C_V = \frac{q_V}{\Delta T} = \left(\frac{\partial U}{\partial T}\right)_V \tag{6.10}$$

and

$$C_p = \frac{q_p}{\Delta T} = \left(\frac{\partial H}{\partial T}\right)_p. \tag{6.11}$$

Since, on boiling, $\Delta T = 0$ (i.e., heat is input without temperature rise at the boiling point), $C_V = C_p = \infty$ at the boiling point, T_b. This is depicted in Figure 6.2.

Because ΔH is the integration of C_p over the temperature (Equation 6.4), H increases slightly with temperature over the liquid region, then has a step (integrating the infinite heat capacity) at the boiling point, followed by a further gentle increase as the temperature increases in the gas phase. This is shown in Figure 6.2. An infinite heat capacity is not limited to the case of a gas → liquid phase transition, as we shall see later.

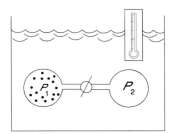

Figure 6.3. Joule's experiment. The gas at pressure P_1 was allowed to expand to the other bulb, where $P_2 = 0$, while the temperature of the water bath was recorded.

Before turning to the heat content of a nonideal gas, it is worth stating a few basic concepts concerning ideal gases. An **ideal gas** is a theoretical gas for which

$$PV = nRT \tag{6.12}$$

is the equation of state for n moles under all circumstances. The equation of state for an ideal gas can be derived from the kinetic theory of gases, with two main assumptions: (1) the molecules of the gas have negligible volume in comparison with the volume of the container, and (2) the molecules exert no attraction or repulsion on each other. All gases approach ideality as $P \rightarrow 0$, because these two conditions are most closely met at low pressures.

Joule's Experiment

Joule[5] carried out an important experiment concerning ideal gases, which led to the discovery of an important property: the internal energy of an ideal gas depends only on temperature.

In Joule's experiment, two connected gas bulbs were placed in a water bath. One bulb was evacuated and the other contained gas, as shown in Figure 6.3.

Joule opened the stopcock between the two bulbs and measured the resulting temperature change in the water bath, ΔT_{bath}. He found

$$\lim P_1 \rightarrow 0 \qquad \Delta T_{bath} = 0 \tag{6.13}$$

where P_1 is the pressure of the gas and it expands into the vessel at $P_2 = 0$.

It can be shown, as follows, that this result leads to $\Delta U = 0$ for the gas in the limit of $P \rightarrow 0$ (i.e., approaching ideality). By the **first law of thermodynamics,** the change in internal energy, ΔU, is

$$\Delta U = q + w \tag{6.14}$$

---

[5] James Prescott Joule (1818–1889) was an English scientist noted for the establishment of the mechanical theory of heat, and honoured by the SI energy unit. He carried out his work on heat while he was in his early 20s, and then made other contributions to the field of thermodynamics, including the first estimate of the speed of gas molecules (1848). Lord Kelvin wrote of Joule, "His boldness in making such large conclusions from such very small observational effects is almost as noteworthy and admirable as his skill in extorting accuracy from them."

and work, w, is given by

$$w = -\int P_{ext}\, dV \tag{6.15}$$

where P_{ext} is the external pressure. In this experiment, the gas expands within a fixed volume while doing no external work, so $w = 0$ here. This shows that expansion into a vacuum takes no work. Consideration of the heat, q,

$$q = C\Delta T \tag{6.16}$$

and Joule's finding that $\Delta T_{bath} = 0$ leads to $q = 0$. Since $w = 0$ and $q = 0$, it follows from Equation 6.14 that $\Delta U = 0$.

In other words, when a low-pressure (ideal) gas is expanded, it does so without a change in internal energy, U. Since U is most simply expressed as a function of T and V (just as H is a function of T and P), an infinitesimal change in U, written as dU, can be expressed in terms of infinitesimal change in T, written as dT, and infinitesimal change in V, written as dV:

$$dU = \left(\frac{\partial U}{\partial T}\right)_V dT + \left(\frac{\partial U}{\partial V}\right)_T dV. \tag{6.17}$$

From Joule's experiment, $dU = 0$ and $dT = 0$, so $(\partial U/\partial V)_T\, dV = 0$, but $dV \neq 0$ so $(\partial U/\partial V)_T = 0$ for an ideal gas. This shows that U is independent of volume for an ideal gas at constant temperature, and this is because the molecules are neither attracted nor repelled in an ideal gas. Therefore, for an ideal gas, U is a function of temperature only, in accordance with the equipartition theorem (i.e., U depends only on T, not on V or P).

Joule–Thomson Experiment

Joule teamed up with Thomson[6] to carry out an important experiment that has allowed quantification of nonideality of a gas. This experimental apparatus can be considered as an insulated double piston, as shown schematically in Figure 6.4.

Initially, the gas was all in the left chamber. When the left piston was pushed in, the gas moved through a porous plug (said to have been a linen handkerchief) into the right chamber. The gas state functions were transformed from initial values (on the left $P = P_L$, $T = T_L$, $V = V_L$; on the right $P = P_R$, $T = T_R$, $V = V_R = 0$) to final values (on the left $P = P_L'$, $T = T_L'$, $V = V_L' = 0$; on the right $P = P_R'$, $T = T_R'$, $V = V_R'$).

Since the piston was made of insulating material, there was no extraneous heat exchange with the surroundings, i.e., the experiment was carried out under **adiabatic** conditions: $q = 0$.

---

[6] William Thomson (1824–1907) was a Belfast-born physicist later made Baron Kelvin of Largs and known as Lord Kelvin; the temperature unit is named after him. He graduated from Glasgow University at age 10 and was made Professor of Natural Philosophy there at age 22. Thomson made major contributions to many areas: electrodynamics, thermodynamics including the absolute temperature scale and the second law (on his own and in collaboration with Joule), and electromagnetism (with Faraday). He was knighted for his contribution to the laying of the first Atlantic cable and made many other practical advances including the invention of a tide predictor and an improved mariner's compass. He was the acknowledged leader in physical sciences in the British Isles during his lifetime.

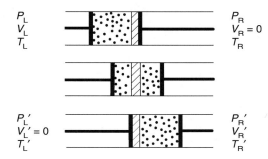

Figure 6.4. The Joule–Thomson double-piston experiment, shown as a function of time with the initial conditions at the top. In this experiment, gas was pushed through a porous plug in the center of a piston and taken from an initial set of pressure, volume, and temperature conditions on the left side of the piston (denoted by subscript L) to a final set of conditions (denoted by a prime) on the right side of the plug (denoted by subscript R).

This information can be used to show that the overall process is **isenthalpic** (i.e., at constant enthalpy, $\Delta H = 0$), as follows. Since H is defined as

$$H = U + PV \tag{6.18}$$

it follows that

$$\Delta H = \Delta U + \Delta(PV). \tag{6.19}$$

Making use of the first law of thermodynamics (Equation 6.14) and the change in PV for this process leads to

$$\Delta H = q + w + P'_R V'_R - P_L V_L \tag{6.20}$$

and here $q = 0$. From Equation 6.15,

$$w = -\int_{V_L}^{0} P_L \, dV - \int_{0}^{V'_P} P'_R \, dV \tag{6.21}$$

so

$$w = P_L V_L - P'_R V'_R \tag{6.22}$$

and substitution of Equation 6.22 and $q = 0$ into Equation 6.20 leads to

$$\Delta H = P_L V_L - P'_R V'_R + P'_R V'_R - P_L V_L = 0 \tag{6.23}$$

which is to say that the process takes place at constant enthalpy.

The Joule–Thomson experiment was carried out to determine the change in temperature of the gas with respect to pressure changes at constant enthalpy. The Joule–Thomson coefficient, μ_{JT}, so determined, is defined:

$$\mu_{JT} = \left(\frac{\partial T}{\partial P}\right)_H. \tag{6.24}$$

Experimentally it is found that μ_{JT} for a given gas is positive at relatively low temperatures, negative at relatively high temperature, and zero at some intermediate temperature (the Joule–Thomson inversion temperature; the exact value of this temperature depends on the gas). This temperature dependence can be understood as follows.

Unlike an ideal gas in which the forces between molecules can be ignored, the molecules in real gases do "feel" each other. At infinitely large separations the force between molecules

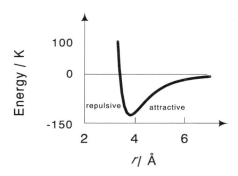

Figure 6.5. Intermolecular interaction energy as a function of intermolecular separation, r. The curve shown is for the interaction between two Ar atoms, and energy is expressed in units equivalent to thermal energy (=kT at a temperature T).

is vanishingly small, but as the intermolecular separation, r, diminishes, the molecules begin to feel attraction for each other. It is this attraction that eventually leads to liquefaction of gases when they are compressed. This attraction lowers the energy of the system. At extremely short intermolecular distances, the intermolecular interaction becomes repulsive, increasing the energy of the system. This is shown schematically in Figure 6.5.

At relatively low temperatures, where $\mu_{JT} > 0$, the definition of μ_{JT} implies that as the pressure decreases, so does the temperature: the gas cools on expansion. This is because of intermolecular forces. When the gas is expanded (P decreased), some work is required to overcome the attractive intermolecular forces. This work comes at the expense of the energy (or temperature) of the gas and causes the gas to cool on expansion.

At very high temperatures, where $\mu_{JT} < 0$, the opposite situation exists. At these very high temperatures, the temperature increases when the pressure decreases (gas is expanded). This is because high temperatures allow for many repulsive intermolecular interactions, so when the gas is expanded, energy is released, and this is manifested in an increase in the temperature of the gas.

At the Joule–Thomson inversion temperature, there is an accidental cancellation of the attractive and repulsive forces, and there is no change in temperature on expansion of the gas: $\mu_{JT} = 0$ at this temperature.

An ideal gas, with its negligible intermolecular interactions, also has $\mu_{JT} = 0$. Therefore, μ_{JT} is a direct way to quantify the nonideality of a gas.

Most gases (exceptions noted below) are still below their Joule–Thomson inversion temperature at room temperature. Because these gases have $\mu_{JT} > 0$, they cool on expansion and can be used as refrigerants. Examples are NH_3, freons (chlorofluorocarbons), and CO_2. In fact, CO_2 can cool so much on expansion that an attachment can be placed on a CO_2 gas cylinder to produce dry ice, i.e., CO_2 in its solid form, directly from expansion of the gas from the cylinder. Freons were originally developed to be nonreactive gases that would efficiently cool on expansion, for use as refrigerants. We know now that freons are rather unreactive on earth but quite reactive with ozone in the upper atmosphere,[7] and the search is on for replacement materials with suitable refrigeration capacity.

Low-boiling gases are already above their Joule–Thomson inversion temperature at room temperature. Examples are H_2, He, and Ne. To use these gases as refrigerants that expand on cooling, they must first be precooled below their Joule–Thomson inversion temperatures.

---

[7] The 1995 Nobel Prize in Chemistry was awarded to Paul Crutzen (1933– , Max-Planck-Institute for Chemistry, Mainz, Germany), Mario Molina (1943– , MIT), and F. Sherwood Rowland (1927–, University of California, Irvine) for their work in atmospheric chemistry, particularly the formation and decomposition of ozone.

6.4 HEAT CAPACITIES OF SOLIDS

In 1819 Dulong[8] and Petit[9] found experimentally that for many nonmetallic solids at room temperature, the molar heat capacity at constant volume, $C_{V,m} \approx 3R = 25$ J K^{-1} mol^{-1}. Originally known as the **Law of Atomic Heats,** this is now known as the **Dulong–Petit Law.**

The Dulong–Petit Law can be understood for monatomic solids by use of equipartition theory. For each atom in the monatomic solid, there are three degrees of freedom, and each atom's three degrees of freedom can be expressed in terms of vibrational motion of the atoms at a point of the lattice. From equipartition this gives, per atom,

$$U = 3 \times 2\frac{kT}{2} = 3kT \tag{6.25}$$

so

$$C_V = 3k \tag{6.26}$$

and, on a molar basis, with $kN_A = R$,

$$C_{V,m} = 3R \tag{6.27}$$

as found experimentally.

What the Dulong–Petit Law does not address is the further experimental fact that all heat capacities decrease as the temperature is lowered, i.e., $C_{V,m} \to 0$ as $T \to 0$ K. In fact, this is not explainable in terms of classical theories and requires quantum mechanics.

The **Einstein**[10] **model of heat capacity of a solid** (published by Einstein in 1906) was, historically, one of the first successes of quantum mechanics. In this model, thermal depopulation of vibrational energy levels is used to explain why $C_{V,m} \to 0$ as $T \to 0$ K.

---

[8] Pierre Louis Dulong (1785–1838) was a French physicist who, with Petit, discovered the Law of Atomic Heats, now known more commonly as the Dulong–Petit Law. This work was particularly important in establishing atomic masses. Dulong's work in his later life was hampered by inadequate funding (unlike most of his contemporaries, he had no funding from industrial connections) and the loss of an eye from his 1811 discovery of nitrogen chloride.

[9] Alexis Thérèse Petit (1791–1820) was a French physicist who, with Dulong, discovered the Law of Atomic Heats (Dulong–Petit Law). In 1810, at the age of 19, he was appointed Professor of Lycée Bonaparte in Paris. With Dulong, Petit was one of the first physical scientists to reject the caloric theory of heat (which treated heat as a fluid) and espouse the atomic theory of matter in 1819; this stand made little headway with his peers in France at that time.

[10] Albert Einstein (1879–1955) was a German-born theoretical physicist. Einstein's early life did not predict great success: his entry to university was delayed because of inadequacy in mathematics and when he completed his studies in 1901 he was unable to get a teaching post, so he took a junior position at a patent office in Berne. While working there he managed to publish three important papers: one on Brownian motion which led to direct evidence for the existence of molecules; one connecting quantum mechanics and thermodynamics proving that radiation consisted of particles (photons), each carrying a discrete amount of energy, which gives rise to the photoelectric effect; and a third concerning the special theory of relativity. In 1932 Einstein left Germany for a tour of America, but decided, as a Jew, that it was not safe to return to Germany, so he remained in the United States. Einstein received the Nobel Prize in Physics in 1922 for his work on the photoelectric effect.

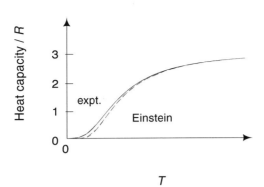

Figure 6.6. The Einstein heat capacity for a monatomic solid, compared with experiment.

Einstein considered each atom in the solid to be sitting on a lattice site, vibrating at a frequency ν. For N atoms, this led to the following expression for the heat capacity:

$$C_V = 3Nk\left(\frac{h\nu}{kT}\right)^2 \frac{e^{h\nu/kT}}{(e^{h\nu/kT} - 1)^2} \tag{6.28}$$

where h is Planck's constant and k is the Boltzmann constant. The repeating factor $(h\nu/kT)$, which must be unitless because it appears as an exponent, leads to units of s^{-1} (also called Hertz[11]) for ν. Often vibrational excitation information is given in the unit cm^{-1} ($\tilde{\nu}$, wavenumber), i.e.,

$$E = h\nu = \frac{hc}{\lambda} = hc\tilde{\nu} \tag{6.29}$$

where c is the velocity of light.

Alternatively, $h\nu/k$ can be expressed directly in terms of the **Einstein characteristic temperature, θ_E**, defined as

$$\theta_E = \frac{h\nu}{k} = \frac{hc\tilde{\nu}}{k} \ . \tag{6.30}$$

As the term $h\nu/kT$ is the ratio of vibrational energy ($h\nu$) to thermal energy (kT), the use of θ_E makes the comparison even more direct. Written in terms of θ_E, the heat capacity for N atoms in the Einstein model can be expressed:

$$C_V = 3Nk\left(\frac{\theta_E}{T}\right)^2 \frac{e^{\theta_E/T}}{(e^{\theta_E/T} - 1)^2} \ . \tag{6.31}$$

This gives a temperature-dependent molar heat capacity of a solid (Figure 6.6), approaching $3R$ as $T \to \infty$ in agreement with the Dulong–Petit finding, and approaching 0 as $T \to 0$ K in agreement with experiment. However, the Einstein model falls below the experimental heat capacity at low temperatures (Figure 6.6).

---

[11] Heinrich Rudolf Hertz (1857–1894) was a German physicist. Hertz was an experimentalist who studied electrical waves and made important discoveries that led to significant advances in understanding of electricity. Hertz also discovered radio waves.

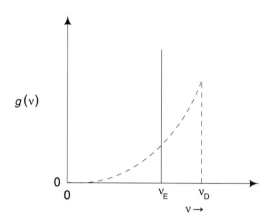

Figure 6.7. Comparison of Debye $(- - -)$ and Einstein (———) distributions of frequencies. The corresponding frequencies, $\nu_D(=\theta_D k/h)$ and $\nu_E(=\theta_E k/h)$ are also shown.

The 1912 **Debye**[12] **model of heat capacity of a solid** considers the atoms on lattice sites to be vibrating with a distribution of frequencies. Debye's model assumed a continuum of frequencies in the distribution, up to a maximum (cut-off) frequency, ν_D, the Debye frequency. The dependence of the distribution of frequencies on the frequency is shown in Figure 6.7, where the (single-frequency) Einstein model also is shown for comparison.

Mathematically, the two models can be compared in terms of the frequency distribution function, $g(\nu)$, where $g(\nu)$ is the number of modes with frequency ν:

$$\textit{Einstein model:} \quad g(\nu)\,d\nu = 0 \qquad \nu \neq \nu_E \tag{6.32}$$

$$= 1 \qquad \nu = \nu_E$$

$$\textit{Debye model:} \quad g(\nu) = a\nu^2 \qquad 0 \leq \nu \leq \nu_D \tag{6.33}$$

$$g(\nu) = 0 \qquad \nu > \nu_D \tag{6.34}$$

where a in Equation 6.33 is a constant, dependent on the material. This leads to the following expression for the Debye heat capacity for N particles:

$$C_V = 9Nk\left(\frac{kT}{h\nu_D}\right)^3 \int_0^{h\nu_D/kT} \frac{(h\nu/kT)^4 e^{h\nu/kT}}{(e^{h\nu/kT}-1)^2}\, d\left(\frac{h\nu}{kT}\right) \tag{6.35}$$

and again introducing a characteristic temperature, this time θ_D, the **Debye characteristic temperature,** where

$$\theta_D = \frac{h\nu_D}{k} \tag{6.36}$$

[12] Peter Joseph William Debye (1884–1966) was a Dutch-born chemical physicist who, at age 27, succeeded Einstein as the Professor of Theoretical Physics at the University of Zurich. He moved to the United States in 1940. Debye's contributions included fields as diverse as the theory of specific heat, the theory of dielectric constants, light scattering, x-ray powder (Debye–Scherrer) analysis, and the Debye–Hückel theory of electrolytes. In 1936 he was awarded the Nobel Prize in Chemistry.

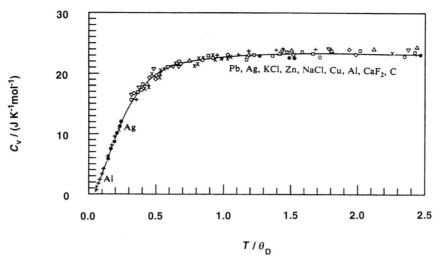

Figure 6.8. Heat capacities of various crystalline solids (each represented by different symbols, as labeled) as a function of temperature scaled to the Debye temperature for the material. The solid line is the Debye curve (Equation 6.37), and the data are for various systems, as listed. Based on data from A.W. Adamson (1973). *A Textbook of Physical Chemistry*, Academic Press.

this leads to a simpler way of writing Equation 6.35:

$$C_V = 9Nk\left(\frac{T}{\theta_D}\right)^3 \int_0^{\theta_D/T} \frac{x^4 e^x}{(e^x - 1)^2}\, dx \tag{6.37}$$

where

$$x = \frac{h\nu}{kT}. \tag{6.38}$$

The Debye heat capacity equation (Equation 6.37) makes C_V look like a universal function of θ_D/T, with θ_D as the scaling function for different materials. This is borne out very well in experimental data, as shown in Figure 6.8, which illustrates C_V as a function of temperature expressed in units of θ_D for many materials. As a scaling function, θ_D reflects the strength of the interatomic interactions, i.e., θ_D is a measure of the force constant between molecules or atoms in the solid. Some typical values of θ_D are given in Table 6.3, where it can readily be seen that materials that are harder to deform have higher values of θ_D. The higher the value of θ_D, the less the vibrational modes are excited at a given temperature, and the lower the heat capacity (in comparison with materials with lower values of θ_D).

Like the Einstein model, the Debye model of heat capacity correctly describes the experimental situation in the temperature extrema ($C_{V,m} \to 3R$ as $T \to \infty$ and $C_{V,m} \to 0$ as $T \to 0$ K). The other major success of the Debye theory is that it correctly predicts the way in which $C_V \to 0$ as $T \to 0$ K for a monatomic solid. In the limit of $T \to 0$, the Debye heat capacity expression becomes, for N atoms,

$$\lim_{T \to 0} \quad C_V = \frac{12}{5}\,\pi^4 Nk\left(\frac{T}{\theta_D}\right)^3 \tag{6.39}$$

TABLE 6.3.
Debye Temperatures for Selected Materials[a]

Material	θ_D/K
Diamond	2230
Gold	225
Neon	75
Mercury (solid)	72

[a] From C. Kittel (1986), *Introduction to Solid State Physics*, 6th Ed. John Wiley & Sons, Ltd. Note that at a given temperature, a material with a lower value of θ_D will have a higher heat capacity than a material with a higher value of θ_D.

COMMENT

DEBYE TEMPERATURE AND STEREO SPEAKERS

Debye temperature, θ_D, and sound velocity, V, are related as

$$\theta_D = \frac{vh}{2\pi k}\left(\frac{6\pi^2 N}{V}\right)^{1/3} \qquad (6.41)$$

where N is the number of particles and V is the volume. The high Debye temperature of diamonds, and consequently high sound velocity of diamonds, led to the use of diamond coatings in diaphragms in tweeter speakers to reduce high-frequency sound distortion. These are currently marketed by JVC.

---

i.e.,

$$for \qquad T < \theta_D/100, \quad C_V \propto T^3 \qquad (6.40)$$

as observed experimentally for nonmetals. This is referred to as the **Debye-T^3 law.**

For a monatomic solid, the Debye model gives a more accurate description than the Einstein model, because the frequency distribution (e.g., as determined by inelastic neutron scattering) is more Debye-like. For a polyatomic molecular solid, the rigid-molecule contribution to the heat capacity will be Debye-like, but the contributions of internal vibrations (which are localized at particular frequencies and can be observed in infrared and/or Raman spectroscopy experiments) are better treated by the Einstein model. This is considered further in Problem 10 at the end of this chapter.

Heat Capacities of Metals

For metals, there is an additional factor beyond the lattice contribution to the heat capacity. The extra degrees of freedom of the conducting electrons, i.e., those with energy above the Fermi energy, must be taken into account. That this is a very small fraction of the valence electrons can be seen as follows.

The experimental molar heat capacity of a monatomic solid metal at high temperatures is usually a little more than $3R$. We can use equipartition theory to determine the contribution of free electrons to the heat capacity. On the basis of vibrations alone, the internal energy of a monatomic solid is

$$U = 3 \times 2 \times \frac{kT}{2} = 3kT \tag{6.42}$$

and the corresponding molar heat capacity is

$$C_{V,m} = 3R. \tag{6.43}$$

If the atoms each had one free valence electron, the total internal energy per atom would be

$$U = 3 \times 2 \times \frac{kT}{2} + 3 \times \frac{kT}{2} = 4.5kT \tag{6.44}$$

where the first term is from atomic vibrations and the second term is from the translation of the free electron. This would lead to

$$C_{V,m} = 4.5R \tag{6.45}$$

which is far in excess of the experimental observation that the molar heat capacity of a monatomic metal is just slightly in excess of $3R$. Since $3R$ is the contribution from the vibrational degrees of freedom, only a small fraction of the valence electrons must be free to translate. This was the first indication that Fermi statistics (Equation 3.1) holds.

It can be shown from $P(E)$, the probability distribution function for electrons (Chapter 3), that the electronic contribution to the molar heat capacity at low temperatures is given by

$$C_{V,m}^{elec} = \gamma T \tag{6.46}$$

where γ depends on the particular metal. Therefore, using Equation 6.40, the total molar heat capacity of a metal at very low temperatures is

$$\lim_{T \to 0} C_{V,m}^{metal} = \gamma T + AT^3 \tag{6.47}$$

where the first term is the electronic contribution and the second term is the lattice contribution.

6.5 HEAT CAPACITIES OF LIQUIDS

As in many of their properties, a description of a liquid in terms of its degrees of freedom falls between solids, where low-frequency vibrations are dominant, and gases, where translation, rotation, and perhaps vibrational degrees of freedom are active. Rigid crystals and liquids can be further distinguished by the excitement of configurational degrees of freedom in the latter. As is usual with other phases of matter, the heat capacity of a liquid usually increases as the temperature increases, due to the increased numbers of excited degrees of freedom at elevated temperatures, requiring more energy to invoke the same temperature rise.

The heat capacity of liquid water, as many other physical properties of water, illustrates a special case. The heat capacity of water near room temperature historically was used to define the calorie.[13] The heat capacity of water can be converted to units of R:

$$C_V = 1 \text{ cal K}^{-1} \text{ g}^{-1} \approx 18 \text{ cal K}^{-1} \text{ mol}^{-1} \approx 9R. \tag{6.48}$$

This value can be used to illustrate a point concerning degrees of freedom in water. As a triatomic molecule, water has a total of $3 \times 3 = 9$ degrees of freedom per molecule. If these were partitioned as three translational, three rotational and three vibrational degrees of freedom then the internal energy per molecule would be given by

$$U = U_{\text{trans}} + U_{\text{rot}} + U_{\text{vib}} = \frac{3}{2}kT + \frac{3}{2}kT + \frac{3}{2} \times 2kT = 6kT \tag{6.49}$$

and the molar heat capacity of water would be

$$C_{V,m} = 6R. \tag{6.50}$$

However, this is not enough to account for the known molar heat capacity of water ($9R$). Since the only way to increase $C_{V,m}$ within the equipartition model is to increase the twice-weighted vibrational degrees of freedom, consider the extreme case of all nine degrees of freedom being vibrational. This leads to the internal energy per molecule of

$$U = U_{\text{vib}} = 9 \times 2\frac{kT}{2} = 9kT \tag{6.51}$$

and molar heat capacity

$$C_{V,m} = 9R \tag{6.52}$$

as observed experimentally. This simplified view of water is rather like that of a solid: all the degrees of freedom would be vibrational. This could arise because of the extended hydrogen-bonding network in water, which prevents free translation and rotation and allows water to look lattice-like in terms of its heat capacity. (It also gave water its color; see Chapter 2.) Although this view of water is oversimplified because at higher temperatures the heat capacity of water exceeds $3R$ due to thermal excitation of configurational degrees of freedom, detailed thermodynamic calculations of water are consistent with the existence of hydrogen-bonded clusters of water molecules in the liquid state.

6.6 HEAT CAPACITIES OF GLASSES

A **glass** can be defined as a rigid supercooled liquid formed by a liquid that has been cooled below its normal freezing point such that it is rigid but not crystalline. A glass is also said to be **amorphous,** which means without shape, showing its lack of periodicity on a molecular scale. A **supercooled liquid** (i.e., a liquid cooled below its normal freezing point) is **metastable**

---

[13] The calorie was defined as the amount of energy required to increase the temperature of exactly 1 g of water from 14.5 to 15.5°C. Since this is a relative measurement and the joule is based on SI concepts, the calorie has been redefined in terms of the joule: 1 cal $\equiv$ 4.184 J.

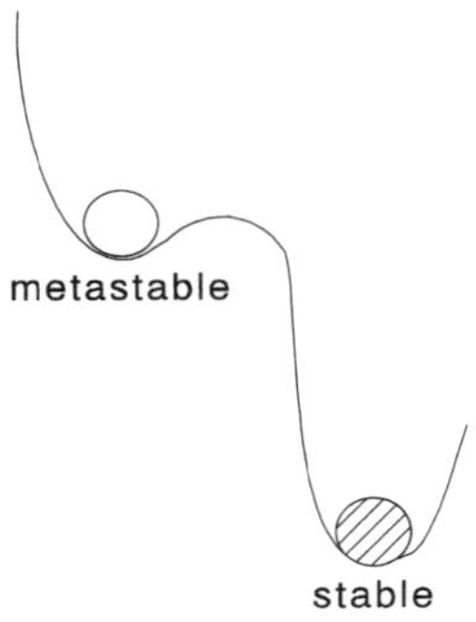

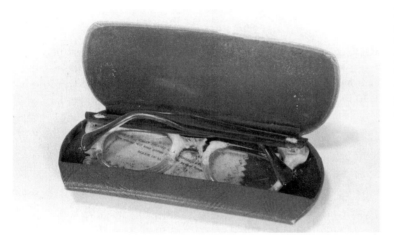

Figure 6.9. The metastable situation is at a local energy minimum, whereas the stable situation is at a global energy minimum.

metastable

stable

Figure 6.10. The frames of these eyeglasses are made of polymers that are predominantly in a glassy form. Over the course of about 30 years, they have begun to convert to a more stable crystalline form, causing destruction of the structure. The conversion is especially prominent at stress points, near the hinges and in the bridge.

with respect to its corresponding crystalline solid. Metastability is defined as a local Gibbs energy minimum, with respect to the global energy minimum being stable (see Figure 6.9). Whereas a supercooled liquid is an equilibrium state (the molecules obey the Boltzmann distribution), a glass, which is obtained by cooling a supercooled liquid, is not in equilibrium with itself. This is because in glasses, molecular configurations (i.e., the relative orientations and packing of the molecules) change slowly (sometimes hardly at all) so that Boltzmann distributions cannot be achieved. Given sufficient time, a glass would eventually convert to a crystalline form because it is more stable; an example of such a conversion is shown in Figure 6.10.

When a glass is heated, it does not have a well-defined melting point, but instead becomes gradually less viscous. Due to the irregularity of the molecular packing in a glass (see Figure 6.11 for a two-dimensional comparison of the arrangement of atoms in crystalline and glassy SiO_2), a glass can flow when a stress is applied.[14]

---

[14] It has been said that window glass in ancient cathedrals is thicker at the bottom than at the top due to the flow of glass over the centuries. More recently this effect has been shown to be insufficient to account for the observed difference in thickness. It is possible that it is due to mounting of panels of glass of irregular thickness: for stability, glaziers may have mounted the glass with the thickest edge to the bottom.

a. 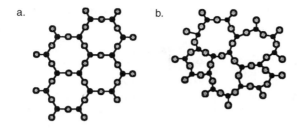 b.

Figure 6.11. A two-dimensional schematic comparison of (a) crystalline SiO_2, which shows periodic atomic arrangement, and (b) glassy SiO_2, in which the atoms are not periodically arranged.

Common glass ($Na_2O \cdot CaO \cdot 6SiO_2$) is also called **soda lime glass.** This glass can be made by the reaction at elevated temperatures of ash with lime and sand. It accounts for about 90% of all glass manufactured today and has been known for about 4500 years. Although this glass is very useful (and has better strength characteristics than many give it credit for), it has a major disadvantage in that it cracks when either heated or cooled rapidly.

The addition of 12% B_2O_3 to soda glass produces a glass with better thermal shock resistance. This glass is commonly known by the brand name **Pyrex®.**

When PbO replaces some of the CaO of soda glass, a denser softer glass, with a high refractive index, is produced. This glass is commonly known as "**lead crystal**" (although this is not technically a correct description of its structure, as it is still glass, not crystalline). The high refractive index of this glass allows its facets to "catch" light and cause its dispersion.

The addition of 5% Al_2O_3 to soda glass gives a strong acid-resistant glass. When fibers of this glass are reinforced with an organic plastic, this is **Fiberglass®.**

The heat capacity of a typical glass is shown in Figure 6.12 in comparison with that of a crystalline solid. From this figure it can be seen that the heat capacity of a typical liquid is higher than that of the corresponding crystalline solid, due to the increased number of degrees of freedom in the liquid. Decreasing in temperature from the liquid, the heat capacity of the supercooled liquid is in line with that of the liquid because no change in state occurred in passing below the melting point to the supercooled liquid state.

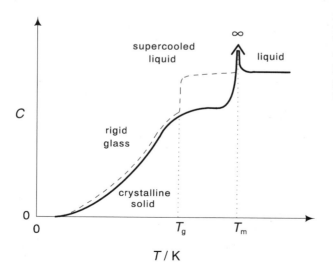

Figure 6.12. Heat capacity as a function of temperature showing typical crystalline solid, liquid, and rigid glass phases. T_m and T_g are the melting and glass transition temperatures, respectively.

As the temperature of the supercooled liquid is lowered further, considerable entropy is removed according to

$$\Delta S = \int \frac{C}{T} \, dT. \tag{6.53}$$

In fact the rate of entropy removal is so high in comparison with the rate of removal in the crystalline solid, because the heat capacity of a supercooled liquid is more than that of the crystalline solid, that continuing at that rate as the temperature was lowered would eventually lead to a negative entropy for the supercooled liquid. This would be an unphysical situation. This is avoided because the heat capacity of the supercooled liquid drops down a step at a temperature characteristic of the material; concurrently rigidity sets in. Below this temperature, known as the **glass transition temperature** T_g, the material is a **rigid glass.** The drop in heat capacity below T_g arises from the change in degrees of freedom from the supercooled liquid (an equilibrium state in which configurational degrees of freedom are active) to the rigid glass (a nonequilibrium state in which the heat capacity is essentially vibrational because the configurational part is frozen). T_g is always less than the normal melting point of the pure crystalline material. For example, for SiO_2, $T_g \approx 1200°C$, and $T_m = 1610°C$.

The heat capacity of the rigid glass is still in excess of that of the crystalline solid, and this can be understood in terms of Debye's heat capacity theory applied to each phase. The fact that the heat capacity of the glass is higher than that of the crystal indicates that the effective Debye characteristic temperature of the glass is lower than that of the crystal. With reference to Figure 6.8, the temperature at which a given heat capacity is achieved in the glass will be lower than the temperature required for the crystal to have this same heat capacity, which implies that $\theta_D^{glass} < \theta_D^{crystal}$. This also means that atomic motion is easier in the glass than in the crystal. This is in line with the results of Table 6.3, which show that materials with lower Debye temperatures are softer. The relative ease of atomic motion in the glass reflects the lack of periodicity in its structure in comparison with the crystalline solid, as can be anticipated from the two-dimensional schematic look at the structures of glassy and crystalline SiO_2 in Figure 6.11.

Many chemical systems, both simple and complex, can exhibit glassy behavior. An appropriate thermal treatment is required to trick the material into forming the metastable supercooled liquid rather than the stable crystalline material. Examples of materials that form glasses include many biological materials, polymers (including plastics and paints), and foodstuffs. The glass transition temperature, T_g, can be very important in the characterization of materials from milk solids to paints, as is illustrated in the Tutorial on Thermal Analysis.

6.7 PHASE STABILITY AND PHASE TRANSITIONS, INCLUDING THEIR ORDER

A number of useful thermodynamic results that can be readily applied to phase stability and phase transitions can be derived from some basic thermodynamic definitions. These will be used here to investigate the relationships among derivatives of Gibbs energy, and this will be used to consider phase stability. Phase stability is very important in materials science because all the properties of a material—optical, thermal, electronic, magnetic, mechanical—depend critically on its phase.

Starting from the definitions of enthalpy, H (Equation 6.18) and Gibbs energy, G:

$$G = H - TS = U + PV - TS \tag{6.54}$$

and Helmholtz[15] energy, A:

$$A = G - PV = H - TS - PV = U - TS, \tag{6.55}$$

for a closed system, infinitesimal changes in H, G, and A can, respectively, be written as

$$dH = dU + PdV + VdP \tag{6.56}$$

$$dG = dU + PdV + VdP - TdS - SdT \tag{6.57}$$

and

$$dA = dU - TdS - SdT. \tag{6.58}$$

If the work is pressure–volume work and it is carried out reversibly then

$$dU = \delta q + \delta w = \delta q_{rev} - PdV = TdS - PdV \tag{6.59}$$

where the **second law of thermodynamics,**

$$dS = \frac{\delta q_{rev}}{T} \tag{6.60}$$

has been used. Equations 6.56 to 6.59 can be used to give

$$dH = TdS + VdP \tag{6.61}$$

$$dG = VdP - SdT \tag{6.62}$$

and

$$dA = -PdV - SdT. \tag{6.63}$$

These expressions for dU, dH, dG, and dA (Equations 6.59, 6.61, 6.62, and 6.63) are known collectively as the **fundamental equations.**

The fundamental equations can be used to derive important thermodynamic relations, known as the **Maxwell[16] Relations.** Consider the fundamental equation for U, Equation 6.59. Since U is a state function of S and V, dU also can be written as

$$dU = \left(\frac{\partial U}{\partial S}\right)_V dS + \left(\frac{\partial U}{\partial V}\right)_S dV \tag{6.64}$$

---

[15] Hermann Ludwig Ferdinand von Helmholtz (1821–1894) was a German physiologist and natural scientist. Helmholtz's thesis on connections between nerve fibers and nerve cells led to his studies of heat in animals, and this led to his theories of conservation of energy. Helmholtz made contributions to a wide variety of subjects: the invention of the ophthalmoscope, the structure and mechanism of the human eye, the role of the bones in the middle ear, electricity and magnetism, and music theory.

[16] James Clerk Maxwell (1831–1879) was a Scottish-born physicist who made many important contributions to electricity, magnetism, color vision, and color photography. His studies of Saturn's rings led to his seminal work on the kinetic theory of gases.

and the coefficients of the two forms of dU (Equation 6.59 and 6.64) can be equated so

$$\left(\frac{\partial U}{\partial S}\right)_V = T \tag{6.65}$$

and

$$\left(\frac{\partial U}{\partial V}\right)_S = -P. \tag{6.66}$$

Since U is a **state function** the order of differentiation for its second derivative does not matter and

$$\frac{\partial^2 U}{\partial V \partial S} = \frac{\partial^2 U}{\partial S \partial V} \tag{6.67}$$

so

$$\left(\frac{\partial T}{\partial V}\right)_S = -\left(\frac{\partial P}{\partial S}\right)_V \tag{6.68}$$

which is the Maxwell Relation from U.

Similarly, from the fundamental equation for G (i.e., Equation 6.62), since G is a state function of P and T, dG also can be written as

$$dG = \left(\frac{\partial G}{\partial P}\right)_T dP + \left(\frac{\partial G}{\partial T}\right)_P dT \tag{6.69}$$

and the coefficients of the two forms of dG (i.e., Equations 6.62 and 6.69) can be equated so

$$\left(\frac{\partial G}{\partial P}\right)_T = V \tag{6.70}$$

and

$$\left(\frac{\partial G}{\partial T}\right)_P = -S. \tag{6.71}$$

Since G is a state function the order of differentiation for its second derivative does not matter and

$$\frac{\partial^2 G}{\partial T \partial P} = \frac{\partial^2 G}{\partial P \partial T} \tag{6.72}$$

so

$$\left(\frac{\partial V}{\partial T}\right)_P = -\left(\frac{\partial S}{\partial P}\right)_T. \tag{6.73}$$

Equation 6.73 is the Maxwell Relation from G.

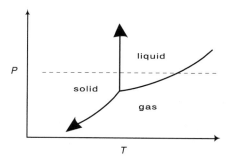

Figure 6.13. The pressure–temperature phase diagram of a typical pure material. The solid lines denote the phase boundaries, and the dashed line indicates an isobaric change in temperature, as discussed in the text.

Two further Maxwell Relations can be derived, one from each of the other fundamental equations. It is left as an exercise for the reader to show that

$$\left(\frac{\partial T}{\partial P}\right)_S = \left(\frac{\partial V}{\partial S}\right)_P \tag{6.74}$$

and

$$\left(\frac{\partial P}{\partial T}\right)_V = \left(\frac{\partial S}{\partial V}\right)_T. \tag{6.75}$$

To consider the stability of a given phase, it is first useful to summarize the pressure–temperature stability of a "typical" material, as given in Figure 6.13.

As the pressure is held constant and the temperature is increased along the dashed line in Figure 6.13, the stable phase goes from solid to liquid to gas. This can be shown in a figure of G as a function of T at constant P (Figure 6.14). This figure shows the driving force for phase changes: at a given temperature and pressure the phase with the lowest G is the most stable. At the equilibrium between two phases (such as the solid–liquid equilibrium at the melting point or the liquid–gas equilibrium at the boiling point), the Gibbs energies of the equilibrium phases are equal ($G_{\text{solid}} = G_{\text{liquid}}$ at the melting point; $G_{\text{liquid}} = G_{\text{gas}}$ at the boiling point).

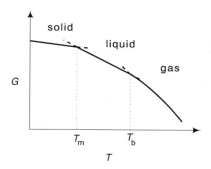

Figure 6.14. Gibbs energy, G, as a function of temperature for a typical material. T_{m} and T_{b} are the melting and boiling temperatures, respectively. Solid lines indicate the stable phase in a given region.

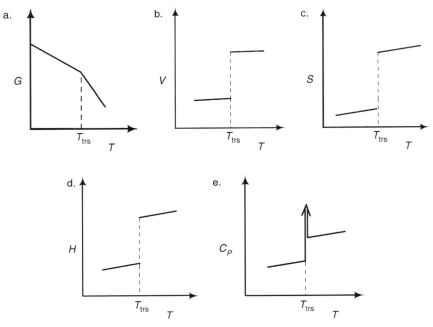

Figure 6.15. (a) $G(T)$, (b) $V(T)$, (c) $S(T)$, (d) $H(T)$, and (e) $C_p(T)$ for a first-order phase transition. T_{trs} designates the transition temperature. Note that C_P of the higher-temperature phase could be lower or higher than that of the lower-temperature phase.

The slope of the curve in the $G(T)$ plot, $(\partial G/\partial T)_P$, is identical to $-S$ (Equation 6.71). Since $S > 0$ (always), all the slopes are negative in Figure 6.14. Since the entropy increases on going from solid to liquid to gas, the slope becomes more negative from solid $\rightarrow$ liquid $\rightarrow$ gas. Even within a given phase the entropy increases as the temperature increases, so the slopes are concave down in Figure 6.14.

Within the constraints of model calculations, G can be calculated for various phases under specific temperature and pressure conditions, and the predicted most stable phase is the one with the lowest G. Comparison with experimental phase stabilities can indicate the validity of the assumed intermolecular interactions.

The **Ehrenfest**[17] **classification of phase transitions** hinges on the behavior of G near the phase transformation. In this classification, phase transitions can be **first-order** (first derivatives of G are discontinuous) or **second-order** (first derivatives of G are continuous but second derivatives of G are discontinuous).

In a **first-order phase transition,** the first derivatives of G [$(\partial G/\partial P)_T = V$ and $(\partial G/\partial T)_P = -S$] are discontinuous. As shown in Figure 6.15, since $G^\alpha = G^\beta$ at the equilibrium of phase α

[17] Paul Ehrenfest (1880–1933) was an Austrian-born theoretical physicist who studied for his Ph.D. under Boltzmann's supervision. As a teacher, Einstein described Ehrenfest as "peerless" and "the best teacher in our profession I have ever known." His students' nickname for him, "Uncle Socrates," showed his probing but personable style. Ehrenfest's contributions to thermodynamics and quantum mechanics stemmed from his ability to ask critical, probing questions. Depression due to the plight of his Jewish colleagues and personal difficulties led Ehrenfest to take his own life.

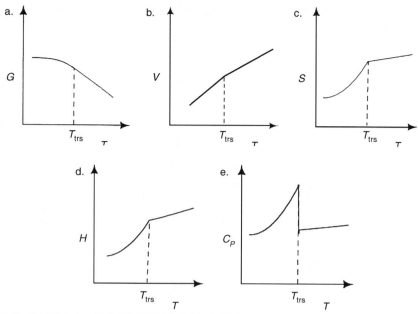

Figure 6.16. (a) G(T), (b) V(T), (c) S(T), (d) H(T), and (e) $C_p(T)$ for a second-order phase transition. T_{trs} is the transition temperature.

with phase β, and given the preceding discussion, $\Delta_{trs}V \neq 0$ and $\Delta_{trs}S \neq 0$ at a first-order phase transition. From $\Delta_{trs}G = 0$ and Equation 6.7 it follows that

$$\Delta_{trs}S = \frac{\Delta_{trs}H}{T} . \qquad (6.76)$$

Since there is a change in the amount of disorder at a transition,

$$\Delta_{trs}S \neq 0 \qquad (6.77)$$

and therefore

$$\Delta_{trs}H \neq 0 \qquad (6.78)$$

for a first-order phase transition. From Equation 6.4 with the integration to give $\Delta_{trs}H$ over the fixed temperature T_{trs}, for $\Delta_{trs}H$ to be nonzero, C_p must be infinite at the transition. This is illustrated in Figure 6.15. Examples of first-order transitions (i.e., transitions with the signatures of $\Delta_{trs}V \neq 0$, $\Delta_{trs}S \neq 0$, $\Delta_{trs}H \neq 0$, $C_p = \infty$) include melting, boiling, sublimation, and some solid–solid transitions (e.g., graphite-to-diamond).

  A **second-order phase transition** does not have a discontinuity in $V[=(\partial G/\partial P)_T]$ or $S[= -(\partial G/\partial T)_p]$, but the second derivatives of G [e.g., $(\partial^2 G/\partial T\partial P) = (\partial V/\partial T)_p$ and $(\partial^2 G/\partial P\partial T) = -(\partial S/\partial P)_T]$ are discontinuous. This is illustrated in Figure 6.16; $\Delta_{trs}V = 0$, $\Delta_{trs}S = 0$, and $\Delta_{trs}H = 0$ at a second-order transition. The third condition leads to the finite heat capacity at a second-order transition, as shown in Figure 6.16. The λ shape of the heat capacity at this transition leads this to be called a λ transition. An example of such a transition (which is much rarer than a first-order transition) is observed in liquid helium (see Chapter 9).

6.8 $(C_P\text{-}C_V)$: AN EXERCISE IN THERMODYNAMIC MANIPULATIONS

The aim here is to derive a useful thermodynamic relationship between C_P and C_V and to look at one special case of its use.

From the definitions of C_P (Equation 6.3) and C_V (Equation 6.1):

$$C_P - C_V = \left(\frac{\partial H}{\partial T}\right)_P - \left(\frac{\partial U}{\partial T}\right)_V \tag{6.79}$$

and $(C_P\text{--}C_V)$ is a measure of the difference in energy required to increase temperature at constant pressure (which allows for $P\text{--}V$ work of expansion on heating) relative to increasing the temperature at constant volume (where there is no expansion work).

The definition of H (Equation 6.18) leads to an expression for dH in terms of dU (Equation 6.56), which can be differentiated with respect to temperature at constant pressure to give

$$\left(\frac{\partial H}{\partial T}\right)_P = \left(\frac{\partial U}{\partial T}\right)_P + P\left(\frac{\partial V}{\partial T}\right)_P \tag{6.80}$$

and substitution of Equation 6.80 into Equation 6.79 leads to

$$C_P - C_V = \left(\frac{\partial U}{\partial T}\right)_P + P\left(\frac{\partial V}{\partial T}\right)_P - \left(\frac{\partial U}{\partial T}\right)_V. \tag{6.81}$$

This expression can be simplified by finding another expression for $(\partial U/\partial T)_P$. Since U is a function of V and T,

$$dU = \left(\frac{\partial U}{\partial V}\right)_T dV + \left(\frac{\partial U}{\partial T}\right)_V dT \tag{6.82}$$

which leads, on differentiation with respect to temperature at constant pressure, to

$$\left(\frac{\partial U}{\partial T}\right)_P = \left(\frac{\partial U}{\partial V}\right)_T \left(\frac{\partial V}{\partial T}\right)_P + \left(\frac{\partial U}{\partial T}\right)_V. \tag{6.83}$$

Substitution of Equation 6.83 into Equation 6.81 leads to

$$C_P - C_V = \left(\frac{\partial V}{\partial T}\right)_P \left[\left(\frac{\partial U}{\partial V}\right)_T + P\right]. \tag{6.84}$$

Although both $(\partial V/\partial T)_P$ and P are directly experimentally accessible, $(\partial U/\partial V)_T$ is not, so it is useful to try to relate it to some other variables. From the fundamental equation for U (Equation 6.59), differentiation with respect to volume at constant temperature leads to

$$\left(\frac{\partial U}{\partial V}\right)_T = T\left(\frac{\partial S}{\partial V}\right)_T - P. \tag{6.85}$$

Substitution of Equation 6.85 into Equation 6.84 gives

$$C_P - C_V = T\left(\frac{\partial V}{\partial T}\right)_P \left(\frac{\partial S}{\partial V}\right)_T. \tag{6.86}$$

Although $(\partial S/\partial V)_T$ is still awkward on its own, a Maxwell Relation (Equation 6.75) allows

$$C_P - C_V = T\left(\frac{\partial V}{\partial T}\right)_P \left(\frac{\partial P}{\partial T}\right)_V \tag{6.87}$$

for all systems, with no approximations. However, Equation 6.87 is usually written in terms of the **coefficient of thermal expansion** (α) and the **isothermal compressibility** (β_T), defined as follows:

$$\alpha = \frac{1}{V}\left(\frac{\partial V}{\partial T}\right)_P \tag{6.88}$$

and

$$\beta_T = \frac{-1}{V}\left(\frac{\partial V}{\partial P}\right)_T. \tag{6.89}$$

From these definitions units of α are K^{-1} and units of β_T are reciprocal pressure units (e.g., Pa^{-1}). The advantage of using α and β_T rather than $(\partial V/\partial T)_P$ and $(\partial V/\partial P)_T$ is that α and β_T are **intensive properties** (i.e., independent of the size of the system), whereas $(\partial V/\partial T)_P$ and $(\partial V/\partial P)_T$ require knowledge of the size of the system (i.e., they are **extensive properties,** dependent upon the size of the system.)

From the definition of α (Equation 6.88) it follows that

$$C_P - C_V = TV\alpha\left(\frac{\partial P}{\partial T}\right)_V. \tag{6.90}$$

To obtain $(\partial P/\partial T)_V$, consider that V is a function of P and T so that

$$dV = \left(\frac{\partial V}{\partial P}\right)_T dP + \left(\frac{\partial V}{\partial T}\right)_P dT \tag{6.91}$$

which leads to

$$\left(\frac{\partial P}{\partial T}\right)_V = -\frac{(\partial V/\partial T)_P}{(\partial V/\partial P)_T} \tag{6.92}$$

and

$$\left(\frac{\partial P}{\partial T}\right)_V = \frac{-V\alpha}{-V\beta_T} = \frac{\alpha}{\beta_T}. \tag{6.93}$$

Substitution of Equation 6.93 into 6.90 leads to the following general expression for $(C_P - C_V)$:

$$C_P - C_V = \frac{TV\alpha^2}{\beta_T} \tag{6.94}$$

for isotropic materials. If the material is anisotropic, α and β_T will depend on the direction, and a form of Equation 6.94 specific to the system's symmetry must be employed.

In general, $(C_P - C_V)$ is small compared with C_P for most solids; it is particularly small at low temperatures (since $\alpha \to 0$ as $T \to 0$; see Chapter 7) and may be <1% of C_P at $T < \theta_D$. $(C_P - C_V)$ typically is a few percent of C_P for a solid at room temperature.

For the special case of ideal gases, $(C_P - C_V)$ has a simple form. An ideal gas is described by the equation of state given by Equation 6.12, so α for an ideal gas can be written

$$\alpha = \frac{1}{V}\left(\frac{\partial V}{\partial T}\right)_P = \frac{P}{nRT}\left[\frac{\partial}{\partial T}\left(\frac{nRT}{P}\right)\right] = \frac{P}{nRT}\left(\frac{nR}{P}\right) = \frac{1}{T} \tag{6.95}$$

and β_T for an ideal gas can be written

$$\beta_T = \frac{-1}{V}\left(\frac{\partial V}{\partial P}\right)_T = -\frac{P}{nRT}\left[\frac{\partial}{\partial P}\left(\frac{nRT}{P}\right)\right] = \frac{P}{nRT}nRT\,P^{-2} = \frac{1}{P} \tag{6.96}$$

and substitution of these into Equation 6.94 leads to

$$C_{p,m} - C_{V,m} = \frac{TV(T^{-2})}{P^{-1}} = \frac{VP}{T} = R \tag{6.97}$$

for an ideal gas, where the result is written in terms of the molar heat capacities, $C_{p,m}$ and $C_{V,m}$.

Heat Storage Materials: A Tutorial

Heat storage materials have many applications, from scavenging heat from hot industrial exhaust to keeping electrical components from overheating during soldering, to storing the energy from solar collectors, to making use of off-peak electricity.

a. Heat storage materials such as rock, brick, water, and concrete absorb energy when heated through what is called **sensible heat storage** (i.e., heat storage due to their heat capacities). These materials are used as "night heaters," for example, taking advantage of off-peak electricity rates. The room temperature **specific heat capacities**[18] (C) of some common materials are given below:

Material	C/J K^{-1} g^{-1}
Common brick	0.920
Balsa wood	2.93
Marble	0.879

Which of these materials can store the most energy per unit mass at this temperature? What additional information is needed to make this decision per unit volume of material? Do you expect this to affect which is the best heat storage material?

---

[18] The term "specific" (e.g., specific heat capacity or specific gravity) is used to denote the value of a physical property per unit mass.

b. Some other materials such as Glauber salts (hydrates of $CaSO_4$ and Na_2SO_4) and paraffins (long chain saturated hydrocarbons) store energy through their phase transitions from solid to liquid. Sketch the following for a material that stores energy through fusion (i.e., melting): (i) the heat capacity through the region from solid to liquid; (ii) the enthalpy through the region from solid to liquid.

c. What do you expect to be the most important criteria for a successful energy storage material that uses fusion? Think of thermodynamics, but also think of other matters.

d. A commercial product, Re-Heater™, contains sodium acetate trihydrate. This salt melts at 58°C, with an enthalpy change of 17 kJ mol⁻¹. Sketch the enthalpy vs. temperature diagram for this salt from 0 to 100°C.

e. Sodium acetate trihydrate can be supercooled fairly easily to room temperature. Using a dashed line, draw the enthalpy curve for the supercooling process on your diagram from (d).

f. A supercooled liquid is an example of a metastable state. In terms of the Gibbs energy, *G,* this state is at a local minimum, and the stable state is at a global minimum. Sketch a Gibbs energy curve that shows both these states.

g. When a supercooled liquid is perturbed, for example when a seed crystal is added, the barrier between the stable and the metastable states is lowered, and the stable state is achieved. This is a *physical* change. What do we usually call this effect for *chemical* changes?

h. How does a seed crystal lower the Gibbs energy path between the metastable supercooled liquid state and the stable crystalline state?

i. If you were able to lower the Gibbs energy path between the supercooled liquid state of sodium acetate trihydrate and its solid state, from the enthalpy diagram of part (d) and (e), what would happen energetically?

j. In a closed system it is not easy to add a seed crystal. How else could you induce crystallization in supercooled liquid sodium acetate trihydrate?

k. This application gives off heat when activated. Does the spontaneous crystallization process always give off heat? Can you think of a similar principle that takes up heat when activated?

Thermal Analysis: A Tutorial

There are three main techniques in thermal analysis: TGA is **t**hermo**g**ravimetric **a**nalysis, DTA is **d**ifferential **t**hermal **a**nalysis, and DSC is **d**ifferential **s**canning **c**alorimetry.

a. **Thermogravimetric analysis** (TGA) is an analytical tool to carry out "gravimetry" (i.e., mass determination) of a sample as the temperature is changed. This tool is straightforward: the sample is heated and mass is determined as a function of temperature.

A typical plot is mass vs. temperature. Suppose that a commercial sample of mass 10.0 mg is heated from room temperature and experiences a 3.0% mass loss at 80°C. Sketch the resulting thermogravimetric curve.

b. Thermogravimetry can be used to characterize pure samples and blends, and it is a very useful technique. An even more useful thermoanalytical tool is **differential thermal analysis.** In this technique, a sample and a reference (e.g., an empty sample cell) are both monitored as a function of temperature. The experiment can be carried out either in the heating or cooling mode, but the rate of sample heating (or cooling) is usually kept constant. In this experiment, the same power is input to the sample and the reference, and the temperature difference between the sample and the reference is measured as a function of temperature. If the heat capacities of the sample and reference are exactly the same throughout the temperature range of the measurement, the signal is "flat" as shown in Figure 6.17. What would the signal look like if the sample container held a solid that melted during the temperature range of the scan?

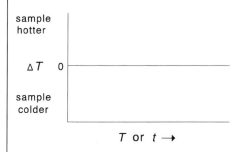

Figure 6.17. A flat signal from DTA when the sample and reference are perfectly matched in heat capacity.

c. Why do you suppose this technique is called **differential** thermal analysis?

d. What would the signal look like if a rigid glass was heated to well into the region where it is a supercooled liquid?

e. **Differential scanning calorimetry** is similar to DTA in several respects. Again, it is a differential technique, with a sample and a reference, and the set-up is heated (or cooled) at a constant rate. What distinguishes DSC from DTA is that in DTA the temperature difference between the sample and the reference is monitored, and in DSC the difference in power to the sample and reference to keep heating (or cooling) them both at the same rate is measured. Sketch and label the DSC curve for the melting of a crystalline solid.

f. It would be very useful in a technique such as DTA or DSC if the area "under" peaks could be converted to enthalpy of the corresponding process. Of DTA and DSC only one has the area scaling to enthalpy. Which one? *Hint:* Consider what the peak area units are in each case.

g. How could the area under the peak be scaled to determine the enthalpy?

h. As an example of an amorphous solid, consider the following. Sugar can be crystalline (e.g., big sugar crystals that can be grown out of sugar/water solutions; also white sugar—this is particulary obvious under a microscope). It also can be amorphous (or glassy or **vitreous;** all mean the same thing). If sugar is heated to a high temperature (above the melting point) and then cooled very quickly, for example by dropping it into cold water, it is transformed to a rigid glassy state. What is the confectionery name for this state of sugar?

i. Why is a candy thermometer needed to make this confectionery?

j. Is the rigid glassy state of sugar stable? What happens to this state of sugar if left for a long time?

k. If the melted sugar is allowed to cool more slowly, will it form the same rigid glassy state?

l. What are the white specks that sometimes form on the surface of chocolates that have been around for a long time? Should you eat these chocolates?

m. Why would a chocolate factory own a differential scanning calorimeter?

n. Although DSC can be used to determine heat capacity of a sample, the precision of the results can be considerably greater than their accuracy. [The most accurate technique to determine heat capacities near and below room temperature is adiabatic calorimetry, in which the heat capacity is determined by the energy input and temperature rise (Equations 6.10 and 6.11) under adiabatic conditions.] Suggest some possible systematic errors in DSC determination of heat capacity.

o. What does Figure 6.18 indicate about the relative enthalpy changes associated with the two stages of setting of cement?

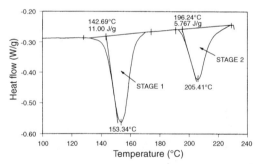

Figure 6.18. DSC scan on heating of cement showing the successive transformations: $CaSO_4 \cdot 2H_2O \rightarrow CaSO_4 \cdot \frac{1}{2}H_2O + \frac{3}{2}H_2O$ and $CaSO_4 \cdot \frac{1}{2}H_2O \rightarrow CaSO_4 + \frac{1}{2}H_2O$. Diagram supplied by Roger Blaine.

FURTHER READING

General References

Many introductory physical chemistry and thermodynamics textbooks discuss the concepts presented here (heat capacity, Maxwell Relations, phase transitions, etc.). In addition, the following address many of the general points raised in this chapter:

A. Bondi (1968). *Physical Properties of Molecular Crystals, Liquids and Glasses.* John Wiley & Sons, Inc.

M. de Podesta (1996). *Understanding the Properties of Matter.* Taylor & Francis.

A. Guinier and R. Jullien (1989). *The Solid State.* Oxford University Press.

H.M. Rosenberg (1989). *The Solid State.* Oxford University Press.

A.J. Walton (1983). *Three Phases of Matter,* 2nd Ed. Oxford University Press.

Glass and Amorphous Matter

R.H. Brill (1963). Ancient Glass. *Scientific American,* November 1963, 120.

R.J. Charles (1967). The Nature of Glasses. *Scientific American,* September 1967, 126.

Y.-T. Cheng and W.L. Johnson (1987). Disordered Materials: A Survey of Amorphous Solids. *Science,* **235,** 977.

W.S. Ellis (1993). Glass: Capturing the Dance of Light. *National Geographic,* **184,** 37.

C.H. Greene (1961). Glass. *Scientific American,* January 1961, 92.

R.C. Plumb (1989). Antique Windowpanes and the Flow of Supercooled Liquids. *Journal of Chemical Education,* **66,** 994.

Liquid Crystals

A.J. Leadbetter (1990). Solid Liquids and Liquid Crystals. *Proceedings of the Royal Institute of Great Britain,* **62,** 61.

Thermal Analysis

R. Blaine (1995). Determination of Calcium Sulfate Hydrates in Building Materials Using Thermal Analysis. *American Laboratory,* September 1995, 24.

M.A. White (1996). Thermal Analysis and Calorimetry Methods. Chapter 4 in Volume 8 of *Comprehensive Supramolecular Chemistry.* J.-M. Lehn, J.L. Atwood, D.D. MacNicol, J.E.D. Davies, and F. Vögtle, Eds. Pergamon Press.

Thermodynamics of Water

S.W. Benson and E.D. Siebert (1992). A Simple Two-Structure Model for Liquid Water. *Journal of the American Chemical Society,* **114,** 4269.

PROBLEMS

1. The ideal gas law is a limiting law.
   a. Derive the relationship between pressure and density for an ideal gas.
   b. Confirm the relationship (as a limiting law) on the basis of the following data for dimethyl ether at $T = -13.6°C$.

P/atm	1.000	0.7000	0.5000	0.3000	0.2000	0.1000
$\rho/(g\ L^{-1})$	2.4110	1.4684	1.0615	0.6635	0.4501	0.2247

   Use a plot to answer this question. It may prove useful to know that the molecular formula of dimethyl ether is CH_3OCH_3.

2. The Einstein function for heat capacity is given by Equation 6.28.
   a. Calculate the value of C_V from the Einstein function for 1 mol of a monatomic material with $\tilde{v} = 100$ cm^{-1} at the following temperatures: 5.00 K; 10.00 K; 25.00 K; 50.00 K; 75.00 K; 100.0 K; 150.0 K; 200.0 K; 250.0 K; 300.0 K. Give your answer for $C_{V,m}$ in terms of R, the gas constant.
   b. What is the percentage difference between the calculated value of $C_{V,m}$ at $T = 300.0$ K compared with the Dulong–Petit Law?

3. Derive the Maxwell Relations from the fundamental equations for (a) H and (b) A.

4. Derive a plot of G as a function of pressure [analogous to $G(T)$ given in Figure 6.14] for a material that goes from gas to liquid to solid as the pressure is increased. Show your reasoning as fully as possible.

5. A new molecular species was investigated with a view to using it as a commercial refrigerant. The temperature drops for a series of initial pressures, P_i, expanding into exactly 1 atm pressure were measured at 0°C, with the following results:

P_i/atm	32	24	18	11	8	5
$-\Delta T$/K	22	18	15	10	7.4	4.6

   a. What is the value of the Joule–Thomson coefficient for this gas at these conditions? *Hint:* Consider the definition of the Joule–Thomson coefficient and the definition of the partial derivative function.
   b. Would another gas that would be a better refrigerant have a lower or higher Joule–Thomson coefficient? Explain.

6. Calculate $(C_p - C_v)$ in J K^{-1} mol^{-1} for (a) liquid water and (b) solid Zn, given the following information for 25°C:

	α/K^{-1}	β_T/atm^{-1}	V_m/(cm^3 mol^{-1})
H_2O (l)	4.8×10^{-4}	4.6×10^{-5}	18
Zn (s)	8.93×10^{-5}	1.5×10^{-6}	7.1

and compare the differences with the C_p values: 75 J K^{-1} mol^{-1} for water and 23 J K^{-1} mol^{-1} for Zn.

7. Sketch how the differential scanning calorimeter trace for water would look on cooling from 15 to -10°C. Label the temperature axis (including values and units), but just include the label and the units (i.e., no values) on the other axis.

8. A new gaseous species has been made in the laboratory, and it is found to be the triatomic molecule ABC. It is not known if ABC is linear or nonlinear. Explain how heat capacity measurements could differentiate between these possibilities. Be as explicit as possible, including stating expected values of the heat capacities and the conditions under which you would carry out the measurements.

9. Very careful calorimetry can be used to carry out quality control tests on pacemakers to determine the reliability of their power supplies. Suggest the physical basis for these tests.

Figure 6.19. A seedling is surrounded by cylinders of water. This is a top view.

10. Although the Debye model is better able to describe the heat capacity of a monatomic solid, the Einstein model is useful in molecular solids with more than one atom per molecule. In this case, there are vibrational motions of the whole molecule (**acoustic phonons**) that are described well by the Debye model, and motions due to internal vibrations (**optic phonons**) that are described well by the Einstein model. In a unit cell with N atoms, there are three acoustic (Debye) modes and $3N - 3$ optic (Einstein) modes. Calculate $C_{V,m}$ for a diatomic molecular solid with two atoms per unit cell and a Debye characteristic temperature of 300 K and all Einstein modes at 200 cm^{-1}. Plot your result as a function of temperature from $T = 0$ to 500 K.

11. A recent advertisement claimed that a new product could allow planting of outdoor crops earlier than would otherwise be possible in cold climates. It might even be possible, using this device, to plant before all danger of frost has passed. This product is a plastic container composed of vertical columns filled with water, which then surrounds the seedling with a cylinder of water. A top view is shown schematically in Figure 6.19.
    a. How does this device work? Explain in thermodynamic terms.
    b. Would this device be as effective if it were filled with another liquid, e.g., oil? Explain.

12. A gas at room temperature and at a constant pressure of 1 atm is going to be used to heat a material that is at a cooler temperature. You are offered the option of using 1 mol of argon (Ar) gas or 1 mol of nitrogen (N_2) gas for this purpose. Which will be more efficient? Explain your reasoning.

13. a. Which form of carbon do you expect to have the higher value of θ_D: diamond or graphite? Justify your answer.
     b. Sketch the molar heat capacity from $T = 0$ K to $T = 300$ K for graphite and diamond. Put both on the same graph, and put a numeric scale on the heat capacity axis.

14. A consumer product made of a synthetic material (which looks very much like leather) is supple at room temperature but found to become brittle in the outdoors in the Canadian winter. It is thought that this is because the material is not very crystalline (i.e., it is highly amorphous), and it undergoes a transformation to a rigid glass at a temperature of about –10°C. Sketch the expected differential scanning calorimetry trace for such a material from –30 to 20°C.

15. When pine ages in a well-aired place over tens of years, there is an increase in the ratio of the crystalline-to-amorphous portion of the wood. This is a desired feature in pine that is to be used for construction of musical instruments, as crystalline materials are less permeable to moisture than amorphous materials. Sketch the DSC curves of two pieces of wood, one well-aged and the other more recently cut.

16. A bowl of frozen vegetables is to be cooked in a microwave oven. When the bowl is covered during the cooking process, 2 minutes is sufficient cooking time to produce steaming hot vegetables. If the bowl is not covered, after 2 minutes cooking the vegetables are still cold. Why?

17. When a bicycle tire is inflated with a hand pump the valve stem is noticeably warmed (temperature increase of about 10 K). Why?

18. For silver, $\theta_D = 225$ K and $\gamma = 6.46 \times 10^{-4}$ J K^{-2} mol^{-1}. Calculate C_V of silver at 0.1 K and at 1 K. Comment on the relative contributions of the lattice (Debye) heat capacity and the electronic heat capacity at each temperature.

19. Air in a container is at a pressure of about 1 atm and room temperature. Inside the container there is a sensitive thermometer and when the air is pumped out of the container with a vacuum pump the temperature is seen to fall. Explain why the temperature falls and what this indicates about the gas.

20. The fuels on the space shuttle Challenger leaked out during the flight, causing the shuttle to explode. The fuels were meant to be held in place by O-rings, i.e,. soft rubbery materials that could be compressed to make tight seals. However, at the cold launch temperature (near 0°C), the soft rubbery material had become a rigid glass. On this basis, sketch the DSC (differential scanning calorimetric) curve for this rubber from 0 to 25°C and indicate any features.

21. Polymers have been proposed as the basis of new electrooptic devices, but one of the problems is that the required molecular alignment of amorphous polymers can decay over time. For a polymer to be useful in an electrooptic device should T_g be very high or very low? Explain.

22. The velocity of sound varies considerably with depth in the earth's crust. In the upper mantle, it is 8 km s^{-1}, and it increases rapidly with depth to about 12 km s^{-1} in the lower mantle.
    a. How does the Debye characteristic temperature vary with depth?
    b. What does this variation of the sound velocity indicate about the interactions within the crustal materials as a function of depth? Does this make sense?
    c. Suggest an application of the variation of sound velocity with depth in the earth's crust.

23. Fast-cooled slag from smelters is glassy and therefore less reactive with water from the environment than slow-cooled slag (which is more crystalline). Therefore, fast-cooled slag is preferred because it leads to less run-off. Explain how DSC could be used for quality control of the slag-cooling process.

7

THERMAL

EXPANSION

7.1 INTRODUCTION

Almost all materials expand when heated, regardless of the phase of matter. This **thermal expansion** can be related directly to the forces between the molecules. Different thermal expansions of materials used together in applications can lead to mechanical problems (loose fits or stress). We begin by considering the basis of thermal expansion in gases (ideal and nonideal) and then move on to the solid state.

7.2 COMPRESSIBILITY AND THERMAL EXPANSION OF GASES

The defining equation (also called equation of state) for an ideal gas is

$$PV_m = RT \tag{7.1}$$

where V_m is the molar volume ($=V/n$). From Equation 7.1 it can be readily seen that as the temperature is increased at constant pressure for an ideal gas, the volume increases. In other words, gases expand on isobaric warming (i.e., they undergo thermal expansion).

For a nonideal gas, corrections must be introduced to Equation 7.1 to account for the shortcomings of the ideal gas model. The **compressibility factor, Z,** is defined as

$$Z = \frac{PV_m}{RT} \tag{7.2}$$

and clearly $Z = 1$ for an ideal gas and $Z \neq 1$ for a nonideal gas. All gases approach ideality as $P \to 0$, i.e., $Z \to 1$ in the limit $P \to 0$.

For a nonideal gas in which $Z < 1$, at a given temperature and pressure the molar volume

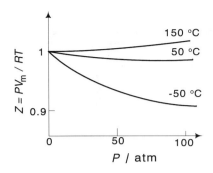

Figure 7.1. Variation of compressibility factor, Z, with temperature and pressure for N_2.

is less than the molar volume of an ideal gas. This means that the gas is more compressed than if it had been ideal, which implies that the intermolecular forces are dominantly attractive under these conditions (e.g., very low temperature).

The opposite would be true if $Z > 1$: at a given temperature and pressure the molar volume would be greater than the ideal molar volume. This reflects the dominance of repulsive intermolecular interactions under these conditions (e.g., very high temperature).

For a real gas, Z varies with both temperature and pressure, and with the type of gas. An example is shown in Figure 7.1 for N_2.

It is worth considering how the value of Z might be qualitatively related to the Joule–Thompson coefficient, as both can be used to quantify nonideality in a gas. This is left as an exercise for the reader.

As it turns out, many gases can be placed together on a series of plots such as shown in Figure 7.1 by using appropriate scaling factors. The scaling factors are related to the exact conditions under which that particular liquid is no longer distinguishable from its gas phase; this point is called the **critical point** (see Figure 7.2) and the temperature, pressure, and volume at

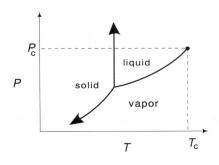

Figure 7.2. Stability of phases of matter for a typical pure material as functions of temperature and pressure, showing the critical pressure, P_c, and critical temperature, T_c, beyond which gases and liquids are no longer distinguishable.

TABLE 7.1.
Critical Temperatures, Pressures, and Volumes for Some Gases[a]

Gas	Boiling point/K	T_c/K	P_c/atm	V_c/(cm³ mol⁻¹)
He	4.2	5.2	2.25	61.55
Ne	27.3	44.75	26.86	44.30
H_2	20.7	33.2	12.8	69.69
O_2	90.2	154.28	49.713	74.42
N_2	77.4	125.97	33.49	90.03
CO_2	194.7	304.16	72.83	94.23
H_2O	373.2	647.3	218.5	55.44
NH_3	239.8	405.5	112.2	72.02
CH_4	109.2	190.25	45.6	98.77

[a] Data are from A. W. Adamson (1973), *A Textbook of Physical Chemistry*, Academic Press.

this point are referred to as the **critical temperature** (T_c), **critical pressure** (P_c), and **critical volume** (V_c), respectively. T_c, P_c, and V_c depend on the gas, and some typical values are listed in Table 7.1.

When the temperature is considered in units of T_c, this defines the **reduced temperature, T_r**:

$$T_r = \frac{T}{T_c} \qquad (7.3)$$

and the **reduced pressure, P_r**, and **reduced volume, V_r**, are similarly defined:

$$P_r = \frac{P}{P_c} \qquad (7.4)$$

and

$$V_r = \frac{V}{V_c}. \qquad (7.5)$$

A plot of compressibility, Z, as a function of P_r for various values of T_r (see Figure 7.3) is very nearly universal for all gases. This so-called **Hougen–Watson plot** is particularly useful as it can be used to derive P,V,T relationships for a gas in conditions in which the ideal equation of state does not hold.

The near universality of the Hougen–Watson plot suggests that there could be a relatively simple expression relating P,V, and T at high pressures where $PV_m \neq RT$. There are many high-pressure equations of state and we now consider one: the **van der Waals**[1] **equation of state.** This equation of state simply accounts for the major shortcomings in the ideal gas model.

First, real molecules do take up space. The ideal gas model assumed this to be negligible,

---

[1] Johannes Diderik van der Waals (1837–1923) was a Dutch physicist. Although his career got off to a slow start [his Ph.D. thesis (the subject was the equation of state that now carries his name) was not published until he was 35 years old], van der Waals' accomplishments concerning equations of state for real gases were widely acclaimed. He was awarded the 1910 Nobel Prize in Chemistry for his work on states of matter.

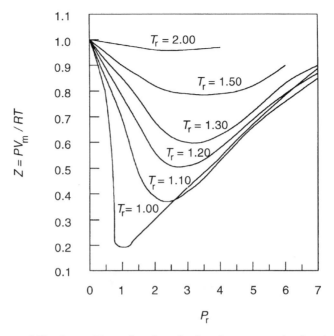

Figure 7.3. Compressibility factor, Z, as a function of reduced pressure and reduced temperature, for a number of gases. Data from O.A. Hougen, K.M. Watson, and R.A. Ragatz (1959), *Chemical Process Principles II Thermodynamics,* 2nd Ed., John Wiley & Sons.

which is why it breaks down at higher pressures. Since the real volume is greater than the ideal volume, the van der Waals equation of state replaces V_m^{ideal} with $(V_m^{real} - b) = (V_m - b)$ where b is the effective volume occupied by the molecules in 1 mol of gas; it follows that b depends on the type of gas.

Second, the molecules in real gases feel intermolecular forces, whereas the ideal gas model assumed there were no intermolecular forces. At long range the forces are dominantly attractive, so the observed pressure is less than the ideal pressure. van der Waals proposed, on the basis of experiments, that the following substitution could be made in the ideal equation:

$$P^{ideal} = P^{obs} + \frac{a}{V^2} \tag{7.6}$$

where P^{obs} is the observed pressure ($=P$), and a depends on the gas.

With these two substitutions, the ideal equation of state (Equation 7.1) can be modified to produce the van der Waals equation of state:

$$\left(P + \frac{a}{V_m^2}\right)(V_m - b) = RT. \tag{7.7}$$

A list of the van der Waals constants (a and b) for some gases is given in Table 7.2.

TABLE 7.2.
van der Waals Constants for Selected Gases[a]

Gas	$a/(\text{L}^2 \text{ atm mol}^{-2})$	$b/(\text{L mol}^{-1})$
He	0.03412	0.02370
Ne	0.2107	0.01709
H_2	0.2444	0.02661
O_2	1.360	0.03183
N_2	1.390	0.03913
CO_2	3.592	0.04267
H_2O	5.464	0.03049
NH_3	4.170	0.03707
CH_4	2.253	0.04278

[a] Data are from A.W. Adamson (1973), *A Textbook of Physical Chemistry,* Academic Press.

The van der Waals equation of state is still too inaccurate to describe every gas in every circumstance. Other more advanced equations of state (including those taking anisotropy of molecular shape into account) are also available, and the interested reader is referred to the Further Reading section at the end of this chapter for references to these. However, it is worth mentioning that Kamerlingh Onnes[2] suggested in 1901 that the equation of state for a gas could be expressed as a power series (also called the **virial equation**) such as

$$Z = \frac{PV_m}{RT} = 1 + B'(T)P + C'(T)P^2 + D'(T)P^3 + \cdots \tag{7.8}$$

or, more accurately, as a power series in V_m:

$$Z = \frac{PV_m}{RT} = 1 + \frac{B(T)}{V_m} + \frac{C(T)}{V_m^2} + \frac{D(T)}{V_m^3} + \cdots \tag{7.9}$$

where the coefficients $B'(T)$, $C'(T)$, and $D'(T)$ [or $B(T)$, $C(T)$, and $D(T)$] are called the second, third, and fourth **virial coefficients,** respectively. Virial coefficients can be directly related to intermolecular forces.

In these considerations of nonideality, it may be interesting to note that although gases deviate from nonideality at pressures as low as a few atmospheres, the gas in the sun (at pressures of billions of atmospheres and extremely high temperatures) can be expressed well by the ideal gas law,[3] presumably due to accidental cancellation of attractive and repulsive forces at these extreme conditions.

[2] Heike Kamerlingh Onnes (1853–1926) was a Dutch physicist. In his inaugural lecture as Professor of Experimental Physics at Leiden University, Kamerlingh Onnes stated: "In my opinion it is necessary that in the experimental study of physics the striving for quantitative research, which means for the tracing of measure relations in the phenomena, must be in the foreground. I should like to write 'Door meten tot weten' ['Through measuring to knowing'] as a motto above each physics laboratory." This outlook led him to build a laboratory for the first low-temperature research, including the first liquefaction of H_2 (20.4 K) in 1906 and He (4.2 K) in 1908. Kamerlingh Onnes was awarded the 1913 Nobel Prize in Physics for low-temperature research.

[3] D.B. Clark. (1989) *Journal of Chemical Education,* **65,** 826.

The thermal expansion, α, of a gas can be calculated for any equation of state, ideal or nonideal. When it is not possible to manipulate the equation of state to derive thermal expansion analytically, α can be evaluated numerically by calculation of the volume at two close temperatures and use of the definition of α from Equation 6.88; see Problem 2 at the end of this chapter.

7.3 THERMAL EXPANSION OF SOLIDS

To a first approximation a solid can be described as atoms (or molecules) interacting with their neighbours as if they are connected with springs. For two atoms separated by a distance x away from the equilibrium distance, the force between them, F, can be expressed, to a first (harmonic) approximation, by **Hooke's[4] law:**

$$F = -k'x \tag{7.10}$$

where k' is the **Hooke's law force constant.**

The potential energy, $V(x)$, can be derived from the force, $F(x)$, by the general relation

$$F(x) = -\frac{dV(x)}{dx} \tag{7.11}$$

i.e.,

$$V(x) = -\int F(x)\,dx \tag{7.12}$$

where Equation 7.11 is a form of Equation 5.2. For a force given by Equation 7.10, the potential energy is

$$V(x) = \frac{k'}{2}x^2 = cx^2 \tag{7.13}$$

where $V = 0$ at $x = 0$. Since the potential function is a parabolic shape (see Figure 7.4 for a one-dimensional representation), the average separation of the atoms will be independent of temperature in the **harmonic approximation.** Although the atoms will be nearer the bottom of the well at low temperatures and higher up the well at higher temperatures, the average value of x, represented by the symbol $<x>$ (where $<\ >$ denotes the average over statistical fluctuations), would be independent of temperature, due to the symmetric shape of the parabola. Of course we know that most solids do expand as the temperature increases, and we must look further to see how this comes about.

A true intermolecular potential is not quite parabolic in shape; as shown in Figure 7.5 it is parabolic only at the very bottom of the well. A consequence of this deviation from the harmonic potential of Equation 7.13 is the fact that bonds can be dissociated at large separation.

---

[4] Robert Hooke (1635–1703) was an English scientist and inventor. Although in his lifetime his scientific reputation suffered due to conflicts with his contemporaries over scientific priority, Hooke is now considered second only to Newton as the seventeenth century's most brilliant English scientist. Hooke's law was published in 1676 as *Ut tensio sic vis,* Latin for "as the tension, so the force."

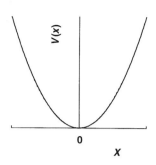

Figure 7.4. A harmonic potential.

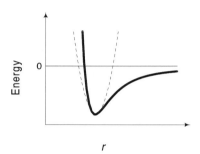

Figure 7.5. A comparison between a true intermolecular potential (——) and a harmonic potential (– – –) as a function of intermolecular separation, r.

Higher up the well, the two sides are no longer symmetric. A better representation of the intermolecular potential in terms of x, the displacement from minimum-energy separation, is given by the **anharmonic potential:**

$$V(x) = cx^2 - gx^3 - fx^4. \tag{7.14}$$

The first term in Equation 7.14 represents the harmonic potential (as in Equation 7.13), the second term represents the asymmetry of the mutual repulsion, and the third term represents the softening of the vibrations at large amplitude. The second term leads to $<x>$ increasing as the temperature increases, since at higher temperatures the asymmetry of the potential leads to a larger value of the intermolecular separation distance. This is shown schematically in Figure 7.6. This increase in $<x>$ is the source of physical expansion in solids, as will now be quantified.

For any property generalized as L, where the displacement of the atoms from their equilibrium position is one example of L, the value of L averaged over statistical fluctuations, represented by $<L>$, will be given by

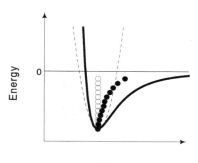

Figure 7.6. Thermal expansion for a harmonic and an anharmonic potential. The asymmetry in the anharmonic intermolecular potential leads to increase in <x> (displacement from the minimum-energy separation) as the temperature is increased. (○) With a harmonic (parabolic) intermolecular potential, <x> is independent of the energy level occupied; (●) with a true (anharmonic) intermolecular potential, <x> depends on the energy of the system and at higher energies (i.e., higher temperatures), <x> is increased.

$$<L> = \frac{\sum_i L_i n_i}{\sum_i n_i} \tag{7.15}$$

where L_i is the value of L for state i, and n_i is the population of state i, and the sum is over all states.[5] Since the number of possible states for the position of an atom in a solid is extremely large, the sum can be replaced with an integral. Furthermore, n_i can be replaced with a **Boltzmann factor** to weight each state:

$$n_i = ne^{-V_i/kT} \tag{7.16}$$

where V_i is the energy of state i and n is a normalization factor. Therefore, Equation 7.15 can be replaced with

$$<L> = \frac{\int_{-\infty}^{\infty} Le^{-V(L)/kT}\, dL}{\int_{-\infty}^{\infty} e^{-V(L)/kT}\, dL} \tag{7.17}$$

which for the average value of the displacement from minimum-energy separation, <x>, becomes

$$<x> = \frac{\int_{-\infty}^{\infty} xe^{-V(x)/kT}\, dx}{\int_{-\infty}^{\infty} e^{-V(x)/kT}\, dx}. \tag{7.18}$$

From the potential energy (Equation 7.14), the exponentials in Equation 7.18 can be written:

$$e^{-V(x)/kT} = e^{-(cx^2-gx^3-fx^4)/kT} = e^{-cx^2/kT}e^{(gx^3+fx^4)/kT} \tag{7.19}$$

---

[5] It may help to consider an example: Consider <L> to be the average mark on a term test, where n_i is the number of students with mark L_i, and the sum is over all possible marks.

and since, for small values of x,

$$e^{(gx^3+fx^4)/kT} = 1 + \frac{gx^3}{kT} + \frac{fx^4}{kT} + \cdots \tag{7.20}$$

where the higher-order terms can be neglected, Equation 7.18 can be written as

$$<x> = \frac{\int_{-\infty}^{\infty} e^{-cx^2/kT}[x+(gx^4/kT)+(fx^5/kT)]\,dx}{\int_{-\infty}^{\infty} e^{-cx^2/kT}[1+(gx^3/kT)+(fx^4/kT)]\,dx}. \tag{7.21}$$

Since the second and third terms in the denominator are $\ll 1$, Equation 7.21 can be approximated as

$$<x> = \frac{\int_{-\infty}^{\infty} e^{-cx^2/kT}[x+(gx^4/kT)+(fx^5/kT)]\,dx}{\int_{-\infty}^{\infty} e^{-cx^2/kT}\,dx}. \tag{7.22}$$

Both the numerator and the denominator are standard integrals, and they can be solved analytically to give

$$<x> = \frac{(3\sqrt{\pi}/4)(g/c^{5/2})(kT)^{3/2}}{(\pi kT/c)^{1/2}} \tag{7.23}$$

which simplifies to

$$<x> = \frac{3gkT}{4c^2}. \tag{7.24}$$

This predicts that the dimensions of a material (represented by the $T = 0$ K dimension plus $<x>$ multiplied by a factor to account for the total number of atoms in the system) will increase by a factor that is linear in temperature. Experimental results show that the dimensions of a material increase almost linearly in temperature above a certain temperature. An example is shown in Figure 7.7 for the lattice constant of solid Ar.

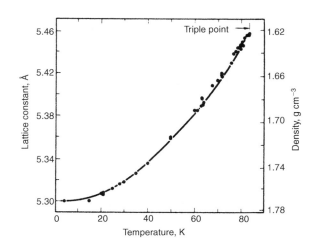

Figure 7.7. The lattice constant, a, as a function of temperature for solid argon. From C. Kittel (1995). *Introduction to Solid State Physics,* 7th Ed. Copyright 1995 John Wiley & Sons, Inc. Reprinted by permission of John Wiley & Sons, Inc.

TABLE 7.3.
Values of Thermal Expansion of Selected Materials at Various Temperatures

Material	T/K	α_a/K^{-1}
Al	50	3.8×10^{-6}
Al	300	2.32×10^{-5}
Diamond	50	4×10^{-9}
Diamond	300	1.0×10^{-6}
Cu	50	3.8×10^{-6}
Cu	300	1.68×10^{-5}
Ag	50	8×10^{-6}
Ag	300	1.93×10^{-5}
Quartz(cr)	50	-3.3×10^{-7}
Quartz(cr)	300	7.6×10^{-6}
Ice	100	1.27×10^{-5}
Ice	200	3.76×10^{-5}
Pyrex	50	5.62×10^{-5}
Pyrex	300	-2.3×10^{-6}

Thermal expansion, α, was defined previously (Equation 6.88) in terms of volume:

$$\alpha = \alpha_v = \frac{1}{V}\frac{dV}{dT} \tag{7.25}$$

but α can also be defined in a particular direction in a solid. For a noncubic solid, the thermal expansion will usually be different in the different directions, and it will be important to take account of this in considering the overall thermal expansion. If the unit cell of a solid has one side of dimension a, the thermal expansion in this direction, α_a, is defined as

$$\alpha_a = \frac{1}{a}\frac{da}{dT} \tag{7.26}$$

and in the case of Ar, α_a at any temperature would correspond to $1/a$ times the slope of the curve shown in Figure 7.7. For all materials, because the dimensions approach a constant value as the temperature approaches absolute zero, $\alpha \to 0$ as $T \to 0$. This is a consequence of the Maxwell relation (Equation 6.73) and the fact that $S \to 0$ as $T \to 0$.

Typical values of coefficients of thermal expansion for a variety of materials at different temperatures are given in Table 7.3. Note that $\alpha > 0$ for most materials at most temperatures, in accord with Equation 7.24, but in some instances, e.g., crystalline quartz at 50 K and Pyrex at room temperature, $\alpha < 0$.

Thermal expansion of a crystalline solid can be measured by scattering X-rays from a material and determining the lattice spacing as a function of temperature, with **Bragg's**[6] **law:**

$$n\lambda = 2d \sin\theta \tag{7.27}$$

where this is the condition for constructive interference of the X-ray beam from successive layers of the solid; n is an integer, λ is the wavelength of the X-ray beam (of the order of the

---

[6] Sir William Henry Bragg (1862–1942), English scientist, was the founder of the science of crystal structure determination by X-ray diffraction. With his son (William Lawrence Bragg) he was awarded the Nobel Prize in Physics in 1915 for seminal work showing the positions of atoms in simple materials such as diamond, copper, and potassium chloride.

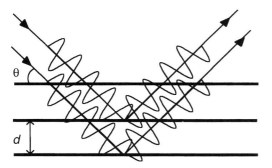

Figure 7.8. The Bragg diffraction condition for X-rays or neutrons scattered from successive layers of a solid. Only some angles and wavelengths allow the scattered beam to be enhanced by constructive interference; if the scattered beams are not completely in phase then less intense scattering is observed. This diagram shows the diffracted waves from successive layers to be in phase, resulting in constructive interference.

lattice spacing), d is the lattice spacing, and θ is the angle of incidence, as defined in Figure 7.8. In some instances it is preferable to scatter neutrons rather than X-rays; the principles are much the same, although X-rays are more readily available at laboratory sources, whereas neutrons require a nuclear reactor for their source. However, since X-rays scatter from the electrons of the atoms involved and neutrons scatter from the nuclei (and also interact magnetically, especially with unpaired electrons), the two techniques can be quite complementary.

Another way to determine thermal expansion is to use the sample to prop apart the plates of a capacitor and measure the capacitance as a function of temperature. The capacitance depends directly on the distance between the plates, so this can lead to a very accurate assessment of the dimensions of the sample as a function of temperature, and, hence, the thermal expansion in the **capacitance dilatometer.**

As can be seen from Equation 7.24, thermal expansion reflects both c (which characterizes the strength of the interaction potential) and g (which represents the asymmetry of the intermolecular potential). More asymmetric potentials (larger g values) lead to larger thermal expansion, which makes sense considering that a harmonic potential, which is perfectly symmetric, has no thermal expansion, i.e., $\alpha^{\text{harmonic}} = 0$. Larger values of c, which imply stiffer springs and steeper walls to the potential, lead to lower coefficients of thermal expansion. Therefore, measurement of thermal expansion of a solid can directly provide information concerning the intermolecular potential.

A MATERIAL WITH NO THERMAL EXPANSION: INVAR

Although most materials increase in dimension as the temperature is increased, one rather curious material, an alloy of 35 atomic% nickel and 65 atomic% iron, has virtually zero thermal expansion throughout the temperature range from −50 to 150°C. The discovery of this material was the subject of the 1920 Nobel Prize in Physics, to Charles E. Guillaume of France. This material, commonly called **Invar** (to represent its invariant thermal expansion), has many practical uses: it is used in laser optics laboratories where dimensional stabilities the order of the wavelength of light are required; it was used to determine the diurnal (day–night) height variations of the Eiffel tower; and hulls of supertankers are made of Invar to avoid strains due to thermal expansion associated with welding processes. The unusual absence of thermal expansion in Invar is related to its magnetic properties, but it is not yet fully understood. Another low thermal expansion material is Zerodur (Schott Glass Technologies, $\alpha = 0 \pm 0.10 \times 10^{-6}$ K^{-1}, 273 K < T < 323 K).

Thermal expansion can be an important property of a material, as it may have some influence over other properties such as mechanics of fitting.[7] We have previously seen the effect of thermal expansion on the color of a liquid crystal as in a mood ring (a piece of costume jewelry that changes color with temperature and could be used to indicate mood). Thermal expansion also may lead to stress in a material and this can greatly affect mechanical stability.

SOME MATERIALS CONTRACT WHEN HEATED

Most but not all materials expand when heated (see Table 7.3). A few materials shrink in one dimension as they are heated, but generally they expand in another dimension such that the overall thermal expansion is positive. However, there are some materials that contract when heated, such as quartz at very low temperature and Pyrex at room temperature. A newly discovered material, of approximate composition $ZrVPO_7$, is found to have the unusual property of contracting when heated, in the temperature range from 300 to 900°C. Even more surprising, zirconium tungstate, ZrW_2O_8, has the unusual property of shrinking uniformly in all three dimensions as the temperature is increased from 0.5 K to 1050 K! When the units within this material are heated, their thermal agitation leads to a more closely packed structure, and negative thermal expansion results. A familiar material with anomalous thermal expansion is H_2O: when liquid water is heated from 0 to 4°C it contracts. Ice also shows negative thermal expansion at very low temperatures.

---

[7] Woodwind musical instruments have the correct pitch only when they are at the correct temperature; if they are cold, they are too short, and sound flat. This is not due to contraction of the cold instrument, as the shorter cold instrument would sound sharp. The important factor here is the contraction of the cold gas. The velocity of sound in the gas decreases as the gas density increases (i.e., as the temperature decreases), and this decreases the frequency of the sound as temperature decreases, making a cold woodwind instrument sound flat. In this case, thermal expansion of the air (not that of the instrument) is the important factor.

Examples of Thermal Expansion: A Tutorial

Typical thermal expansion of a solid at room temperature is of the order 10^{-5} K^{-1}.

a. For a rod of material 1 m long, how much does the length of the rod change for 3 K increase in temperature?

b. In some applications it is important not to have changes in length as the temperature changes. One such example is the pendulum of a clock, where the accuracy of the time depends on the stability of the length of the pendulum. Many of the decorative metals have large values of thermal expansion near room temperature. Given that there will be variations in the temperature from time to time in the vicinity of the clock, it has been suggested that a bimetallic arrangement could be used for the pendulum, with a decorative metal (large thermal expansion) on the outside face, securely attached to a less attractive metal (smaller thermal expansion) on the inside face. In the case of extreme temperature fluctuations, will the pendulum remain linear?

c. Can the stress induced by differential thermal expansion in a bimetallic strip be used to amplify thermal expansion effects? Consider the change in the pointer position in the thermostat shown in Figure 7.9, in comparison with the change in dimension found in (a).

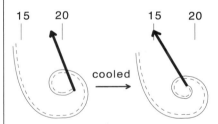

Figure 7.9. The bimetallic strip inside a thermostat. As the temperature is reduced, the coil tightens since the thermal expansion coefficient of the outer material (solid line) is less than that of the inner material (dashed line).

d. Suggest an application for thermal expansion of a material.

FURTHER READING

General References

A.W. Adamson (1973). *A Textbook of Physical Chemistry*. Academic Press.

J.A. Barker and D. Henderson (1981). The Fluid Phases of Matter. *Scientific American,* November 1981, 130.

R.S. Berry, S.A. Rice, and J. Ross (1980). *Physical Chemistry*. John Wiley & Sons Ltd.

A. Bondi (1968). *Physical Properties of Molecular Crystals, Liquids and Glasses.* John Wiley & Sons Ltd.

M. de Podesta (1996). *Understanding the Properties of Matter*. Taylor & Francis.

A. Guinier and R. Jullien (1989). *The Solid State.* Oxford University Press.

C. Kittel (1996). *Introduction to Solid State Physics.* John Wiley & Sons Ltd.

Experimental Methods

G.K. White (1979). *Experimental Techniques in Low-Temperature Physics.* Clarendon Press.

Materials with Unusual Thermal Expansion

A.N. Aleksandrovskii, V.B. Esel'son, V.G. Manzhelii, G.B. Udovidchenko, A.V. Soldatov, and B. Sundqvist (1997). *Low Temperature Physics,* **23,** 943.

S.W. Benson and E.D. Siebert (1992). A Simple Two-Structure Model for Liquid Water. *Journal of the American Chemical Society,* **114,** 4269.

T.A. Mary, J.S.O. Evans, T. Vogt, and A.W. Sleight (1996). Negative Thermal Expansion from 0.3 to 1050 Kelvin in ZrW_2O_8. *Science,* **272,** 90.

D.G. Rancourt (1989). The Invar Problem. *Physics in Canada,* **45,** 3.

PROBLEMS

1. A total of 44.0 g of CO_2 is contained in a 1.00-L vessel at 31°C.
   a. What is the pressure of the gas (in atm) if it behaves ideally?
   b. What is the pressure of the gas if it obeys the van der Waals equation of state? See Table 7.2 for data. Give your answer in atm.
   c. Estimate the pressure of the gas (in atm) based on the Hougen–Watson plot (Figure 7.3) and the values of the critical constants for CO_2 given in Table 7.1.

2. Based on the data given in Problem 1, calculate the volume coefficient of thermal expansion, α, for CO_2 at 31°C and $P = 1.00$ atm, for the following cases: (a) the gas is ideal; (b) the gas obeys the van der Waals equation of state.

3. An apparatus consisting of a rod, made of material A, placed inside a tube, made of material B, as shown in Figure 7.10, is being constructed. The aim is to have the rod fit tightly in the tube at room temperature. One way to achieve this is to cool both the rod and the tube (e.g., in liquid nitrogen) and put them together when they are cold and the fit is loose. As they warm to room temperature, the fit will be tight. This procedure is called **cold welding.** It relies on different thermal expansions of the two materials for the fit to be tight at room temperature and loose when much colder. Which material needs to have greater thermal expansion for this to work: A (rod) or B (tube)? Explain.

4. Boiling water is poured into two glasses; the only difference between the glasses is that one is made of thick glass and the other is made of thin glass. Which glass is more likely to break? Why?

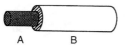

Figure 7.10. A rod, A, is placed inside a tube, B.

A   B

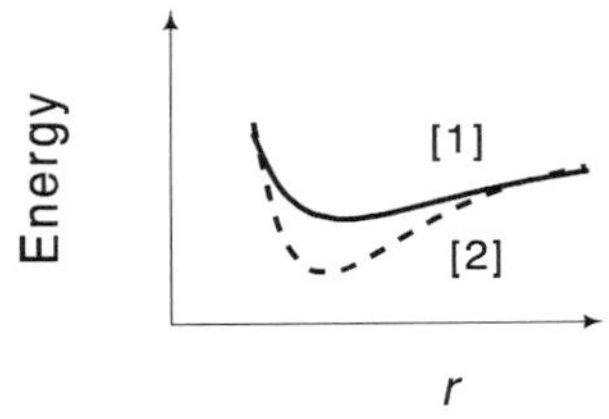

Figure 7.11. Two intermolecular potentials.

5. Consider the curves for intermolecular potential as a function of molecular separation, r, shown in Figure 7.11.
   a. Which case, [1] or [2], has the stronger intermolecular potential? Explain briefly.
   b. Which case, [1] or [2], is more anharmonic? Explain briefly.
   c. Which case, [1] or [2], has greater thermal expansion? Explain briefly.

6. Owens-Corning Fiberglas Corp. has introduced a new form of glass fiber for insulation. The fiber is composed of two types of glass fused together. At the cooling stage in the production, the components separate so that one side of the fiber is made of one component and the other side has the other composition. This can be seen when looked at in cross section under a microscope. The key to the applications of this material is that the fibers are coiled, not straight as for regular glass fibers: the coils in the new fibers allow them to spring into place, giving them excellent insulating properties. What causes the fiber to be coiled?

7. When a polymer is cooled below its glassy phase transition temperature, T_g, there is an abrupt drop in its coefficient of thermal expansion. Given that the configurational degrees of freedom are frozen below T_g and become active above T_g, what does this imply about the harmonicity of the dominant intermolecular interactions in the rigid glass compared with the more flexible state?

8. The linear thermal expansion coefficient for a borosilicate crown glass is zero at $T = 0$ K, gradually rises to 80×10^{-5} K^{-1} at $T = 600$ K, and then rapidly rises to 500×10^{-5} K^{-1} at $T = 800$ K. How will this increase in thermal expansion coefficient affect the high-temperature applications of this material?

9. On the basis of a tree's structure, would you expect thermal expansion of wood to be isotropic or anisotropic? Explain how this could affect the design of a wooden musical instrument.

10. A gas-filled piston is central to the operation of a pneumatic window opener that opens and closes according to the temperature. The piston is attached to the window frame such that its extension determines the opening of the window.
    a. Explain the principle on which this works.
    b. If the gas were ideal, would it be better (all other factors being equal) to have it operate at low pressure or high pressure? Explain.
    c. If the gas were not ideal, would it be preferable to have $Z > 1$ or $Z < 1$? Explain.

11. Invar was used on a space flight to determine the outside temperature of a space capsule during its voyage. Given that Invar does not change dimensions in the expected temperature range (–50 to 150°C), why was it chosen?

12. For a particular application it is necessary to introduce a piece of wire into a glass tube. This will be done by heating the glass and inserting the wire. If you have a choice of platinum (linear expansion coefficient 8.6×10^{-6} K^{-1}) or copper (linear expansion coefficient 1.7×10^{-5} K^{-1}) to insert in glass with a linear thermal expansion coefficient of 8.6×10^{-6} K^{-1}, which metal should you choose? Explain.

13. At some sinks, when the hot water tap is opened only slightly, the water first flows, and then slows until only a trickle, finally shutting off completely. A second opening of the hot water tap does not usually lead to a repeat of this behavior, and the cold water tap does not behave this way. Give an explanation for this phenomenon.

14. The coefficient of thermal expansion, α, of a sample could be measured by placing the sample between parallel circular copper plates to form a capacitor. Since the capacitance of the capacitor is directly proportional to the area of one of the plates and inversely proportional to the distance between the plates, α can be determined by monitoring the capacitance as a function of temperature. However, the sample is usually placed outside the capacitor plates (which are in vacuum) in such a way that it moves one of the capacitor plates when the temperature is changed. Why would it be better to determine the capacitance of the cell in this geometry than with the sample between the plates?

15. The speed of sound in the ocean varies with the depth, depending mostly on the temperature of the water (i.e., salinity is of secondary importance). The speed is highest in warmer water, and can vary by a few percent from its average value of 1500 m s^{-1}, depending on the depth.
    a. Explain why the speed of sound varies as it does with temperature.
    b. Under certain conditions, such as shallow ocean water in the summer, the water temperature first decreases with increasing depth, reaches a minimum at a certain depth (approximately 50 m), and then increases again to a nearly constant value at depths greater than 150 m.
        i. Sketch the sound speed (x-axis) as a function of depth (y-axis) under these conditions.
        ii. Like light, sound waves bend (refract) toward the lower velocity. Use this information to show how a sound wave can be guided by the ocean to travel great distances without loss (similar to fiber optics acting as a **wave guide;** see Chapter 4).

16. a. Which would be expected to have the higher volume thermal expansion coefficient, diamond or graphite? Explain.
    b. Graphite is an anisotropic material due to its layered structure. Would you expect the coefficient of thermal expansion to be higher along the layer directions or perpendicular to the layers? Explain.

17. Near room temperature, polymers have higher thermal expansion than metals, which have higher thermal expansion than ceramics. Explain this order. Consider intermolecular forces.

18. Rubies can exhibit thermochromism. If a ruby of composition 10% Cr_2O_3, 90% Al_2O_3 is heated from room temperature to 500°C, it will turn from red to green. The color is due to absorption in both cases. Suggest a reason for the change in color.

8

THERMAL
CONDUCTIVITY

8.1 INTRODUCTION

Thermal conductivity determines whether or not a material feels cold to the touch, and why some materials are good insulators while others efficiently conduct heat. In this chapter we consider thermal conductivities of gases and then of solids.

The symbol used to represent thermal conductivity is κ. [Some sources use λ, but we prefer to reserve this for wavelength (already introduced) and mean free path (introduced later in this chapter).] The higher the value of κ the better the material conducts heat. Like electrical conductivity, the values of thermal conductivity can vary by orders of magnitude from one material to another, and even within a given material when the temperature is changed.

8.2 THERMAL CONDUCTIVITY OF GASES

The formal derivation of a simple theory of thermal conductivity of gases, attributed to Peter Debye, is based on kinetic theory of gases.

The model of a gas for kinetic theory is a swarm of molecules, each of mass m, all in continual random motion. The molecules are considered to have negligible size with respect to the average distance they travel between collisions. The average distance traveled by a gas molecule between collisions is called the **mean free path** and is denoted λ. In this model, the molecules within the gas do not interact except when they collide; in other words, long-range attractive forces are neglected.

The kinetic theory of gases has many important successes. For example, it leads to the equation of state for an ideal gas:

$$PV_\text{m} = RT. \tag{8.1}$$

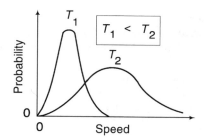

Figure 8.1. Distributions of molecular speeds in a gas at two different temperatures.

The kinetic theory of gases also leads to the distribution of molecular velocities within a gas, as shown in Figure 8.1. This figure shows that the distribution shifts to higher velocities as the temperature is increased. In fact, it leads to a relationship between the weighted average speed, $\bar{v}$ and the temperature of the gas:

$$\bar{v}^2 = const \times T \tag{8.2}$$

which follows from consideration of the translational kinetic energy, KE:

$$KE = \frac{1}{2}m\bar{v}^2 = const \times kT \tag{8.3}$$

where the latter proportionality comes from equipartition theory (Chapter 6).

Thermal conductivity is a **transport property,** involving motion of heat (energy) from hot to cold. Other transport properties are **effusion** (mass is transported out of a small orifice), **diffusion** (general mass transport), **viscosity** (momentum transfer), and **electrical conductivity** (charge transport). Much of the formalism of the derivation of various transport properties follows similar lines, so the derivation of the relationship between thermal conductivity of a gas and other physical parameters will be presented here in some detail.

The **flux,** $J(i)$, of some physical property, i, can be defined as the amount of i flowing per unit time per unit area. For example, in a box with a temperature gradient (one end hot, one end cold, as shown in Figure 8.2), $J(energy)$ is the energy flux, or the amount of energy flowing

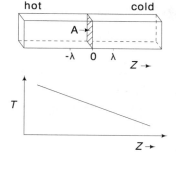

Figure 8.2. A box of gas with a temperature gradient, used to consider the thermal conductivity of a gas. T is the temperature of the gas as a function of position z.

per unit area per unit time. SI units of $J(energy)$ are J m^{-2} s^{-1} ≡ W m^{-2}. The way the box of Figure 8.2 has been designed, the temperature gradient is parallel to the z-axis, and this can be considered to be a one-dimensional problem in which the fluxes can be represented by the flux in the z-direction, J_z.

The energy flux along the z-direction can be written:

$$J_z(energy) = -\kappa \left(\frac{dT}{dz} \right) \tag{8.4}$$

where κ is the **coefficient of thermal conductivity** or, often, simply **thermal conductivity**; SI units of κ are W m^{-1} K^{-1}. Since $(dT/dz) < 0$ here, $J_z(energy) > 0$, and energy flows from decreasing z value (Figure 8.2), i.e., from hot to cold.

The purpose now is to express κ in terms of physically meaningful parameters. To do so it is useful to start by expressing the energy flux in terms of the matter flux. If each molecule moving under the temperature gradient carries energy $\varepsilon(z)$, where z is its position, we can calculate the net flux of molecules per unit area per unit time. In the box in Figure 8.2, consider the area A to be defined as exactly 1 unit area. To calculate $J_z(energy)$, we need to know the net number of molecules passing through A per second.

The **mean free path, λ,** is defined as the average distance that a molecule in a gas can travel before colliding with another molecule. On average, molecules that hit area A from the left have traveled one mean free path (λ) since their last collision. Other molecules will be too far away (and will bump into other molecules before reaching A) or too close (and will pass through A). The number of molecules hitting A from the left per unit time is then $(N\bar{v}/6)$, where $\bar{v}$ is the average speed of the molecules and N is the number density. The factor of 1/6 accounts for the fact that of the six directions (up, down, left, right, in, out) only one (to the right) leads to A. The expression $(N\bar{v}/6)$ can be seen to lead to units of molecules m^{-2} s^{-1} for the matter flux to A from the left, as one would expect. With an energy of $\varepsilon(-\lambda)$ for each molecule at a distance $-\lambda$ from A (see Figure 8.2), this leads to left-to-right energy flux, $J_z(energy)^{L \rightarrow R}$:

$$J_z(energy)^{L \rightarrow R} = \frac{N\bar{v}}{6} \varepsilon(-\lambda). \tag{8.5}$$

There is also a flux of molecules to A from the right, of $(N\bar{v})/6$, giving right-to-left energy flux, $J_z(energy)^{R \rightarrow L}$:

$$J_z(energy)^{R \rightarrow L} = \frac{N\bar{v}}{6} \varepsilon(\lambda). \tag{8.6}$$

The net energy flux at A will be the difference between these two fluxes:

$$J_z(energy) = \frac{1}{6} N\bar{v} [\varepsilon(-\lambda) - \varepsilon(\lambda)] = \frac{1}{6} N\bar{v} \Delta \varepsilon \tag{8.7}$$

where $\Delta \varepsilon$ is the difference in energy per molecule at a distance λ to the left of A (i.e., at $z = -\lambda$) to a distance λ to the right of A (i.e., at $z = \lambda$), i.e., $\Delta \varepsilon = \varepsilon(-\lambda) - \varepsilon(\lambda)$. $\Delta \varepsilon$ can be written in terms of the energy gradient, $d\varepsilon/dz$:

$$\Delta \varepsilon = \left(\frac{d\varepsilon}{dz} \right) \Delta z = -\left(\frac{d\varepsilon}{dz} \right) 2\lambda \tag{8.8}$$

where $\Delta z = -2\lambda$. From Equations 8.7 and 8.8, the energy flux can be written:

$$J_z(energy) = -\frac{1}{6}N\bar{v}\left(\frac{d\varepsilon}{dz}\right)2\lambda = -\frac{1}{3}N\bar{v}\lambda\left(\frac{d\varepsilon}{dz}\right).$$

(8.9)

Letting E be the total energy per unit volume, it follows that

$$\varepsilon = \frac{E}{N}$$

(8.10)

and

$$\frac{d\varepsilon}{dz} = \frac{1}{N}\frac{dE}{dz}.$$

(8.11)

Further, dE is related to dT by the molar heat capacity at constant volume, $C_{V,m}$, the number of moles, n, and the volume, V:

$$dE = \frac{C_{V,m}n}{V}dT$$

(8.12)

so the energy gradient, dE/dz, can be expressed in terms of the temperature gradient, dT/dz:

$$\frac{dE}{dz} = \frac{C_{V,m}n}{V}\frac{dT}{dz}.$$

(8.13)

Substitution of Equations 8.13 and 8.11 into Equation 8.9 leads to:

$$J_z(energy) = -\frac{1}{3}\bar{v}\lambda C\left(\frac{dT}{dz}\right)$$

(8.14)

where $C = C_{V,m}n/V$ is the heat capacity per unit volume (SI units of C are J K^{-1} m^{-3}). Comparison of Equations 8.4 and 8.14 leads to

$$\kappa = \frac{1}{3}C\bar{v}\lambda.$$

(8.15)

Equation 8.15 is known as the **Debye equation for thermal conductivity,** and it will prove to be very useful for both gases and for solids.

Equation 8.15 shows that κ for a gas is governed by three factors: its heat capacity per unit volume, the mean free path of the molecules that carry the heat, and the average speed of these molecules. Each of these factors is temperature dependent, and since their temperature dependences do not cancel, it follows that κ for a gas also is temperature dependent.

What about the pressure dependence of κ? From the kinetic theory of gases it can be shown that the mean free path, λ, can be written as

$$\lambda = \frac{V}{\sqrt{\pi}d^2nN_A}$$

(8.16)

where V is the volume, n is the amount of gas in moles, and d is the diameter of the gas molecule under consideration. From Equation 8.15 it follows that

TABLE 8.1.
Thermal Conductivities of Selected Materials at $T = 300$ K

Material	κ / (W m^{-1} K^{-1})
Al_2O_3	36.0
Argon (gas)	0.018
Boron	2.76
Copper	398
Diamond	2310
Graphite	2000 along c-axis
	9.5 along a-axis
Helium (gas)	0.15
Iodine	0.449
MgO	60.0
Nitrogen (gas)	0.025
Black phosphorus	12.1
White phosphorus	0.235
Sapphire	46
SiO_2 (crystalline)	10.4 along c-axis
	6.2 along a-axis
SiO_2 (amorphous)	1.38

$$\kappa = \frac{1}{3}\,\bar{v}\,\frac{V}{\sqrt{2}\pi d^2 n N_A}\,\frac{n}{V}\,C_{V,m} = \frac{\bar{v}C_{V,m}}{3\sqrt{2}\pi d^2 N_A} \qquad (8.17)$$

which shows thermal conductivity, κ, to be independent of the pressure of the gas. However, Equation 8.17 shows that κ depends on $C_{V,m}$ and d, which both depend on the nature of the gas, so, at a given temperature, the thermal conductivity of a gas will depend on its composition. This can be used to detect different gases, and is used, for example, in thermal conductivity detectors in gas chromatography.

The physical interpretation of the pressure independence of the thermal conductivity is as follows. As the pressure is increased, more molecules are available to transport the heat, and this would increase κ, but the presence of more molecules decreases the mean free path, and this decreases κ. These equal and opposite effects lead to κ being independent of pressure.

However, if the pressure is extremely low, as in the case of a Thermos™ bottle, pressure can be important. The mean free path, λ, increases as the pressure decreases, but eventually λ is limited by the dimensions of the container and it stays at this constant value no matter how low the pressure. As the number density of molecules has been able to fall faster than the mean free path has been able to grow at very low pressure, this leads to very poor thermal conduction at these low pressures, and use of vacuum flasks for insulation.

Typical values of thermal conductivities of some gases and solids are presented in Table 8.1.[1]

8.3 THERMAL CONDUCTIVITIES OF SOLIDS

In a gas, molecules carry the heat. What carries the heat in a nonmetallic solid? The answer to this question was first considered by Debye: vibrations of the atoms in the solid, known as **lat-**

---

[1] Data for solids are taken from Y.S. Touloukian, R.W. Powell, C.Y. Ho, and P.G. Klemens (1970), *Thermal Conductivity: Nonmetallic Solids,* Plenum. Data for gases are from A.W. Adamson (1973), *A Textbook of Physical Chemistry,* Academic Press.

Figure 8.3. Depiction of a lattice wave (phonon) in a two-dimensional solid. At $T = 0$ K, the atoms would be arranged on a regular grid, but this instantaneous picture shows the atoms displaced from their equilibrium positions.

tice waves (also known as **phonons**) carry the heat. A phonon is a quantum of crystal wave energy, and it travels at the speed of sound in the medium. The presence of phonons in a solid is a manifestation of the available thermal energy (kT); the more heat, the greater the number of phonons (i.e., more excited lattice waves). A depiction of a lattice wave is shown in Figure 8.3.

The thermal conductivity of a solid can be measured by heating the material (with power $\dot{q}$, this power being applied at one end and lost at the same rate at the other end, in order to achieve steady-state conditions; see Figure 8.4) while determining the temperature gradient, dT/dx. The thermal conductivity, κ, is given by

$$\kappa = \frac{\dot{q}}{A}\frac{dx}{dT} \tag{8.18}$$

where A is the cross-sectional area of the material.

By analogy with the derivation of the thermal conductivity in gases (Equation 8.15), for nonmetallic solids thermal conductivity, κ, is given by

$$\kappa_{\text{solid}} = \frac{1}{3}C\bar{v}\lambda \tag{8.19}$$

where C is the heat capacity per unit volume (as before), but now $\bar{v}$ is the **mean phonon velocity** and λ is the **mean free path of the phonons.**

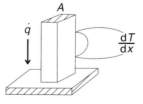

Figure 8.4. The thermal conductivity, κ, of a single crystal (or well-defined block) of a material can be determined by direct measurement. For a crystal of cross-sectional area A, with a temperature gradient along its face of dT/dx, the thermal conductivity is given by $\kappa = \dot{q}(dx/dT)A^{-1}$, where $\dot{q}$ is the power supplied to the top of the crystal. The measurement would be carried out at steady-state conditions, i.e., the rate of power introduced to the top of the crystal would be equal to the loss at the bottom, such that the temperature of the crystal remains constant. In addition, for high-accuracy measurements, care would need to be taken to reduce extraneous heat exchange with the surroundings; for example, the crystal and its platform would be in a vacuum, and the crystal could be surrounded with a temperature-matched heat shield.

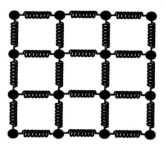

Figure 8.5. A solid can be considered to be a network of atoms, each connected to other atoms by springs.

To understand the thermal conductivity of nonmetallic solids, we must first return to the simple harmonic oscillator. If all the atoms in a solid are connected to each other by spring-like forces (with all the force constants given by Hooke's law, Equation 7.10), as shown in Figure 7.4, then the potential in all directions is shaped like a three-dimensional parabola; a schematic view of a solid is given in Figure 8.5.

As the lattice is heated, there is an increased probability of exciting high-energy phonons, and more of the atoms will be vibrated away from their equilibrium lattice sites. (Incidentally, the following will indicate just how far off their lattice site they could be. Vibration increases as temperature increases and in 1912 Lindemann[2] showed that at the melting point, the vibrations result in about a 10% increase in volume over the volume at $T = 0$ K. This corresponds to about a 3% increase in the lattice parameters over the $T = 0$ K values.)

If the lattice were perfectly harmonic, the phonons would carry heat perfectly, with no resistance to heat flow, and the thermal conductivity would then be infinite. Clearly, non-metallic solids must have some heat-flow resistance mechanism taking place, since observed thermal conductivities are not infinite.

This problem was solved by Peierls,[3] when he showed that phonon–phonon collisions can lead to "turning back" ("Umklapp" in German) processes, as shown in Figure 8.6. This requires the presence of anharmonicity in the intermolecular potential. Recall that anharmonicity also was responsible for thermal expansion, and now we see that it also leads to thermal resistance.

The thermal conductivity for a "typical" simple crystalline nonmetallic solid is shown in Figure 8.7, and its shape as a function of temperature can be understood simply in terms of the Debye equation (Equation 8.19) as follows.

At high temperatures $(T > \theta_D)$, C is approximately constant and $\bar{v}$ (which depends on the material as it reflects the force constant of interactions within the solid; the stiffer the lattice, the higher $\bar{v}$) is nearly independent of temperature. The mean free path, λ, depends on temperature such that as the temperature is increased, λ decreases due to the increased probability of phonon–phonon collisions. Therefore, from Equation 8.19, at high temperatures, as T increases κ decreases.

---

[2] Frederick Alexander Lindemann (1886–1957) was a German-born physicist who spent most of his career at Oxford University in England. In addition to his melting point theory and his theory of unimolecular reactions, Lindemann was personal scientific advisor to Winston Churchill during World War II.

[3] Sir Rudolf Ernst Peierls (1907–1995) was a German-born theoretical physicist. His 1929 theory of heat conduction (predicting an exponential increase in thermal conductivity as a perfect crystal is cooled) was finally verified experimentally in 1951. Peierls also has made important contributions to nuclear science (including work on the Manhattan Project), quantum mechanics, and other aspects of solid-state theory.

a.

b.

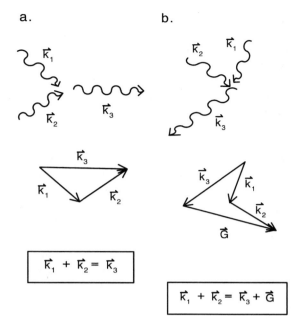

$$\vec{K}_1 + \vec{K}_2 = \vec{K}_3$$

$$\vec{K}_1 + \vec{K}_2 = \vec{K}_3 + \vec{G}$$

Figure 8.6. (a) Two incident phonons (each shown as a wave) can interact to give a resultant phonon with a motion in the same general direction as the incident phonons. This process is known as a **"normal"** or **"N" process;** it does not resist heat flow. The vector view shows $\hat{\mathbf{k}}_1$ and $\hat{\mathbf{k}}_2$ as incident momentum (wave) vectors, and $\hat{\mathbf{k}}_3$ as the resultant wavevector. (b) Two incident phonons (each shown as a wave) can interact to give a resultant phonon with one component of the motion in a direction opposite to the general motion of the incident phonons. This process, since it "turns back" heat flow, is known as an **"Umklapp" process.** It is these sorts of processes that lead to thermal resistance (reducing thermal conductivity). The vector view shows $\hat{\mathbf{k}}_1$ and $\hat{\mathbf{k}}_2$ as incident wavevectors, and $\hat{\mathbf{k}}_3$ is the resultant wavevector. $\hat{\mathbf{G}}$ represents a momentum transfer to the lattice.

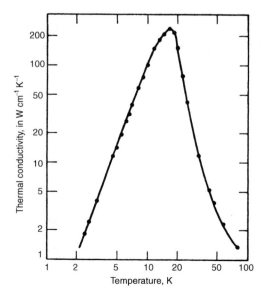

Figure 8.7. The temperature dependence of the thermal conductivity of a "typical" simple insulating crystalline solid, highly purified NaF. From C. Kittel (1995). *Introduction to Solid State Physics,* 7th Ed. Copyright 1995 John Wiley & Sons, Inc. Reprinted by permission of John Wiley & Sons, Inc.

As the temperature is decreased, κ increases because the mean free path, λ, becomes longer. This is because at low temperatures the phonons can travel for longer distances without colliding, since there are fewer phonons around at lower temperatures. Eventually, λ will have grown so long that it is limited either by the size of the crystal (if the crystal is perfect) or by the distance between defects (such as packing faults or even other isotopes) if the crystal is not perfect. Below this temperature, λ cannot increase further. However, as the temperature is lowered below the Debye characteristic temperature, θ_D, the heat capacity, C, falls. Given that the average phonon speed, $\bar{v}$, is approximately independent of temperature, the Debye equation (Equation 8.19) shows that κ will decrease with decreasing temperatures at low temperatures. A typical thermal conductivity–temperature profile for a simple insulating crystalline solid is shown in Figure 8.7.

8.4 THERMAL CONDUCTIVITIES OF METALS

In metals there are two mechanisms to carry heat, the phonons and the free electrons, which is why metals are good thermal conductors. For a metal the total thermal conductivity, κ, can be written as the sum of the phononic contribution, κ_{phonon}, and the electronic contribution, κ_{elec}:

$$\kappa = \kappa_{phonon} + \kappa_{elec} \tag{8.20}$$

where κ_{phonon} is as given by Equation 8.19 and, by analogy with Equation 8.15, κ_{elec} is given by

$$\kappa_{elec} = \frac{1}{3}C_{elec}\bar{v}_{elec}\lambda_{elec} \tag{8.21}$$

where C_{elec} is the electronic contribution to the heat capacity per unit volume, $\bar{v}_{elec}$ is the mean speed of the conducting electrons, and λ_{elec} is the mean free path of the conducting electrons.

Since, typically, κ for a metal is 10 to 100 times κ for a nonmetal, it is clear that most of the heat in a metal is carried by conducting electrons not phonons.

Nevertheless, for parallel reasons, the thermal conductivity of a metal has a temperature dependence similar to that of a nonmetal. The temperature dependence of the thermal conductivity of Cu is shown in Figure 8.8 as a typical example.

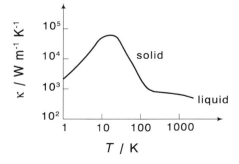

Figure 8.8. Temperature dependence of the thermal conductivity of a "typical" metal. The data shown are for Cu.

IS MIRACLE THAW™ A MIRACLE?

A commercial thawing tray advertises that it thaws frozen food using a "super-conducting" material. In reality, the thawing tray is a sheet of aluminium, painted black. Since aluminium is a metal, it is a good conductor of heat. The black paint ensures efficient "capture" of ambient thermal energy, which is then transferred to the food (Figure 8.9).

Figure 8.9. Ice melts rapidly on the food-thawing tray.

SPACE SHUTTLE TILES

Thermally insulating tiles play an important role in protecting the Space Shuttle as it passes through the atmosphere. Most of the tiles are made of amorphous silica fibers. The fact that they are amorphous means that they are poor conductors of heat. The fiber structure cuts down further on heat conduction and reduces the density to about 0.1 g cm^{-3} (total mass is a major fuel consumption factor on space flights). In some areas of the Space Shuttle, tiles have added alumina-borosilicate content, which adds mechanical strength. The tiles are prepared by making a water slurry of the fibers, casting it into soft blocks to which colloidal silica binder has been added, and then sintering at high temperature to form a rigid block that can be machined to exact dimensions.

Thermal expansion is an important factor in placement of the tiles; if they buckle against one another during heating, the tiles could fall off. For this reason, gaps of about 0.1 mm are left between tiles.

COMMENT

POPPING CORN

Popcorn "pops" when it is heated because the inner moisture is vaporized, and this breaks open the hull. Recent measurements of the thermal conductivities of various types of corn show that popcorn hull is two to three times better as a thermal conductor than nonpopping corn kernels. This indicates that popcorn husks have more crystalline (less amorphous) cellulose than other corn kernels, and this has been confirmed by optical experiments. The high thermal conductivity of popping corn ensures efficient heat transfer into the kernel, while the high mechanical strength associated with its more organized crystalline structure means that the kernel can sustain a high pressure before popping.

Thermal Conductivities of Materials: A Tutorial

a. Consider Equation 8.19, which describes the thermal conductivity of an insulating solid. Consider the effect of the perfection of a crystal on the phonon mean free path. Show what the presence of lattice imperfections would do to the thermal conductivity by sketching $\kappa(T)$ both for a perfect crystal and for the same crystal with imperfections (e.g. impurities) of a few percent.

b. Sketch the mean free path for a perfect crystal as a function of temperature. On the same diagram, sketch the mean free path for the same composition of material, now in an amorphous state.

c. Considering the same equation for thermal conductivity, and the microscopic picture of an amorphous material (compared with a perfect crystal), sketch how the thermal conductivity of an amorphous material could be expected to look as a function of temperature.

d. **Ceramics** are inorganic materials such as MgO, $BaTiO_3$, SiO_2, and SiC. Many ceramics are amorphous materials with particularly useful properties. Why does coffee stay warm longer in a ceramic mug than in a metal cup?

e. Based on Equation 8.19, how would you expect the thermal conductivity of a harder material to compare with a softer material? Does this explain why marble is a good heat conductor (and therefore a poor energy storage material—Tutorial in Chapter 6)? What about diamond?

f. Metals also have high thermal conductivities. Why is diamond so special with regard to its thermal conductivity, in consideration of its other properties?

g. How could the thermal conductivity of diamond be made even higher?

h. An article titled "Hot Rocks" in the business pages of a newspaper reported on the finding that diamonds made from isotopically purified carbon-12 are better conductors of heat than normal diamonds. What do you think of the title of the article? Would these diamonds feel hot or cold to the touch?

FURTHER READING

General References

Many physical chemistry textbooks contain derivations of the kinetic theory of gases. In addition, the following address many of the general points raised in this chapter:

A.W. Adamson (1973). *A Textbook of Physical Chemistry.* Academic Press.

A. Bondi (1968). *Physical Properties of Molecular Crystals, Liquids and Glasses.* John Wiley & Sons.

B.S. Chandrasekhar (1998). *Why Things Are the Way They Are.* Cambridge University Press.

M. de Podesta (1996). *Understanding the Properties of Matter.* Taylor & Francis.

A.B. Ellis, M.J. Geselbracht, B.J. Johnson, G.C. Lisensky, and W.R. Robinson (1993). *Teaching General Chemistry: A Materials Science Companion.* American Chemical Society.

C. Kittel (1996). *Introduction to Solid State Physics,* 7th Ed. John Wiley & Sons Ltd.

H.M. Rosenberg (1989). *The Solid State,* 3rd Ed. Oxford University Press.

R.L. Sproull (1962). Conduction of Heat in Solids. *Scientific American,* December 1962, 92.

A.J. Walton (1983). *Three Phases of Matter,* 2nd Ed. Oxford University Press.

Experimental Techniques

M.E. Bacon, R.M. Wick, and P. Hecking (1995). Heat, Light and Videotapes: Experiments in Heat Conduction Using Liquid Crystal Film. *American Journal of Physics,* **63,** 359.

V.V. Murashov, P.W.R. Bessonette, and M.A. White (1995). A First-principles Approach to Thermal Conductivity Measurements of Solids. *Proceedings of the Nova Scotian Institute of Science,* **40,** 71.

Implications of Thermal Conductivity

W.H. Corkern and L.H. Holmes, Jr. (1991). Why There's Frost on the Pumpkin. *Journal of Chemical Education,* **68,** 825.

Thermal Conductivities of Specific Materials

D.R. Askeland (1994). *The Science and Engineering of Materials,* 3rd Ed. PWS Publishing Company.

W.J. da Silva, B.C. Vidal, M.E.Q. Martins, H. Vargas, A.C. Pereira, M. Zerbetto, and L.C.M. Miranda (1993). What Makes Popcorn Pop. *Nature* **362,** 417.

Y.S. Touloukian, R.W. Powell, C.Y. Ho, and P.G. Klemens (1970). *Thermal Conductivity: Non-metallic Solids,* Plenum.

M.A. White, V. Murashov, and P. Bessonette (1996). Thermal Conductivity of Food-Thawing Trays. *Physics Teacher,* **34,** 4.

K.M. Wong (1990). Space Shuttle Thermal Protection System. *California Engineer,* December 1990, 12.

PROBLEMS

1. a. Calculate the thermal conductivity for He at 25°C and 1 atm, given the following information: at these conditions, $\bar{v}$ of He is 1360 m s^{-1}, and the mean free path is 7.2×10^{-7} m.
    b. Compare this with the following thermal conductivities, at the same conditions (all in W cm^{-1} K^{-1}): air, 2.4×10^{-4}; crystalline SiO$_2$, 0.13; Cu, 5.

2. Several expressions for κ_{gas}, the thermal conductivity of a gas, have been presented in this chapter. One of the expressions was

$$\kappa_{gas} = \frac{\bar{v} C_{V,m}}{3\sqrt{2}\pi d^2 N_A}.\tag{8.22}$$

Use this equation and your knowledge of the temperature dependence of the variables in it to show how κ_{gas} changes as temperature is increased.

3. Figure 8.10 shows the variation of thermal conductivity of a semiconductor after doping with n- or p-type impurities. Explain the origin of the change in thermal conductivity on introduction of the impurity.

4. At a trade show exhibiting new building materials, one of the new products showcased was a window with good thermal insulation. The window had two panes of glass and the space between them contained argon gas. The advertisement said that this window gave much better insulation than could be achieved when the space was filled with air (the more usual case for double-paned windows). Explain why argon is a better thermal insulator than air. Assume both to be at the same pressure. It may be helpful to know that the molecular mass of argon is 39.95 g mol^{-1} and that of air (which is about 80% N$_2$ and 20% O$_2$) is about 29 g mol^{-1}, and to consider all factors in the Debye expression for thermal conductivity. State and justify any assumptions.

5. **Ceramics** are very useful inorganic materials composed of metallic and nonmetallic ions held together by partly ionic and partly covalent bonds. Ceramics are electrical insulators, characterized by their hardness, strength, and heat and chemical resistance. Ceramics can be amorphous or crystalline.

Figure 8.10. The temperature dependence of the thermal conductivity of a semiconductor with n- and p-type impurities at various concentrations.

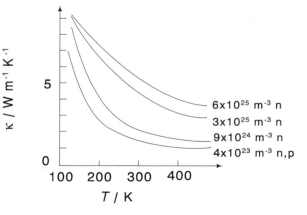

Explain how the microstructure, brought about by small changes in the composition of a ceramic, can affect the thermal conductivity of a ceramic. Take into account the fact that the porosity (and hence density) of a ceramic can be modified by the introduction of impurities.

6. Incandescent light bulbs usually have an argon atmosphere sealed inside them. The purpose of the argon is to prevent the filament from oxidizing (in air) or subliming (in a vacuum). A portion of the energy that is used to light an incandescent bulb is "lost" as heat. Can you suggest another inert gas that could be used in place of argon, but which is a better insulator? Explain your reasoning.

7. Appropriate properties of dental filling materials are the key to their success. A large "silver" filling (made of an amalgam, i.e., mixture, of silver with mercury) can make a tooth very sensitive to hot or cold food. However, this temperature sensitivity is not so dramatic when a white polymeric organic filling material is used. Explain why these two materials, silver amalgam and organic polymer, lead to such different thermal sensitivities. In this case, consider the nerve endings inside the tooth to be sensitive thermometers.

8. Two countertop materials look very similar; one is natural marble and the other is a synthetic material. They can be distinguished easily by their feel: one is cold to the touch and the other is not. Explain which is which and the basis of the difference in their thermal conductivities.

9. **Stainless steel** is iron with added chromium and possibly also nickel. It has good corrosion resistance and can be used for pots for cooking. In this case, the pot bottoms are often made of copper. Why is stainless steel not used on the pot bottoms? Why is copper better for the pot bottoms?

10. Black phosphorus has a density of 2.7 g cm^{-3}, whereas white phosphorus has a density of 1.8 g cm^{-3}. (Both values are at room temperature.) Use this information to explain the difference in thermal conductivities of the two forms of phosphorus (Table 8.1).

9

THERMODYNAMIC ASPECTS
OF STABILITY

9.1 INTRODUCTION

The thermal stability of a material, with respect to phase change and/or chemical change, can be very important in determining uses of a material. In this chapter we examine such properties, first for pure materials, then for two-component materials, and finally for three-component materials.

9.2 PURE GASES

A typical phase diagram for a pure material as a function of temperature and pressure is shown schematically in Figure 9.1, where several isotherms near the critical point are indicated.

The thermodynamic stability of a gas depends on its Gibbs energy being less than a condensed phase (liquid or solid). As previously seen, an ideal gas cannot condense (because it has no intermolecular forces), but we know that real gases do condense to give liquids and/or solids as we compress them, due to attractive intermolecular forces.

Considering the pressure as a function of volume for the isotherms of Figure 9.1 gives a series of curves as shown in Figure 9.2. At T_1, which is greater than the critical temperature, compression of the gas just increases the pressure; it does not lead to condensation. However, at a temperature below the critical temperature, e.g., T_3, a decrease in the volume of the gas leads first to an increase in pressure and then to a sudden drop in volume at constant pressure, corresponding to the drop in volume due to liquefaction of the gas. Further compression at T_3 leads to a rapid increase in pressure since liquids are not very compressible. At a lower temperature, T_4, liquefaction also occurs. At $T_2 = T_c$, the critical temperature, liquid and gas are indistinguishable throughout the whole volume range.

Although the ideal equation of state does not lead to the liquefaction of a gas, the van der Waals equation of state shows instabilities in the pressure–volume curves below the critical

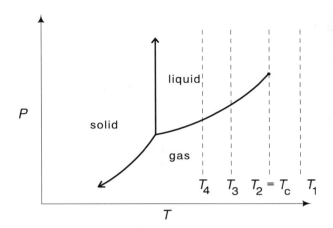

Figure 9.1. The generalized phase diagram for a pure material as a function of temperature (T) and pressure (P). The dashed lines correspond to constant-temperature lines (isotherms) considered further in Figure 9.2.

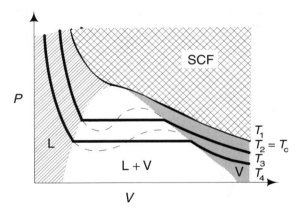

Figure 9.2. The pressure (P) as a function of volume (V) for various isotherms of a real gas. The temperatures correspond to those shown in Figure 9.1. The dashed lines correspond to the solutions to the van der Waals equation of state, and the solid lines correspond to experimental findings. The shading indicates distinctions among the regions, L (liquid), V (vapor), L + V (liquid and vapor) and SCF (supercritical fluid). At the critical temperature, T_2 in Figure 9.1, compression of the gas leads to an inflection point in the P vs. V curve shown here, corresponding to $(\partial P/\partial V)_T = 0$ and $(\partial^2 P/\partial V^2)_T = 0$. This inflection point is the apex of the curve joining the ends of the constant pressure volume liquefaction lines, and this is another way to define the critical point of a gas.

temperature, T_c. The solutions of the van der Waals equation of state lead to the $P(V)$ dashed curves indicated in Figure 9.2; the portion of those curves that leads to decreases in pressure with decreasing volume indicates their unphysical nature in this region. The isobaric condensation of the gas exactly balances the areas of the peak and inverse peaks of the $P(V)$ plot from the van der Waals equation of state (dashed lines in Figure 9.2).

9.3 PHASE EQUILIBRIA IN PURE MATERIALS: THE CLAPEYRON EQUATION

Phase diagrams such as Figure 9.1 can be understood by examining the experimental pressure–temperature boundaries of the phases. In view of this comment, it would be useful to derive an expression for dP/dT along the phase equilibrium lines.

To do so, it is first useful to consider the fact that the Gibbs energies of two single-component phases that are in equilibrium are identical. This is true for liquid–solid equilibrium ($G^{solid} = G^{liquid}$ at the melting point), liquid–vapor equilibrium ($G^{liquid} = G^{vapour}$ at the boiling point), or any equilibrium of phase α with phase β ($G^{\alpha} = G^{\beta}$).

If initial temperature and pressure conditions allow phase α to be in equilibrium with phase β,

$$G^{\alpha}_{initial} = G^{\beta}_{initial} \tag{9.1}$$

and the temperature and pressure are changed (changing G by dG) to final conditions such that the equilibrium of phases α and β is maintained, then

$$G^{\alpha}_{final} = G^{\beta}_{final}. \tag{9.2}$$

Since

$$G^{\alpha}_{final} = G^{\alpha}_{initial} + dG^{\alpha} \tag{9.3}$$

and

$$G^{\beta}_{final} = G^{\beta}_{initial} + dG^{\beta} \tag{9.4}$$

it follows that

$$dG^{\alpha} = dG^{\beta}. \tag{9.5}$$

The fundamental equation for dG (Equation 6.62) can be written for phase α and also for phase β:

$$dG^{\alpha} = V^{\alpha}dP - S^{\alpha}dT = dG^{\beta} = V^{\beta}dP - S^{\beta}dT. \tag{9.6}$$

This can be rearranged to give the quantity we were aiming for, dP/dT:

$$\frac{dP}{dT} = \frac{S^{\alpha} - S^{\beta}}{V^{\alpha} - V^{\beta}} = \frac{\Delta_{trs}S}{\Delta_{trs}V} = \frac{\Delta_{trs}H}{T\Delta_{trs}V} \tag{9.7}$$

where the second equality uses the general form of the transition entropy change ($\Delta_{trs}S$) and transition volume change ($\Delta_{trs}V$), and the third equality makes use of the generalization

$$\Delta G = \Delta H - T\Delta S \tag{9.8}$$

where, due to the equality of G for any two phases in equilibrium,

$$\Delta_{trs}G = 0. \tag{9.9}$$

Equation 9.7 is called the **Clapeyron**[1] **equation.** The Clapeyron equation can be applied to any first-order phase transition, and holds exactly.

To see how the Clapeyron equation can be used to determine the sign of the slope of phase boundaries, we consider two examples.

For the transition from solid to vapor, the Clapeyron equation gives

$$\left(\frac{dP}{dT}\right)_{sol-vap} = \frac{\Delta_{subl}H}{T\Delta_{subl}V} \tag{9.10}$$

where $\Delta_{subl}H$, the enthalpy change on sublimation, is positive (it always takes heat to convert a material from solid to vapor), and the volume change on sublimation, $\Delta_{subl}V = V_{vapor} - V_{solid}$, also is positive. Since T is always positive (on the Kelvin scale), it follows that (dP/dT) on the solid–vapor equilibrium line, i.e., the slope of the solid–vapor equilibrium line, is always positive.

For the transition from solid to liquid, the Clapeyron equation gives

$$\left(\frac{dP}{dT}\right)_{sol-liq} = \frac{\Delta_{fus}H}{T\Delta_{fus}V} . \tag{9.11}$$

Since $\Delta_{fus}H$, the enthalpy change on fusion (melting) is positive, and T is positive, and $\Delta_{fus}V = V_{liquid} - V_{solid}$ is positive for most materials, dP/dT along the solid–liquid line is positive for most materials. However, for some materials $V_{liquid} < V_{solid}$; examples are water (we know this because ice floats in liquid water), Ga, Sb, Bi, Fe, Ge, and diamond. The larger volume for the solid leads to $dP/dT < 0$. We will see this in the experimentally determined water phase diagram later in this chapter.

To generalize, given that it takes heat ($\Delta H > 0$) to proceed from a lower-temperature to a higher-temperature phase (another way to look at this is that the entropy of the higher-temperature phase exceeds that of the lower-temperature phase), the slope of P–T equilibrium lines in the phase diagram of a pure material can indicate, using the Clapeyron equation, the relative densities of the two phases.

Let us turn now to some specific phase diagrams for pure materials.

9.4 PHASE DIAGRAMS OF PURE MATERIALS

In Figure 9.3, the pressure–temperature phase diagram of carbon dioxide is shown. This diagram is an example of the simple P–T phase diagrams discussed earlier. One particularly interesting feature is that at a pressure of 1 atm, the solid converts directly to vapor, at a temperature of 194.7 K; this can be seen in the sublimation of dry ice [$CO_2(s)$] at room temperature, one of the few examples of conversion from solid to vapor at ambient pressure without first passing through the liquid phase.

The relatively moderate critical point of CO_2 ($P_c = 72.8$ atm and $T_c = 304.2$ K) leads to supercritical CO_2 relatively easily. Fluids beyond the critical point are called **supercritical**

[1] Benoit Pierre Émile Clapeyron (1799–1864) was a French engineer who specialized in bridges and locomotives. His only publication in pure science concerned the expression of vapor pressure as a function of temperature.

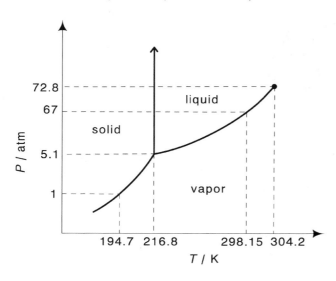

Figure 9.3. The pressure–temperature phase diagram for pure CO_2. The scales are not linear.

fluids, and they are neither gas nor liquid. Supercritical fluids have unusual properties such as much lower viscosity than the liquid and much higher density (and therefore greater solvent power) than the gas. Supercritical CO_2 has diverse uses such as the carrier for some chromatographic separations and the solvent for decaffeination of coffee.

In Figure 9.4, the pressure–temperature phase diagram of sulfur is shown. The solid lines represent the stable phase diagram. The main new feature in the phase diagram is the existence of two solid phases, one of monoclinic structure and the other of orthorhombic structure. [The names of the structures designate the packing arrangements of the smallest repeat unit (**unit cell**) that can replicate the entire crystal. See Appendix 3 for a summary of the types of unit cells.] The existence of more than one solid phase is called **polymorphism,** literally "many shape types." The two polymorphs of sulfur have different temperature and pressure regions of stability. (For elements, polymorphs are also called **allotropes.**)

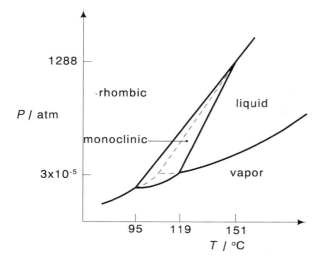

Figure 9.4. The pressure-temperature phase diagram for sulfur. The solid lines indicate the stable phases, and the dashed lines indicate metastability.

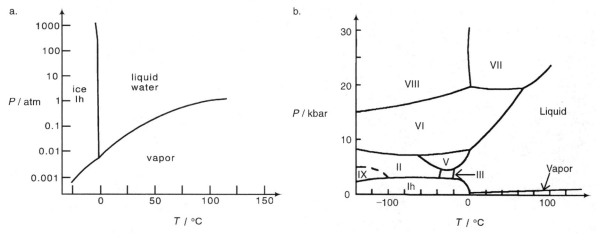

Figure 9.5. The pressure–temperature phase diagram of H_2O (a) at low pressure (note the negative P–T slope) and (b) over a wider pressure range (1 kbar $\approx$ 1000 atm).

Figure 9.4 also shows, by the dashed lines, the metastable phase diagram for sulfur. By appropriate thermal treatment (slow cooling from above the melting point) it is possible to trick sulfur into transforming directly from a supercooled liquid to the rhombic form, without first passing through the monoclinic form. In the region where the monoclinic form is most stable (i.e., monoclinic has the lowest G of monoclinic, liquid, and rhombic), the liquid or rhombic form can exist only metastably. That they can exist is testament to the high activation energy required for the conversion from the metastable form to the monoclinic form in this temperature–pressure region.

Figure 9.5 shows two views of the phase diagram of water. At low pressure (Figure 9.5a), the negative slope of the P–T solid–liquid line is readily apparent. This line indicates, as we have seen earlier in the discussion of the Clapeyron equation, that $H_2O(s)$ is less dense than $H_2O(l)$ under these conditions. The phase of ice in this temperature and pressure region is generally referred to as ice Ih, where the "h" stands for "hexagonal." (The hexagonal structure is the origin of the six points in a snowflake.) The hexagonal form of ice (shown in Figure 9.6) has a very open structure (i.e., low-density structure) with dynamically disordered hydrogen bonds. The structure is not close-packed because of trying to satisfy the hydrogen-bonding requirements of the H_2O molecules. It is the openness of this structure that leads to the low density of ice Ih relative to water and the negative value of dP/dT in this region. It is worth considering that the negative value of dP/dT leads to a decrease in the melting point of ice when pressure is applied. Although this has been cited as the basis for ice skating, calculations[2] show that the temperature drop is insufficient to account for a liquid layer between the skate and blade.

At higher pressures, more polymorphs of ice are observed, as illustrated in Figure 9.5b. The large number of polymorphs for H_2O is due to the many ways in which H_2O molecules can be arranged to satisfy their hydrogen bonds to their neighbors, with different structures

---

[2] L.F. Louks (1986), *Journal of Chemical Education*, **63**, 115.

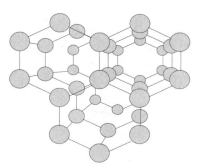

Figure 9.6. The oxygen positions in the hexagonal structure of ice Ih. The hydrogens are not shown because they are disordered. The very open structure makes ice Ih float in liquid water.

having lowest G in different pressure and temperature regions. Note that at high pressures (above about 2 kbar) the values of dP/dT for ice (in phase ice III) have become positive, indicating that the higher-pressure polymorphs of ice are denser than water. This is because the higher-pressure forms are compressed relative to ice Ih.

Figure 9.7 shows the temperature–pressure phase diagram of CH_4. (The solid–liquid–vapor triple point of methane is at 90.69 K.) However, there are at least four solid phases for CH_4, and polymorphism in this case can be attributed to the shape of the molecule. Although we may consider methane to be tetrahedral, as in the ball-and-stick model of Figure 9.8a, the shape of any molecule is really more closely related to the electron density distribution, and this is more nearly represented as a sphere with slight protuberances at tetrahedral locations (Figure 9.8b). The nearly spherical shape of the methane molecule means that it takes very little energy to rotate it even in the solid state. In fact, in Phase I of CH_4, the methane molecules are nearly freely rotating, although they are located on particular lattice sites (Figure 9.9). Another way to express this is to say that they are translationally ordered but rotationally disordered. This phase is often referred to as an **orientationally disordered phase.**

As the temperature is lowered, CH_4 molecules become more ordered, and Phase II has a rather unusual structure with some molecules in the unit cell ordered and others in other positions in the unit cell dynamically disordered (Figure 9.9).

The structures of the other phases of CH_4 are more complicated still and not yet fully sorted out. A complication here is that the CH_4 molecule is so light that Newtonian mechanics cannot properly describe it at such low temperatures, and quantum mechanics must be used to describe the disorder in these solid phases.

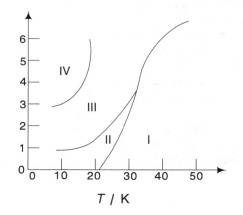

Figure 9.7. The low-temperature phase diagram of CH_4, showing only the solid phases.

a.

b.

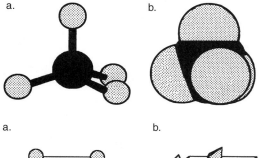

Figure 9.8. Two views of the CH_4 molecule. (a) The ball-and-stick model is useful, but does not truly represent the space-filling shape of the molecule. (b) An electron density distribution view of CH_4, where there is 95% probability of finding the electrons within the surface shown, shows methane to be a more nearly spherical molecule.

a.

b.

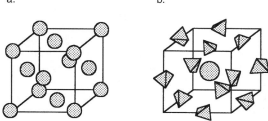

Figure 9.9. Structures of two phases of solid CH_4. (a) Phase I, with the CH_4 molecules orientationally disordered (shown as spheres to indicate rotating tetrahedra) on sites of a face-centered cubic lattice. (b) Phase II, with some CH_4 molecules orientationally ordered (orientationally ordered tetrahedra) and some experiencing very low (practically zero) barrier to rotation (shown as a sphere in the unit cell). More precisely, the latter molecules are in a spherically symmetric ground state at low temperature.

The concept of an orientationally disordered solid also can be described from the point of view of thermodynamics. In particular, the idea of an orientationally disordered solid leads to the picture of a solid with an entropy that is higher than the usual value for a solid. In fact, it has been suggested that an orientationally disordered solid can be described as having an unusually low entropy change on fusion ($=\Delta_{fus}S = S_{liquid} - S_{solid} < 20$ J K^{-1} mol^{-1} for an orientationally disordered solid), given that S_{liquid} is nearly the same for most liquids and S_{solid} is unusually high for an orientationally disordered solid.

An orientationally disordered solid also has been referred to as a "**plastic crystal**" becauseof the ease of deformation of this type of material due to the rotating molecules that make it up. The term "orientationally disordered solid" is now preferred over "plastic crystal," as the latter is sometimes too limiting, and the former term accurately describes the physical picture.

Orientational disorder can be understood further in terms of the intermolecular potential: the barrier to rotation of a nearly spherical molecule (see Figure 9.10) can be quite a bit less than the available thermal energy, kT. This can be contrasted with liquid crystals, where the barrier to reorientation is high due to the molecular shape, but the barrier to translation is relatively low (Figure 9.11). In both orientationally disordered solids and liquid crystals when there is less thermal energy (kT decreases), the material forms an ordered solid.

Methane is not the only example of a molecular solid exhibiting some aspect of orientational disorder: H_2, N_2, O_2, F_2, CCl_4, neopentane, norbornane, cubane, adamantane, and C_{60} are a few of many other examples. Furthermore, portions of large molecules can be orientationally disordered if there is sufficient energy: typical examples are alkyl chains and methyl groups. This flexibility in a large molecule can be responsible for biological activity, and one reason for the intolerance of living organisms to even relatively small temperature changes is the subtle importance of internal degrees of freedom to their biological functions, and the interplay between thermal energy (kT) and activation barriers to orientational motion.

The phase diagram of helium (Figure 9.12 shows the phase diagram for the ^{4}He isotope) shows other interesting and unique features. For example, there are two solid phases, one with hexagonal close packing and the other with body-centered cubic packing (see Appendix 3 for definitions of these structures). Perhaps more interesting is the fact that helium does not exist

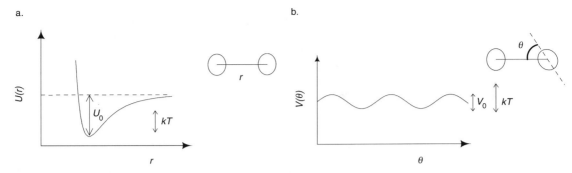

Figure 9.10. The energy of interaction of two nearly spherical molecules that are capable of forming an orientationally disordered solid, as a function of (a) intermolecular separation and (b) angle at fixed separation. If the thermal energy is less than the binding energy [$kT < U_0$ in (a)], the material will be a solid. For a material that can form an orientationally disordered solid, $V_0 \ll U_0$, i.e., the molecules can rotate more easily than they can translate. If the thermal energy (kT) is greater than V_0, as shown here, the material will be orientationally disordered. At lower temperatures, the stable structure will be that of an ordered solid.

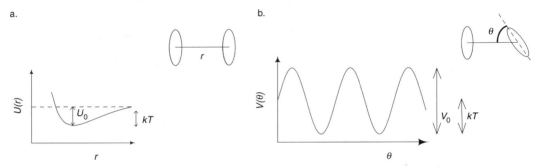

Figure 9.11. The energy of interaction of two rod-shaped molecules that are capable of forming a liquid crystalline phase, as a function of (a) intermolecular separation and (b) angle at fixed separation. For a material that can form a liquid crystal, $U_0 \ll V_0$, i.e., the molecules can translate more easily than they can rotate. If the thermal energy (kT) is nearly equal to U_0, as shown here, then the material will be liquid crystalline. At lower temperatures, the stable structure will be that of an ordered solid.

Figure 9.12. Phase diagram of ^{4}He.

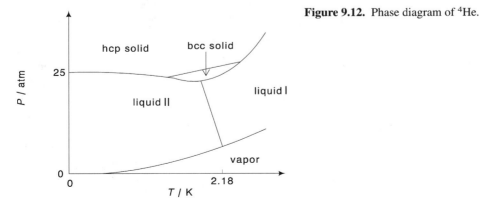

as a solid except under applied pressure; it remains as a liquid down to absolute zero. Furthermore, there are two liquid phases! The higher-temperature liquid form, liquid helium I, is the phase that exists at the critical point ($T_c = 5.2$ K), and this is the form of liquid helium that boils at 4.2 K and has wide use as a **cryogen** (a low-temperature fluid).

Liquid helium II has such fascinating properties that it is called a **superfluid.** For example, a suspended beaker filled with liquid helium II will spontaneously lose its liquid due to creep up the sides of the beaker, followed by dripping until the beaker is empty. Liquid helium II can flow through the narrowest of openings without any resistance (i.e., it has zero viscosity). Furthermore, the thermal conductivity of liquid helium II close to the I–II transition temperature, is about 1000 times greater than that of copper (and remember that a liquid has neither phonons nor free electrons to carry heat!). The ^{3}He isotope has at least three distinct superfluid phases.[3]

The phase transition from liquid helium I to liquid helium II is often referred to as a **λ-transition,** because of the shape of the associated heat capacity anomaly. The source of the unusual properties of liquid helium II is the zero-momentum state of some of the helium atoms in this phase (moving at a speed of about 3×10^{-7} m s^{-1}, which is very slow compared with the speed of the other He atoms, which move about 10^9 faster, i.e., 300 m s^{-1} at absolute zero due to zero-point motion). The helium atoms in this phase are said to have undergone a **Bose–Einstein condensation** (i.e., they are "condensed" into their ground state energy), as suggested first by Fritz London in 1938 and now confirmed experimentally.[4]

COMMENT

POLYMORPHISM IN NATURE

$CaCO_3$ exists in two common polymorphs in nature, calcite and aragonite. Although the two crystal structures are very similar, calcite is the more thermodynamically stable form at room temperature and pressure. The form produced under biological conditions depends on nucleating macromolecules, and possibly on the presence of other ions in solution.

As a final example of an interesting phase diagram of a pure material, we consider polymorphism in carbon (Figure 9.13). Figure 9.13 shows that the low-pressure stable form is graphite while at higher pressures (such as inside the earth) diamond is the stable form. Since we know that diamonds can exist at room temperature and pressure we must conclude that they are metastable and should spontaneously convert to the more thermodynamically stable form, graphite. That diamonds do not convert before our eyes indicates that the activation energy for this conversion is very high. This makes sense in view of the great rearrangement that would be required to go from the three-dimensional diamond lattice (Figure 9.14a) to the two-dimensional layered structure of graphite (Figure 9.14b). The differences between the diamond and

---

[3] The discovery of superfluidity of ^{3}He at 0.002 K was the topic of the 1996 Nobel Prize in Physics, awarded jointly to David M. Lee (1931– ; Professor of Physics, Cornell University), Douglas D. Osheroff (1945– ; Professor of Physics, Stanford University), and Robert C. Richardson (1937– ; Professor of Physics, Cornell University).

[4] See, for example, John Bardeen (1990), Superconductivity and Other Macroscopic Quantum Phenomena, *Physics Today,* December 1990, 25.

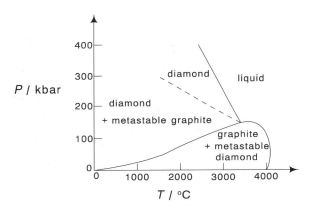

Figure 9.13. The pressure–temperature phase diagram of carbon. Solid lines indicate the regions of stable phases; metastability is indicated with dashed lines.

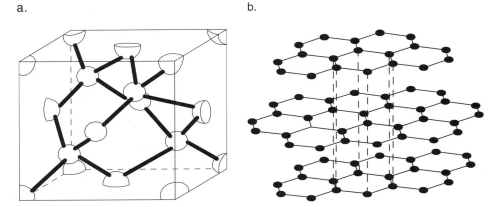

Figure 9.14. Two polymorphs of carbon: (a) diamond and (b) graphite. The three-dimensional bonding network of diamond is one of the reasons why diamond is so hard. On the other hand, the layered structure of graphite, with the layers held together with weak van der Waals interactions, makes graphite useful as a lubricant.

the graphite structures are also responsible for the very large differences in their properties: diamond is hard (due to the three-dimensional structure) whereas graphite is so soft that it is the "lead"[5] in a pencil and also a useful lubricant [due to the weak (van der Waals) interaction holding the layers together].

Other interesting polymorphs of carbon belong to the family of compounds generally known as **Fullerenes.** The archetypal member of this family is the molecule C_{60}, which was first discovered in 1985.[6] The geometry of the C_{60} molecule is that of a soccer ball (see Fig-

---

[5] This is called "lead" because the first deposit of graphite used to make pencils, which was revealed when a tree toppled over, was originally mistaken for the element lead.

[6] This discovery was recognized with the 1996 Nobel Prize in Chemistry, awarded jointly to Robert F. Curl, Jr. (1933– ; Professor of Chemistry, Rice University), Sir Harold W. Kroto (1939– ; Professor of Chemistry, University of Sussex) and Richard E. Smalley (1943; Professor of Chemistry, Rice University).

Figure 9.15. The molecular structure of C_{60}.

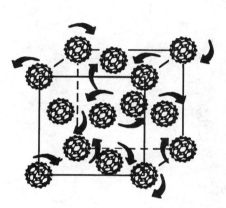

Figure 9.16. The structure of C_{60} at room temperature, showing the C_{60} molecules dynamically orientationally disordered on their face-centered cubic lattice sites.

ure 9.15), with one carbon atom at each vertex. There can be other molecules with similar structures, and the general family is called "Fullerenes" after R. Buckminster Fuller, the architect who designed geodesic domes as architectural structures. Although first synthesized in the laboratory only a few years ago, it is now known that Fullerenes exist naturally on earth and even in interstellar space. C_{60} has been found to occur naturally in a 2-billion-year-old impact crater in Sudbury, Canada. The discovery of C_{60} and its family members has led to a very exciting new area of materials science, as some metal atoms can be trapped inside the cage of Fullerenes, designated by the general formula $M@C_n$. For example, $La@C_{82}$ and $Y@C_{82}$ have been prepared. The combination of metals with the insulating fullerenes can lead to unusual optical and electronic devices: Fullerene derivatives can be made to be **superconducting** (e.g., K_3C_{60}; see Chapter 12 for a discussion of superconductivity). Other molecules with special properties can be added outside or inside the cage, and C_{60} even can be used to produce diamond films.[7] Pure C_{60} is yellow when in the solid form and magenta when dissolved in benzene, indicating that its electronic effects depend on the environment.

The solid form of C_{60} is orientationally disordered at room temperature (see Figure 9.16), with the molecules sitting on the lattice sites of a face-centered cubic lattice. Below a solid–solid phase transition at about $T = 260$ K, the C_{60} molecules become more ordered.

---

[7] For a review, see R.F.Curl and R.E. Smalley (1991), Fullerenes, *Scientific American,* October 1991, 54.

UNUSUAL OPTICAL PROPERTIES OF C_{60}

C_{60} has an unusual optic property: when more intense light shines on it, less is transmitted. Furthermore, C_{60} becomes a better "light limiter" at longer wavelengths, because the absorption cross sections of the ground state and the excited state both change with wavelength, but in opposite directions. This may lead to using C_{60} for protection from light radiation, but first materials scientists have to work out a way to produce stronger films of C_{60} and to keep them from oxidizing (see *Chemistry in Britain,* 887, November 1994).

QUASICRYSTALS

The 14 crystal types shown in Appendix 3 are the smallest cells that can be regularly and repeatedly translated to produce periodic macroscopic crystalline structures. By contrast, glassy (or amorphous) materials do not have periodic atomic arrangements. For many years it was thought that these were the only possibilities for arrangements of atoms in solids—fully periodic or fully disordered. More recently, materials such as alloys of aluminium and manganese have been found that have quasiperiodic structures, and they are called **quasicrystals.** These structures are similar to quasiperiodic tiling patterns worked out by the British mathematician Roger Penrose; an example of a Penrose tiling pattern is shown in Figure 9.17. The main feature of the packing arrangement is that all space is filled, yet the arrangement is never exactly repeated.

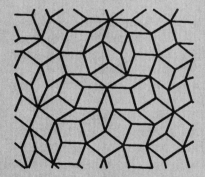

Figure 9.17 A two-dimensional representation of a quasicrystalline structure.

9.5 THE PHASE RULE

There is a very useful rule, called the **Phase Rule** (also known as Gibbs' Phase Rule), that can be used to generalize the relationships between the number of free variables specifying the state of a system, and the number of chemical components present and the number of phases present. The Phase Rule can provide very useful information concerning the phases present in a system.

Before deriving the Phase Rule, it is necessary to introduce a concept that is as important for many-component species as Gibbs energy is for single-component (pure) systems. This concept is **chemical potential,** where the chemical potential for component i, μ_i, is defined as

$$\mu_i = \left(\frac{\partial G}{\partial n_i}\right)_{T,P,n_{j \neq i}} \tag{9.12}$$

where the inequality in the constant-variable subscript indicates that the number of moles[8] of all components except component i are held constant. If the system has only component i (i.e., it is pure species i), then $\mu_i = G_m$, i.e., the molar Gibbs energy and chemical potential of a pure material are identical.

The reason for introducing chemical potential is that it is the multicomponent analogue to G. Just as G is equal for all phases at equilibrium, the chemical potential of a given species is the same in all equilibrated phases in which it is present. For example, if phase α is in equilibrium with phase β, and component i is present in both phases, then

$$\mu_i^\alpha = \mu_i^\beta. \tag{9.13}$$

The symbol F is used to represent the **number of degrees of freedom**[9] of a system, and it is the aim of the Phase Rule to determine F, the number of independent variables needed to specify the system. In other words,

$$F = \text{number of variables} - \text{number of relations.} \tag{9.14}$$

The variables needed to describe the system will involve the composition of each phase, and possibly other variables (e.g., temperature and pressure), and we look at these in some detail below.

We use the symbol c to represent the number of **chemical components** in the system. This is the minimum number of chemical species needed to describe the composition of the system. Alternately, c can be described as the number of independent chemical species. For example, in an aqueous sodium chloride solution, $c = 2$ whether H_2O and NaCl are considered the components or whether the chemical species are considered to be H_2O, NaCl, Na^+, Cl^-, H^+, and OH^-. In the latter case the number of species is six but there are four independent relations among them: $[OH^-] = [H^+]$, $[Na^+] = [Cl^-]$, $H_2O \rightleftarrows H^+ + OH^-$; and NaCl $\rightleftarrows Na^+ + Cl^-$. This leaves $6 - 4 = 2$ independent chemical species, i.e., $c = 2$.

We let p represent the number of types of **phases** within the system. A phase is considered to be a homogeneous region separated from other homogeneous regions by a phase boundary (or surface of discontinuity). For example, an ice-water–salt solution (Figure 9.18a) has $p = 3$; the discontinuities from one phase to the next can be seen by the variation in density of the beaker's contents from the top to the bottom of the beaker (Figure 9.18b).

---

[8] IUPAC recommends the term "amount of substance" in place of "number of moles."

[9] This is distinct from the "degrees of freedom" used in Chapter 6 to designate the number of spatial and angular coordinates required to describe the motion and position of a molecule, but in both cases the term "degrees of freedom" is used to indicate the variables that are necessary to define the system.

a.

b.

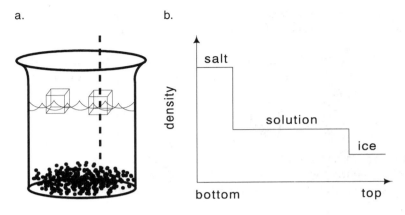

Figure 9.18. An ice-water–salt system. (a) The system itself. (b) The density profile of the system from bottom to top, as indicated by the dashed line through the beaker in (a).

So, using the definition of F (Equation 9.14), we can consider first the number of variables needed to describe the state of the system and then the relations between the variables, to determine a useful relationship for F, i.e., the Phase Rule.

We must know something about the compositions to know the state of the system. Within a given phase, we need to know $(c - 1)$ of the compositions if there are c components present. The value $(c - 1)$ arises because if we know $(c - 1)$ compositions then we know the final composition, i.e., the composition of the cth component, since

$$\sum_i X_i = 1 \tag{9.15}$$

where X_i is the **mole fraction of component** i in the mixture, defined as

$$X_i = \frac{n_i}{n_{\text{tot}}} \tag{9.16}$$

where n_i is the number of moles of component i and n_{tot} is the total number of moles in the system. For p phases, this gives a total of $p(c - 1)$ composition variables that are required to specify the system.

Other variables may also be needed to specify the state of the system. For example, T and P can vary the state of the system and, in some circumstances, electric field (e.g., cholesteric liquid crystals) and/or magnetic field can vary the state of the system. We let n represent these other variables; in our examples often n will be 2 (T and P) or 1 (T or P), but we leave the value n unspecified for the general case. Combined with the composition variables, this leads to $n + p(c - 1)$ variables needed to specify the system. However, not all these variables are independent, and we now address this point.

There are chemical potential relations among the phases that serve to reduce the number of variables needed to specify the state of the system. For each component i in the p equilibrated phases the chemical potentials are all the same, so we can write

$$\mu_i^1 = \mu_i^2 = \ldots = \mu_i^p \tag{9.17}$$

and for the p phases there are $(p - 1)$ such relations for each component i. Therefore, for c components, there are $c(p - 1)$ such relations, each reducing the number of free variables necessary to specify the state of the system.

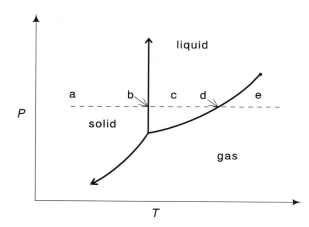

Figure 9.19. Phase diagram for a typical pure material, showing the effect of pressure and temperature on phase stability. The solid lines indicate stable phase boundaries. The dashed line indicates heating at constant pressure (see text).

Putting this together, considering the definition of F (Equation 9.14), leads to

$$F = n + p(c - 1) - c(p - 1) \tag{9.18}$$

which simplifies to

$$F = c - p + n \tag{9.19}$$

where Equation 9.19 is the Phase Rule, first derived in 1875 by J. Willard Gibbs.

The Phase Rule can be used to describe the number of free variables in a system, and, since we have been discussing single-component phase diagrams, we begin there with the use of the Phase Rule.

The typical one-component phase diagram of Figure 9.19, considering that here $c = 1$ and $n = 2$ (T and P can, in principle, vary), leads to $F = 2$ in the region in which only solid exists. This value of F indicates that two variables (T and P) are needed to specify the state of the system. Another way to express this is that P and T can vary independently. Similarly, in the liquid region, $F = 2$ and T and P are independent of each other. In the vapor region $F = 2$ also.

Along the liquid–solid coexistence line, $c = 1$, $p = 2$, so $F = 1 - 2 + 2 = 1$. This shows that although T or P can be varied along this line, they cannot be varied independently; a new T specifies (through the phase diagram or the Clapeyron equation) what the new pressure will be. Similarly, $F = 1$ along the solid–vapor, and along the liquid–vapor line. As long as we are on one of these lines, T and P are not independent of each other; there is only one degree of freedom.

At the **triple point** (the coexistence point of solid, liquid, and vapor), $c = 1$, $p = 3$, and $F = 1 - 3 + 2 = 0$. This is said to be an **invariant point,** since neither the temperature nor the pressure can be arbitrarily changed while remaining at the triple point. For this reason triple points (e.g., $T = 273.16$ K, $P = 610.6$ Pa for H_2O) can be used for calibration of thermometers.

The minimum number of degrees of freedom is 0, as in the invariant triple point. This implies, for a situation in which $n = 2$, that the maximum number of phases in equilibrium in a one-component system is 3. (See Figure 9.4 for sulfur, where there are four phases but no more than three phases coexist at any given temperature and pressure.)

If we consider isobarically warming a sample of the material described by the phase diagram of Figure 9.19 along the dashed line shown in the figure, the Phase Rule can be used to

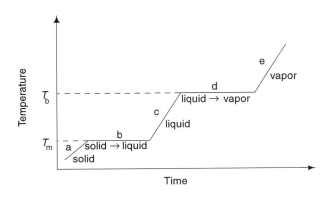

Figure 9.20. The warming curve (temperature–time profile) corresponding to isobaric heating of a pure material from solid to liquid to gas (i.e., along the dashed line of Figure 9.19).

describe the heating curve (temperature as a function of time). As this is a constant pressure situation, $n = 1$. Therefore $F = c - p + n = 1 - p + 1 = 2 - p$ here. In region a, where there is one phase (solid), $F = 2 - 1 = 1$, and the temperature can increase. At point b, $p = 2$ (solid and liquid) so $F = 0$, and the temperature must be invariant throughout the time it takes to pass through b. In c, $p = 1$ and $F = 1$, so the temperature increases again. At point d, $p = 2$ (liquid and vapor) and $F = 0$, so the temperature remains constant until the liquid is fully vaporized. In region e, $p = 1$ (vapor) and $F = 1$, so the temperature can increase again. The corresponding temperature–time profile is shown in Figure 9.20. The horizontal lines in such a warming curve are called **halts** or **arrest points** and they are particularly telling, as they indicate regions where $F = 0$. The determination of such warming curves, as we shall see in particular for multicomponent phase diagrams, can be used to determine phase diagrams. We turn next to phase diagrams consisting of two components.

9.6 LIQUID–LIQUID BINARY PHASE DIAGRAMS

In the world of materials science, mixtures of materials can be somewhat more important than pure materials. The main reason for this is that mixing materials allows one to tailor-make properties, intermediate between those of the pure materials, or sometimes new properties altogether different from those of the pure materials. Although nature has provided many different types of pure materials (over 16,000,000 compounds catalogued in *Chemical Abstracts,* growing at more than 1500 per week!), mixtures can be very important.

When a system consists of more than one chemical component, it is referred to as a **multicomponent system.** In the case of two chemical components, it is called a **binary system.** In our discussion of binary systems, we begin by considering a system prepared by mixing two liquids.

Generally speaking, liquids are considered to be **miscible** (fully soluble, e.g., two liquids with similar polarities, such as water and acetone) or **immiscible** (e.g., two liquids with very different polarities, such as CCl_4 and H_2O). In strict terms, at the lowest concentrations, every liquid will mix with every other liquid, but at higher concentrations there may be a **miscibility gap,** i.e., a range of concentrations for which two phases coexist.

The miscibility of two liquids, like the solubility of a solid in a liquid, can change as the temperature is changed. Liquids with a miscibility gap can either increase or decrease their miscibility as the temperature is increased.

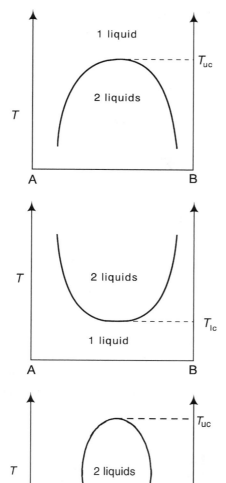

Figure 9.21. An example of a liquid–liquid system with increased solubility at increased temperature, showing an upper consolute temperature, T_{uc}, at a composition intermediate between pure A and pure B. An example of such a phase diagram is water–aniline, with $T_{uc} = 441$ K.

Figure 9.22. An example of a liquid-liquid system with increased solubility at decreased temperature, showing a lower consolute temperature, T_{lc}, at a composition intermediate between pure A and pure B. An example of such a phase diagram is water–triethylamine, with $T_{lc} = 291.6$ K.

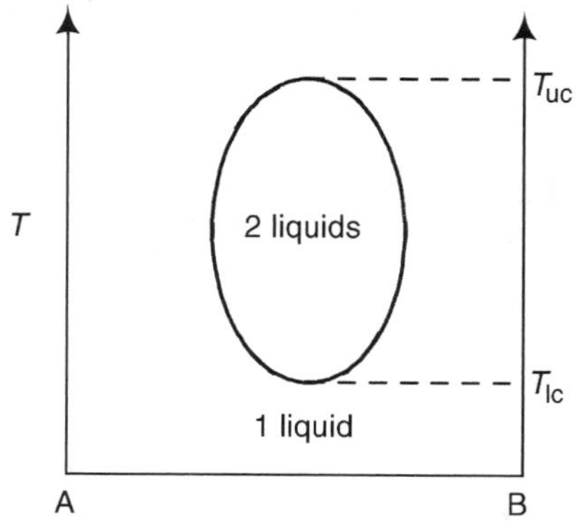

Figure 9.23. A liquid–liquid system showing both an upper consolute temperature and a lower consolute temperature, both at compositions intermediate between pure A and pure B. An example of such a phase diagram is water–nicotine, with $T_{uc} = 480$ K and $T_{lc} = 335$ K.

If the miscibility increases with increasing temperature, this can lead to a temperature above which the two liquids are completely miscible in all proportions. This temperature is referred to as the **upper consolute temperature, T_{uc}.** An example is shown in Figure 9.21.

In contrast, some liquids decrease their miscibility as the temperature is increased, so there is a temperature beneath which the two liquids are miscible in all proportions. This is a **lower consolute temperature, T_{lc}.** An example is shown in Figure 9.22.

There are a few liquid–liquid systems that show both an upper and a lower consolute temperature. One is shown in Figure 9.23.

9.7 LIQUID–VAPOR BINARY PHASE DIAGRAMS

If a liquid is heated sufficiently, it will eventually transform to the vapor phase; this is true for pure liquids and also binary liquid systems. However, the variation of boiling point with composition can reveal interesting features in a binary liquid system. A few of those features are summarized here.

If the components of a binary system are sufficiently similar chemically, they will interact much the same with each other as they do with themselves. In this case, the system is said to be an **ideal solution** and P_A, the vapor pressure of component A, is given by **Raoult's Law**[10]:

$$P_A = X_A P_A^0 \tag{9.20}$$

where X_A is the mole fraction of component A and P_A^0 is the vapor pressure of component A when pure. A similar equation applies to the second component (B), such that the total pressure over an ideal solution, P_{tot}, is given by

$$P_{tot} = X_A P_A^0 + X_B P_B^0 \tag{9.21}$$

where X_B and P_B^0 are the mole fraction of component B and vapor pressure of pure component B, respectively. The resulting vapor pressure as a function of composition for a nearly ideal system is shown in Figure 9.24.

When a binary solution is not ideal, it is because the interactions between the component molecules (i.e., A–B interactions) are different from those between like molecules (i.e., A–A and B–B interactions).

If the A–B type interactions are more repulsive than the A–A and B–B interactions, then the vapor pressure of the solution will be higher than the ideal vapor pressure. An example of this is shown in Figure 9.25, where the increase in vapor pressure has given rise to a maximum

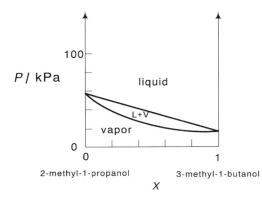

Figure 9.24. The pressure–composition diagram for 2-methyl-1-propanol with 3-methyl-1-butanol, $T = 323.1$ K, which forms a nearly ideal solution. The liquid curve is nearly a straight line, as given by Equation 9.21.

[10] François Marie Raoult (1830–1901) was a leading French experimental physical chemist in the nineteenth century. He made major experimental contributions to electrochemistry and understanding of solutions, including provision of experimental verification of Arrhenius' theory that salts ionize in aqueous solution. His accurate measurements of vapor pressures led to what we now call Raoult's Law.

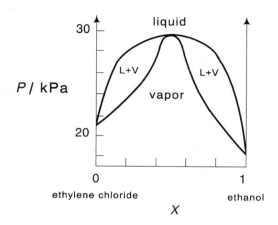

Figure 9.25. The liquid–vapor phase diagram for the system ethylene chloride–ethanol at $T = 313.1$ K, showing pressures greater than ideal pressures, and an azeotrope.

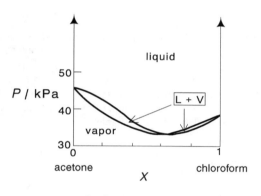

Figure 9.26. A liquid–vapor diagram showing a low vapor pressure azeotrope. The components are acetone and chloroform at $T = 308$ K.

in the vapor-pressure plot. At the composition of this maximum, the liquid boils at constant temperature, just like a pure component. This is called an **azeotrope,** which means "constant boiling mixture."

On the other hand, if the A–B interactions in a binary solution are more attractive than the A–A or B–B interactions, then the vapor pressure will be lower than in the ideal system. An example of this is shown in Figure 9.26, where another azeotrope (this one at low vapor pressure) occurs.

An ideal binary solution, by virtue of its vapor pressure changing monotonically with composition (Equation 9.21), will have a boiling point that is intermediate between the boiling points of the two components. An example is shown in Figure 9.27. Note that at any composition intermediate between the two pure components, there is a temperature range of coexistence between liquid and vapor.

If the deviations from ideality lead to a vapor pressure that is higher than ideal, this corresponds to boiling points that are lower than ideal. (A solution boils when the vapor pressure that it exerts equals the external pressure; if the vapor pressure is high, then a lower temperature will achieve boiling.) An example of a boiling point diagram for such a system is shown in Figure 9.28.

If the deviations from ideality make the vapor pressure lower than the ideal vapor pressure, there will be a maximum in the boiling point diagram. An example is shown in Figure 9.29.

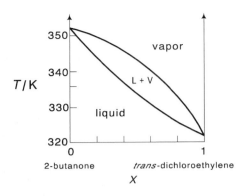

Figure 9.27. Temperature–composition boiling point diagram for 2-butanone with *trans*-dichloroethylene at $P = 101.325$ kPa $= 1.00000$ atm. This system shows nearly ideal behavior.

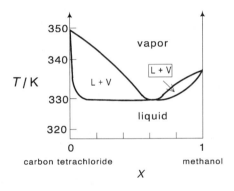

Figure 9.28. The boiling point diagram for carbon tetrachloride and methanol at $P = 101.325$ kPa $= 1.00000$ atm. This system exhibits deviations from ideality that lead to a low-boiling azeotrope. In distillation of a mixture, the **distillate** (i.e., the portion that distils) would have the azeotropic composition, making separation of the components by distillation impossible.

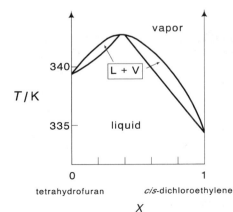

Figure 9.29. Boiling point diagram for the system tetrahydrofuran with *cis*-dichloroethylene at $P = 101.325$ kPa $= 1.00000$ atm. The high-boiling azeotrope means that in distillation of any composition other than the azeotropic composition, the azeotrope will remain behind in the still pot. The azeotropic composition would distil at a fixed temperature, much like a pure component.

9.8 RELATIVE PROPORTIONS OF PHASES: THE LEVER PRINCIPLE

Once we enter the realm of multicomponent systems, beginning with binary systems, the phase rule allows coexistence of two or more phases in fixed proportions at fixed temperature and pressure. In some applications it may be important to know the proportions of the phases that coexist, and it is this problem that is addressed here.

For a binary system at constant temperature and pressure, the number of components is two, the number of other important variables is zero (i.e., $n = 0$), and the minimum number of degrees of freedom (F) is zero. The Phase Rule (Equation 9.19) then shows that the maximum number of phases that can coexist in this circumstance is two.

To consider the relative proportions of these two phases, we can use any phase diagram. Here we make use of the phase diagram for the boiling point of an ideal solution (as in Figure 9.27, redrawn for this purpose in Figure 9.30). At an overall composition X_A' and temperature T^*, as shown in Figure 9.30, the composition of the liquid is given by X_A'' and that of the vapor by Y_A.

With n_A representing the overall number of moles of component A in the system, n_{liq} representing the total number of moles of liquid (including both A and B), n_{vap} representing the total number of moles of vapor (including both A and B), and n_{tot} representing the total number of moles in the system,

$$n_A = X_A' n_{tot} = X_A'(n_{liq} + n_{vap}) = X_A'' n_{liq} + Y_A n_{vap}.$$ (9.22)

To determine the ratio n_{liq}/n_{vap}, Equation 9.22 can be rearranged:

$$\frac{n_{liq}}{n_{vap}} = \frac{Y_A - X_A'}{X_A' - X_A''} = \frac{AB}{BC}$$ (9.23)

where AB and BC refer to the lines in Figure 9.30. This shows that the mole ratio of the phases present is proportional to the "arms" of a lever (called **tie lines**) with the lever fulcrum at the overall composition. This makes sense: the greater the proportion of a given phase, the shorter the "arm" is to it. The proportionality of Equation 9.23 is called the **lever principle.**

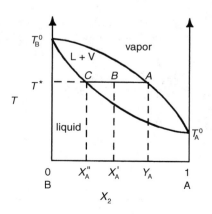

Figure 9.30. A boiling point diagram for an ideal solution, for purposes of illustrating the lever principle.

Since no particular suppositions were made concerning the types of phases present, this lever principle applies to any coexisting two phases, regardless of the types of phases, and we will make use of it in the problems at the end of this chapter. An additional useful feature is that the same general form (ratio of tie line lengths) applies to a mass% diagram to give appropriate mass ratios.

9.9 LIQUID–SOLID BINARY PHASE DIAGRAMS

By far the most important phases in materials science are condensed phases, because these phases [solid (crystalline or amorphous), liquid, liquid crystalline] are materials with structural integrity. Therefore, in this section we concentrate on the phase diagrams of condensed phases of matter that arise when two components are mixed together.

In the case of a single-component (unary) phase diagram, the pressure–temperature profile could be represented in a two-dimensional plot. When a second component is added, showing the pressure–temperature–composition profile would require a three-dimensional plot. Although this can be done, it is more common to show a temperature–composition plot, at constant pressure, as laboratory conditions are most commonly isobaric.

The simplest case of mixing two solids would arise if the solids were so much alike that they dissolved in each other completely in the solid state, giving a **solid solution.** A solution is the solid analogue of a liquid solution: examined microscopically, a solid solution is homogeneous. An example of the formation of a solid solution is given in the copper–nickel phase diagram (Figure 9.31). Although copper melts at a lower temperature than nickel [T_m(Cu) = 1083°C; T_m(Ni) = 1453°C], both Cu and Ni form face-centered cubic solids. Since Cu and Ni are similar in size (the atomic radius is 1.28 Å for Cu and 1.25 Å for Ni), they mix completely in the solid state in all proportions. Note that the melting of a mixture of Cu and Ni begins at a temperature intermediate between the melting point of Cu and that of Ni, but an important difference from either pure Cu or pure Ni is that for the mixture solid and liquid coexist over a temperature range at all compositions intermediate between pure Cu and pure Ni.

Most pairs of solids (like some pairs of liquids) cannot dissolve in one another; in solids

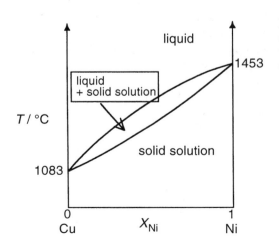

Figure 9.31. Melting point diagram for copper with nickel. Note that only at the pure metal compositions is the melting point at one fixed temperature; at all other compositions, there is a temperature range over which liquid and solid coexist.

Figure 9.32. A photograph of a solid dispersion, called pearlite. This structure is formed in some steels from bands of ferrite and cementite. (See Comment: The Iron Phase Diagram.) The image is 4000× enlarged. Photo supplied by A. Koch.

this is the result incompatibility of crystal structures. When two solids are immiscible, this leads to a **suspension** or **dispersion** of one solid in the other. On a microscopic scale a dispersion is **inhomogeneous** (also called **heterogeneous**). A photograph of a dispersion (Figure 9.32) shows that particles of one solid exist in contact with particles of the other solid. In contrast with a solid solution, which is a single phase, a dispersion contains more than one phase.

A schematic example of a solid system that forms a dispersion is shown in Figure 9.33. The molecular sizes and packing of components A and B make it impossible for them to fit together in one solid phase, so they form a dispersion. Note that as A is added to B, the freezing point is lowered; the same is true when B is added to A. This is the familiar **freezing point depression.** (*Note:* The freezing point of Cu is not depressed by adding Ni in the Cu–Ni phase diagram because one basic premise in the derivation of freezing point depression equations is that the pure solid is formed on freezing the solution, and this is not the case in Cu–Ni because a solid solution is formed.) At low temperatures, the solids formed in the phase diagram of Figure 9.33 are pure solid A and pure B; a dispersion of one in the other would be observed microscopically. Examples of real phase diagrams that are similar to Figure 9.33 include ethylene glycol/water (in its metastable phase diagram; ethylene glycol/water also forms a solid hydrate in its stable phase diagram), and several alloys (As/Au, As/Pb, Au/Si, and Bi/Cd all show almost completely immiscible solids).

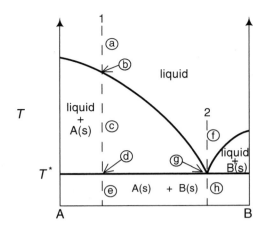

Figure 9.33. A freezing point diagram showing completely immiscible solid phases. The constant-composition slices shown correspond to the cooling curves shown in Figures 9.34 and 9.35.

Consideration of the numbers of degrees of freedom in a few constant-composition slices of the phase diagram in Figure 9.33 allows plotting temperature–time profiles and shows how cooling (or heating) curves can be used to deduce phase diagrams. From the general form of the Phase Rule (Equation 9.19) and the fact that $c = 2$ here and $n = 1$ (T can vary but P is fixed), we find $F = 2 - p + 1 = 3 - p$. Slice 1 (Figure 9.33) gives the phases and degrees of freedom summarized in Table 9.1. The phase(s) in each region are determined by considering what regions an imaginary horizontal line (the tie line) touches. The change from 2 degrees of freedom to 1 at point b leads to a change in slope in the cooling curve; the cooling is not so fast because the precipitation of A is exothermic. This change in slope in a cooling curve is called a **break point,** and it is characteristic of a change in the number of degrees of freedom, with $F \neq 0$. At point d, $F = 0$ so temperature is invariant, and there is a **halt point** (also called **arrest**) in the cooling curve corresponding to this temperature. The cooling curve is shown in Figure 9.34.

Similar consideration of Slice 2 leads to the phases and degrees of freedom summarized in Table 9.2. The corresponding cooling curve is shown in Figure 9.35.

Although cooling curves might provide more reliable experimental information than warming curves (cooling from the melt will generally ensure equilibrium, whereas the solid might contain metastable states), we can consider the effects of warming an equilibrated solid. On warming along the line of Slice 1, the solid (which is a dispersion of solid A in solid B) first begins to melt at T^*. The composition of the first liquid that appears, considering the tie

TABLE 9.1.
The Phases and Degrees of Freedom in the Various Regions
of the Phase Diagram along Slice 1, shown in Figure 9.33.[a]

Region	Phase(s)	p	F
a	Liquid	1	2
b	Liquid + A(s)	2	1 (break)
c	Liquid + A(s)	2	1
d	Liquid + A(s) + B(s)	3	0 (halt)
e	A(s) + B(s)	2	1

[a] P = 1 atm; $F = 3 - p$.

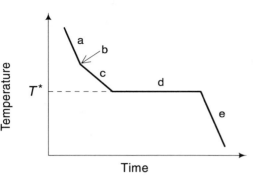

Figure 9.34. The cooling curve (temperature as a function of time) along Slice 1 of the phase diagram shown in Figure 9.33. Note the break at b and halt at d.

TABLE 9.2.
The Phases and Degrees of Freedom in the Various Regions of the Phase Diagram along Slice 2, shown in Figure 9.33.[a]

Region	Phase(s)	p	F
f	Liquid	1	2
g	Liquid + A(s) + B(s)	3	0 (halt)
h	A(s) + B(s)	2	1

[a] P = 1 atm; $F = 3 - p$.

line, will be the composition of point g. Note that this is not the same as the overall composition of the system. As the sample is heated further, the composition of the liquid will change, becoming richer in A as the temperature is increased. The liquid and solid continue to coexist throughout the range from point d to point b, and above b only liquid exists.

Note that the point g on Slice 2 is a special point. At any temperature higher than this point, the system is completely liquid. At any temperature lower than this point, the system is completely solid. Therefore, on heating at this composition, the system melts to give a liquid of the same composition as the solid. This special composition is called the **eutectic composition,** which means that it melts at a single temperature without change in composition. The corresponding temperature (T^* in Figure 9.33) is called the **eutectic temperature.**

The existence of a eutectic can be useful in choosing a material of a particular property. For example, although both tin and lead melt at a temperature too high to make them useful as

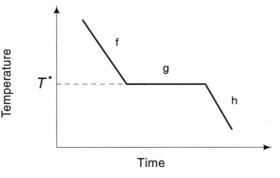

Figure 9.35. The cooling curve (temperature as a function of time) along Slice 2 of the phase diagram in Figure 9.33 at a constant pressure of 1 atm. Note the halt at g.

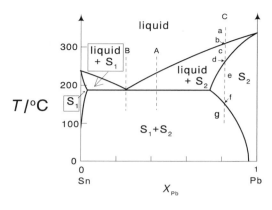

Figure 9.36. Melting point diagram of tin with lead. Note the eutectic at 26 atomic% Pb, which corresponds to tin–lead solder. S_1, also expressed as (Sn), is solid rich in tin; S_2, also written (Pb), is solid rich in lead.

solders for everyday applications, a 26 atomic% Pb mixture of tin and lead (see Figure 9.36) can be useful as a solder because it melts at a fixed temperature without change in composition. Some other useful low-melting alloys are listed in Table 9.3.

Just as liquids can show partial miscibility, so can some solids. The tin–lead phase diagram (Figure 9.36) shows **partial miscibility.** The two solids formed, S_1 and S_2, are **solid solutions** of variable composition, rich in tin and lead, respectively. The composition of S_1 or S_2 can be read from the phase diagram, e.g., S_2 is 93 atomic% lead at $T = 100°C$.

Cooling of Slice A or Slice B (Figure 9.36) of the Sn/Pb phase diagram will lead to cooling curves such as in the Figure 9.33 phase diagram, but Slice C will be slightly different. Again, $F = c - p + n = 2 - p + 1 = 3 - p$ here, which gives the degrees of freedom listed in Table 9.4. The corresponding cooling curve is shown in Figure 9.37. Note that in this slice there is no invariant point ($F = 0$ point) and there is no halt in the cooling curve.

TABLE 9.3.
Compositions and Melting Points of Eutectic Solders

Melting point/°C	Composition/mass %				
	Bi	Cd	Pb	Sn	Hg
60	53.5	—	17	19	10.5
70	45.3	12.3	17.9	24.5	—
80	35.3	9.5	35.1	20.1	—
91.5	51.6	8.1	40.2	—	—
96	52.5	—	32	15.5	—
100	50	—	32.2	17.8	—
124	55.5	—	44.5	—	—
138	57	—	—	43	—
154	14	—	43	43	—
177	—	32	—	68	—

TABLE 9.4.
**Degrees of Freedom in Slice C of the Tin–Lead
Phase Diagram (Figure 9.36).[a]**

Region	Phase(s)	p	F
a	Liquid	1	2
b	S_2 + liquid	2	1 (break)
c	S_2 + liquid	2	1
d	S_2 + liquid	2	1 (break)
e	S_2	1	2
f	$S_1 + S_2$	2	1 (break)
g	$S_1 + S_2$	2	1

[a] $F = 3 - p$ here.

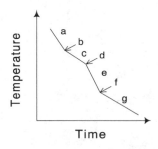

Figure 9.37. Cooling curve (temperature as a function of time) for Slice C of the tin–lead phase diagram (Figure 9.36). Note the breaks at b, d, and f.

PHASE SEPARATION LEADS TO BETTER FIBER OPTICS

Fiber optics with radial refractive index gradients can be prepared by heating a polymethylmethacrylate tube that is filled with methylmethacrylate monomer and higher refractive index dopants. As polymerization proceeds from the outside, the dopants, which are not very soluble in the solid polymer, are pushed to the center in a smooth gradient. These fibers achieve a bandwidth (megabytes per second transmitted through 100 m of cable) of 2500, compared with 100 for copper wire or stepped-index optical fibers, i.e., those produced by coating one material with another.

THE IRON PHASE DIAGRAM

The stable form of pure iron at room temperature has a body-centered cubic (bcc) structure (see Appendix 3 for lattice types) and it is called **α-iron.** Below 760°C it is **ferromagnetic** (i.e., permanently magnetized by an applied magnetic field; see Chapter 13); above 760°C there is too much thermal energy for ferromagnetism to be maintained, but the structure remains bcc until 906°C when it transforms
(continued)

THE IRON PHASE DIAGRAM (continued)

to a face-centered cubic (fcc) structure (γ-**iron**). At 1401°C, pure iron transforms to another body-centered cubic phase, δ-**iron,** which is stable up to the melting point of 1530°C.

Additions of small amounts (< 2 mass%) of carbon to iron leads to the formation of **steel.** Higher concentrations of carbon give **cast iron.** For this reason, the Fe–C phase diagram, a portion of which is shown in Figure 9.38, is among the most important phase diagrams known. As can be seen from the diagram, below 2 mass% carbon heating can lead to a solid solution (γ-iron, which can be hot-rolled, pressed, drawn, or otherwise shaped), but above 2 mass% carbon (i.e., cast iron) heating does not achieve a malleable solid state and therefore shaping is done from the melt by casting.

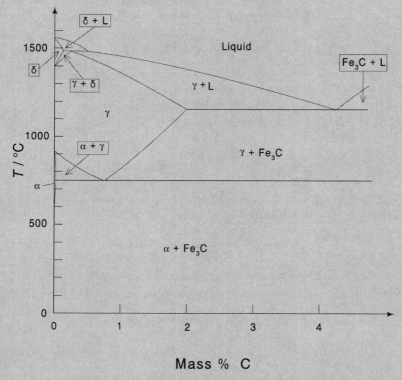

Figure 9.38. A portion of the Fe-C phase diagram showing the Fe-rich region.

The Fe–C phase diagram shows a number of features, including the eutectic at about 1150°C and 4.2 mass% carbon (liquid stable at higher temperatures) and the **eutectoid** (this is similar to a eutectic but with a solid solution stable at higher temperatures) at 723°C and about 1 mass% carbon. The formation of the com-

(continued)

THE IRON PHASE DIAGRAM (continued)

pound Fe_3C plays an important role in the mechanical properties of steel and cast iron, and this can be controlled by heat treatment. The presence of Fe_3C leads to brittle iron that can be cast into its final form. However, Fe_3C is metastable with respect to graphite and in some treatments the presence of graphite allows the production of malleable irons that can be worked into shape.

Figure 9.38 shows that α-iron dissolves only a little carbon (0.05 mass % carbon), whereas γ-iron dissolves up to 2 mass% carbon. This is because there is more room for carbon at **interstitial octahedral sites** (i.e., spaces within the lattice that have octahedrally placed neighbors) in the fcc structure of γ-iron than in the **tetrahedral interstitial sites** (spaces in the lattice with tetrahedral neighbours) of body-centered cubic α-iron.

The solid solution of carbon in γ-iron is called **austentite;** that for α-iron is called **α-ferrite.** If austentite is cooled slowly below the eutectoid at 723°C, it phase separates to α-ferrite and Fe_3C (**cementite**); this phase separation gives a heterogeneous material called **pearlite** (see Figure 9.32).

If austentite is cooled rapidly it forms a metastable structure called **martensite** in which carbon atoms are at the interstitial sites of a bcc lattice, but the crystal structure is tetragonal (expanded along one axis but with 90° angles retained). Martensite is the hardest component in quenched (rapidly cooled) steels because the tetragonal structure of the crystallites does not pack well in the cubic arrangements of their neighboring materials. This is one way in which steel is hardened by heat treatment.

9.10 COMPOUND FORMATION

So far, in considering binary systems, we have taken into account only physical not chemical interactions. If there is a reaction between the two components this can lead to compound formation, and, in terms of their phase diagrams, compounds can be distinguished as congruently melting or incongruently melting.

A **congruently melting compound** is one in which the liquid formed on melting has the same composition as the solid compound. An example of this is shown in Figure 9.39, where AB is the 1:1 compound of components A and B. This diagram can be considered to be the sum of two other simple phase diagrams—the AB/B diagram (right side of Figure 9.39) and the A/AB diagram (left side of Figure 9.39).[11] When looked at this way, the features are not

---

[11] The only difference from being two phase diagrams side by side is that the slope of the melting point around the composition of AB is zero, whereas it is nonzero for the pure components, A and B. The reason for this is that the slope is nonzero only when the liquid is the same chemical species as the solid, and A melts to give liquid A, and B melts to give liquid B, but AB melts to give A(liq) miscible with B(liq), not AB(liq).

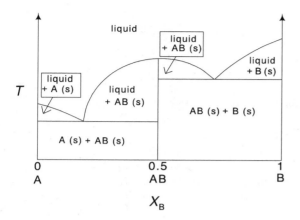

Figure 9.39. The melting point diagram in the binary A-B system, at constant pressure. The compound AB melts congruently.

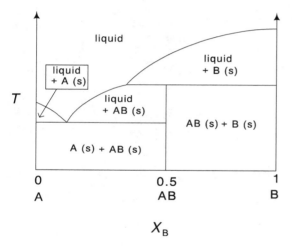

Figure 9.40. Temperature–composition phase diagram for the system A–B at constant pressure showing the incongruently melting compound AB.

new. The compound AB melts to give a liquid of composition 50 mol% A, 50 mol% B, the same as the composition of the solid.

An **incongruently melting compound** gives, on melting, both a liquid and a solid with different compositions from the compound. This is shown schematically for compound AB in Figure 9.40. An example of this is KNa_2, which decomposes on melting to give Na(s) and a liquid that is depleted in Na relative to KNa_2. The Na–K phase diagram is shown in Figure 9.41.

The K–Na phase diagram shows that the composition at 54.0 mass% Na (which corresponds to KNa_2) begins to melt at 7°C, giving a solid solution rich in Na, designated (Na), and a liquid depleted in Na relative to KNa_2. As the temperature is increased, the proportion of liquid (which contains K and Na) and solid (Na-rich solid solution) changes, yielding more liquid as the temperature goes up. This liquid becomes richer in Na as the temperature is increased, because it is forming from (Na). Eventually all the solid will become liquid. A major distinguishing feature between congruent and incongruent melting is that for congruent melting the vertical line at the compound composition directly reaches the liquid region; for incongruent melting the vertical line at the compound composition stops short of the liquid region.

The K–Na phase diagram shows a eutectic (point ① in Figure 9.41) and also another spe-

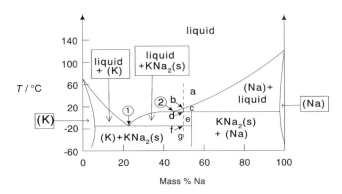

Figure 9.41. The K–Na temperature–composition phase diagram at constant pressure. The compound KNa_2 melts incongruently at 7°C. (K) is solid rich in potassium and (Na) is solid rich in sodium.

cial point (point ② in Figure 9.41), called a **peritectic point.** The peritectic point corresponds to the liquid composition that is formed at the temperature of melting of an incongruently melting compound.

Cooling at constant composition along the dashed line in Figure 9.41 leads to the phases and degrees of freedom (from $F = c - p + n = 2 - p + 1 = 3 - p$), as noted in Table 9.5. The corresponding cooling curve is shown in Figure 9.42. The lower-temperature halt, from the phase diagram, is a **eutectic halt,** whereas the higher-temperature halt is a **peritectic halt.** It is not possible to determine whether a halt is eutectic or peritectic from one cooling diagram alone; several cooling curves are necessary to determine the phase diagram. Indeed, this is how phase diagrams are determined experimentally.

It can be important to know whether a compound melts congruently or incongruently, particularly if there is some probability that the compound will melt in an application. This could be worrisome in the case of an incongruently melting material as it may not be recoverable

TABLE 9.5.
Phases and Degrees of Freedom in the K–Na system.[a]

Region	Phase(s)	p	F
a	Liquid	1	2
b	Liquid+ $(Na)^b$	2	1 (break)
c	Liquid + $(Na)^b$	2	1
d	Liquid + $(Na)^b$ + $KNa_2(s)$	3	0 (peritectic halt)
e	Liquid + $KNa_2(s)$	2	1
f	Liquid + $KNa_2(s)$ + $(K)^b$	3	0 (eutectic halt)
g	$(K)^b$ + $KNa_2(s)$	2	1

[a] Regions correspond to those labeled in Figure 9.41. $F = 3 - p$.

[b] (X) indicates solid rich in X, but not pure.

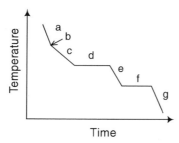

Figure 9.42. The cooling curve (temperature as a function of time) for cooling at constant composition in the K–Na phase diagram shown in Figure 9.41. Note the break at b, the peritectic halt at d, and the eutectic halt at f.

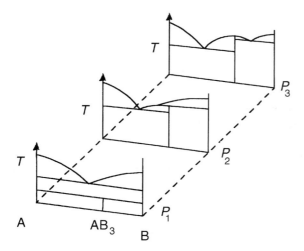

Figure 9.43. Various isobaric temperature–composition diagrams at different pressures. The effect of increasing pressure is to increase the stability of compound AB_3 such that it transforms to A(s) and B(s) at low pressure (P_1); melts incongruently at intermediate pressure (P_2); and melts congruently at high pressure (P_3).

from solidification of the corresponding liquid, since cooling this liquid will first form another solid [(Na) in the K/Na case] that may not be able to rearrange itself on further cooling to provide the desired compound (KNa_2, in the K/Na case).

The main factor determining whether a compound melts congruently or incongruently is the stability of the liquid phase relative to the pure solid components at the temperature at which there is too much thermal energy for the lattice to hold together (i.e., at the melting point). If it would lower the Gibbs energy to produce one of the solid components rather than just liquid, melting will be incongruent (to give the liquid and the more stable solid). Of course, the phase stability could change with a change in another parameter such as pressure. This is shown schematically in Figure 9.43.

9.11 THREE-COMPONENT (TERNARY) PHASE DIAGRAMS

In the single-component (also known as unary) phase diagrams, we were able to plot the phase stability region as a function of temperature and pressure in a two-dimensional plot. For a two-component (binary) phase diagram, to add the additional dimension of composition, the phase

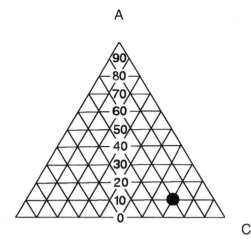

A

B

C

Figure 9.44. Triangular coordinates to represent the constant-temperature, constant-pressure phase diagram of the A–B–C ternary system. Each vertex represents the pure component (A, B, or C, as labeled), and each side presents a binary phase diagram (A–B, opposite vertex C; A–C, opposite vertex B; B–C, opposite vertex A). The interior of the triangle represents nonzero compositions of all three components. The numbers shown correspond to A concentration. The point marked ● represents 10% A, 20% B, 70% C.

diagram was either three-dimensional (*P, T,* composition; see Figure 9.43) or two-dimensional (*P* and composition at constant temperature, e.g., Figure 9.25, or *T* and composition at constant pressure, e.g., Figure 9.28).

For three components, there are two independent composition variables (the third, X_3, the mole fraction of component 3, being given by the relation $X_1 + X_2 + X_3 = 1$) as well as pressure and temperature to consider. Although it is possible to include a diagram that shows composition and temperature or composition and pressure, it is most common in three-component systems to show composition diagrams at constant temperature and constant pressure. The usual way to do this is using triangular coordinates, as shown in Figure 9.44.

In a triangular coordinate system, each vertex represents a pure component, A, B, or C, as shown in Figure 9.44. The advantage of triangular coordinates is that the three components are considered in a symmetric way. The side AB is as far as possible from C, so it contains no C, and varying A–B compositions from 100% A at point A to 100% B at point B. Similarly, the side AC represents the A–C binary phase diagram (no B), and the side B–C represents the binary B–C phase diagram (no A).

The interior of the triangle presents intermediate compositions containing A, B and C. For example, the composition 10% A is found along the "10" line parallel to side BC. (These lines start at 0% A on BC and work up to 100% A at point A.) If the B composition is 20%, this is on the 20% line parallel to side AC. The composition 10% A, 20% B, is at the point where the 10% A line intersects the 20% B line, as shown in Figure 9.44. This meeting point also crosses the 70% C line, as it must since %C = 100% − %A − %B = 100% − 10% − 20% = 70% here. This is shown in Figure 9.44, and the additivity of the overall composition to 100% serves as a useful check in plotting points on ternary graph paper.

A simple example of a ternary phase diagram at constant pressure and temperature is the ethanol–methanol–water phase diagram. At *T* = 25°C and *p* = 1 atm, ethanol, methanol, and water are soluble in one another in all proportions, so the entire interior of the triangle represents the same phase, a liquid. This phase diagram is shown in Figure 9.45.

When water, acetone, and phenol are mixed at room temperature and pressure, the situation is not quite so simple. Water and acetone are miscible in all proportions, so there are no phase boundaries on the water–acetone axis. Similarly, acetone and phenol are miscible in all

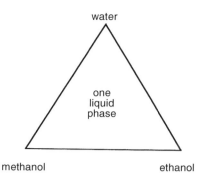

Figure 9.45. Ethanol–methanol–water phase diagram at $T = 25°C$, $P = 1$ atm. All three liquid components are miscible in all proportions.

proportions, so there are no phase boundaries on the acetone–phenol axis. However, water and phenol exhibit only limited miscibility. When only a little phenol is added to water, the two mix, but beyond a certain amount the solution separates into two phases, one rich in water and the other rich in phenol. Similarly, if a little water is added to phenol, it will dissolve, but if more water is added so that the miscibility is exceeded, the solution will separate into two layers, one rich in phenol and the other rich in water. (Note that neither layer is pure as the other component will still dissolve to some extent.) The corresponding water–phenol–acetone phase diagram is shown in Figure 9.46, where the temperature dependence (showing an upper consolute temperature) is included.

The 25°C water–K_2CO_3–methanol phase diagram is shown in Figure 9.47. If sufficient K_2CO_3 is added to water, its solubility would be exceeded, and the result will be a saturated solution with excess K_2CO_3 as the hydrate $K_2CO_3 \cdot 3/2H_2O$ (solid). Similarly, K_2CO_3 has only limited solubility in methanol. These solubility limits are represented by the positions at which the phase boundary lines cross the H_2O–K_2CO_3 and CH_3OH–K_2CO_3 axes. At regions close to the H_2O–CH_3OH axis, there will be one solution, since these liquids are soluble in each other in all proportions when there is nothing else present. However, starting with a 1:1 (by mass)

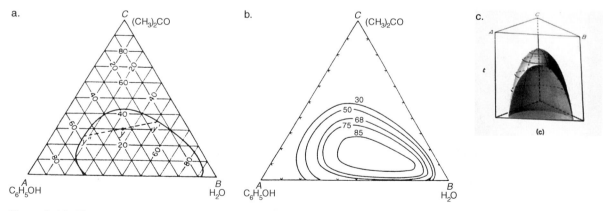

Figure 9.46. The ternary system, phenol-water-acetone: (a) at room temperature; (b) as a function of temperature; (c) showing temperature as an additional dimension (vertical axis), clearly indicating the upper consolute temperature (temperature above which the system gives rise to only one liquid phase, regardless of the proportions). Reproduced, with permission, from W.J. Moore (1962) *Physical Chemistry*, 3rd Edition, Prentice-Hall.

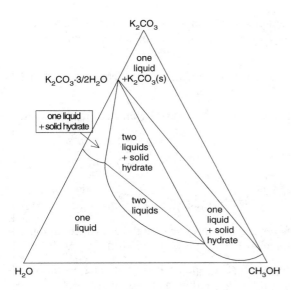

Figure 9.47. The H_2O–K_2CO_3–CH_3OH system at $T = 25°C$. (G.B. Frankforter and F.C. Frary (1913), *Journal of Physical Chemistry*, **17,** 402.)

mixture of H_2O and CH_3OH, and adding a salt such as K_2CO_3 can have a rather startling effect: addition of sufficient salt causes the formation of two separable liquids. One of these liquids is rich in water and the other is rich in methanol; they separate because the salt (K_2CO_3) ties up water in ion hydration spheres, such that there is less water available to dissolve the methanol. Furthermore, because the ionic strength of the aqueous solution is now higher, methanol is no longer miscible. This process is used to cause separation of otherwise miscible solvents by the addition of a salt, and it is termed **salting out.** In the phase diagram it is represented by a region in which two immiscible liquid phases coexist. At higher K_2CO_3 concentrations (i.e., closer to the K_2CO_3 vertex), there are two liquids and excess K_2CO_3 is in the hydrate form, $K_2CO_3\cdot3/2H_2O$ (solid).

A Delicious Phase Diagram: A Tutorial

The phase diagram of Figure 9.48 has been proposed as the mechanism by which soft caramel centered chocolates are made. The idea is that the appropriate composition of chocolate and

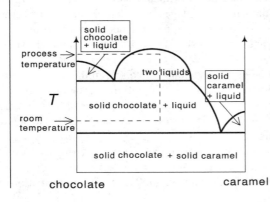

Figure 9.48. Proposed chocolate-caramel binary phase diagram.

caramel is heated to a processing temperature, as noted on the diagram, and then cooled to room temperature (see Figure 9.48). This would cause phase separation into solid chocolate and a liquid phase rich in caramel.

a. Does this proposed phase diagram seem to provide a realistic production mechanism? Why or why not?

b, Speculate on how a liquid caramel phase could be successfully created inside a solid chocolate structure.

Applications of Supercritical Fluids: A Tutorial

Supercritical fluids are fluids at temperatures and pressures beyond the critical conditions. In this region, liquid and gas are not distinguishable.

a. Describe some expected physical properties of a supercritical fluid, such as density (compared with ordinary gas) and viscosity (compared with the liquid).

b. How could the critical point (temperature, pressure, molar volume) for a given compound be determined experimentally?

c. Would you expect the solubility of a species in a supercritical fluid to be the same as in the same solvent in the liquid state? How could this be used to advantage? As an example, consider the extraction of caffeine from green coffee beans by supercritical CO_2 ($P_c = 72.8$ atm, $T_c = 304.2$K), reducing the caffeine from its initial range of 1 to 3% to as low as 0.02% without removing other flavor and aroma components. Following caffeine extraction, the solvent is removed. This is now used as a commercial process to decaffeinate millions of kilograms of coffee beans annually.

d. Some mixtures of high boiling materials cannot easily be separated by fractional distillation. Can supercritical fluids be used to affect a suitable extraction procedure? Consider that supercritical fluids can be used as the carrier in chromatographic separations.

e. In some cases, such as SiO_2 dissolved in water, the solubility is very low—only a few percent even when the water is in its supercritical phase ($P_c = 218.5$ atm, $T_c = 647.3$ K). However, the viscosity of the solution changes considerably at the critical temperature: for SiO_2 in H_2O it is about 10^{-2} P (poise) at room temperature and near 10^{-4} P in the supercritical region. This change in viscosity is very important in the growth of SiO_2 crystals from supercritical water.

   i. Why can SiO_2 not be grown as single crystals from the melt? (See problem 9.)

   ii. Why is the viscosity an important consideration in growth of rather insoluble materials from solution? Consider that large single crystals of SiO_2 are required (e.g., for the electronics industry), and slow growth of crystals can lead to dendrites rather than large single crystals.

   iii. What laboratory conditions would be necessary to grow SiO_2 crystals hydrothermally (i.e., from supercritical water)?

   iv. Can these conditions be achieved in nature? Consider the formation of a geode when a hydrothermal solution saturated with silicon dioxide fills a cavity in a rock matrix, as shown in Figure 9.49.

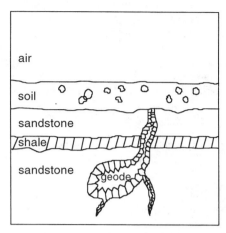

Figure 9.49. In certain geological conditions pockets of supercritical aqueous solutions allow the growth of quartz crystals in **geodes.**

FURTHER READING

General References

M.F. Ashby and D.R.H. Jones (1986). *Engineering Materials 2: An Introduction to Microstructures, Processing and Design.* Pergamon Press.

D.R. Askeland (1994). *The Science and Engineering of Materials.* PWS Publishing Company.

H. Baker, Ed. (1990). *ASM Handbook.* Vol. 3: *Alloy Phase Diagrams.* ASM International.

R.J. Borg and G.J. Dienes (1992). *The Physical Chemistry of Solids.* Academic Press.

D.G. Clerc and D.A. Cleary (1992). Spinodal Composition and an Interesting Example of the Application of Several Thermodynamic Principles. *Journal of Chemical Education,* **72,** 112.

M. de Podesta (1996). *Understanding the Properties of Matter.* Taylor & Francis.

E.G. Ehlers (1972). *The Interpretation of Geological Phase Diagrams.* Dover Publications.

A. Findlay, revised by A.N. Campbell (1938). *The Phase Rule and Its Applications.* Longmans, Green and Co.

J.M. Honig (1982). *Thermodynamics: Principles Characterizing Physical and Chemical Processes.* Elsevier.

H. Jue-ming (1983). The Mass Production of Iron Castings in Ancient China. *Scientific American,* January 1983, 121.

W.J. Moore (1967). *Seven Solid States.* W.A. Benjamin, Inc.

N. Mott (1967). The Solid State. *Scientific American,* September 1967, 80.

A. Navrotsky (1997). Thermochemistry of New, Technologically Important Inorganic Materials. *MRS Bulletin,* May 1997, 35.

C.N.R. Rao and J. Gopalakrishnan (1997). *New Directions in Solid State Chemistry.* Cambridge University Press.

J.F. Shackelford (1988). *Introduction to Materials Science for Engineers,* 2nd Ed. Macmillan.

N.O. Smith (1997). The Gibbs Energy Basis and Construction of Melting Point Diagrams in Binary Systems. *Journal of Chemical Education,* **74,** 1080.

W.F. Smith (1990). *Principles of Materials Science and Engineering.* McGraw-Hill.

R. Trautman, B.J. Griffin and D. Scharf (1998). Microdiamonds. *Scientific American,* August 1998, 82.

R.S. Treptow (1993). Phase Diagrams for Aqueous Systems. *Journal of Chemical Education,* **70,** 616.

L.H. Van Vlack (1989). *Elements of Materials Science and Engineering,* 6th Ed. Addison-Wesley.

Fullerenes

R. Baum (1994). Endohedral Fullerenes Purified by Porphyrin-Silica. *Chemical & Engineering News,* October 10, 1994, 37.

T.W. Ebbesen (1996). Carbon Nanotubes. *Physics Today,* June 1996, 26.

A. Hildebrand, U. Hilgers, R. Blume, and D. Wiechoczek (1996). Playing with the Soccer Ball—an Experimental Introduction to Fullerene Chemistry. *Journal of Chemical Education,* **73,** 1066.

D.R. Huffman (1991). Solid C_{60}. *Physics Today,* November 1991, 22.

Helium

B. Bertman and R.A. Grujer (1967). Solid Helium. *Scientific American,* August 1967, 65.

R.J. Donnelly (1988). Superfluid Turbulence. *Scientific American,* November 1988, 100.

Orientationally Disordered Solids

N.G. Parsonage and L.A.K. Staveley (1978). *Disorder in Solids.* Oxford University Press.

Quasicrystals

MRS Bulletin (1997). Special Issue on Quasicrystals, November 1997.

R.A. Dunlap (1990). Periodicity and Aperiodicity in Mathematics and Crystallography. *Scientific Progress Oxford,* **74,** 311.

D.R. Nelson (1986). Quasicrystals. *Scientific American,* August 1986, 42.

P.W. Stephens and A.I Goldman (1991). The Structure of Quasicrystals. *Scientific American,* April 1991, 44.

H.C. von Baeyer (1990). Impossible Crystals. *Discover,* February 1990, 69.

Supercritical Fluids

T. Clifford and K. Bartle (1993). Chemistry Goes Supercritical. *Chemistry in Britain,* June 1993, 499.

F. Hensel (1988). Critical Point Phenomena of Fluid Metals. *Chemistry in Britain,* May 1988, 457.

R.A. Laudise (1987). Hydrothermal Synthesis of Crystals. *Chemical and Engineering News,* September 26, 1987, 30.

C.L. Phelps, N.G. Smart, and C.M. Wai (1996). Past, Present, and Possible Future Applications of Supercritical Fluid Extraction Technology. *Journal of Chemical Education,* **73,** 1163.

M. Poliakoff and S. Howdie (1995). Supercritical Chemistry: Synthesis with a Spanner. *Chemistry in Britain,* February 1995, 118.

R.W. Shaw, T.B. Brill, A.A. Clifford, C.A. Eckert, and E.U. Franck (1991). Supercritical Water. *Chemical & Engineering News,* December 23, 1991, 26.

K. Zosel (1978). Separation with Supercritical Gases: Practical Applications. *Angewandte Chemie* (*International Edition in English*), **17,** 702.

PROBLEMS

1. At 20°C the vapor pressure of benzene is 100 Torr and that of octane is 20 Torr. If 1.00 mol of octane is dissolved in 4.00 mol of benzene and the resulting solution is ideal, calculate (a) the total vapor pressure; (b) the composition of the vapor; (c) the new vapor composition if the vapor is collected, recondensed and then allowed to reequilibrate.

2. A sample of methane is known to be contaminated with some $N_2(g)$. To determine the extent of the contamination, the sample is cooled in liquid nitrogen to the temperature of the normal boiling point of N_2 (i.e., its boiling point at $P = 1$ atm). At this temperature the methane is fully condensed and may be considered to have negligible vapor pressure. The observed pressure is found to be 11 Torr. What is the purity of the original sample (in mol% methane)?

3. CBr_4 forms four solid phases. Phases II and III coexist at $T = 350$ K, $P = 0.1$ GPa and also at $T = 375$ K, $P = 0.2$ GPa, with phase II stable at lower temperatures. Which phase is denser: II or III? Explain, using the Clapeyron equation.

4. Glycerol and *m*-toluidine form a binary phase diagram that has an upper and lower consolute temperature. Mixtures of various mass% *m*-toluidine (*W*) were found on warming to separate into two phases at T_1 (i.e., become turbid) and lose their turbidity at T_2 as follows:

W	T_1 /°C	T_2 /°C
18	48	53
20	18	90
40	8	120
60	10	118
80	19	83
85	25	53

a. Plot the phase diagram (*T* vs. mass% *m*-toluidine) and find the upper and lower consolute temperatures.

b. Use the phase diagram to explain what happens when *m*-toluidine is added dropwise to glycerol at 60°C.

c. At 40 mass% *m*-toluidine at *T*=70°C, what phases exist and in what proportion (by mass)?

5. From the Clapeyron equation, derive the **Clausius[12]-Clapeyron equation:**

$$\ln\left(\frac{P_2}{P_1}\right) = \left(\frac{\Delta_{vap}H}{R}\right)\left(\frac{1}{T_1} - \frac{1}{T_2}\right). \tag{9.24}$$

Clearly state and justify all assumptions made in the derivation. The Clausius–Clapeyron equation is useful in determining the variation of vapor pressure with temperature.

6. The following data were obtained by cooling solutions of Mg and Ni.

X_{Ni}	0.00	0.044	0.139	0.202	0.383	0.669	0.752	1.00
T_{break}/°C	—	608	—	—	1050	—	—	—
T_{arrest}/°C	651	510	510	510 and 770	770	1180	1080	1450

It is found that two compounds, Mg_2Ni and $MgNi_2$, form in this system. Plot and label the phase diagram for the Mg–Ni system.

7. In the thermal decomposition of $BaCO_3(s)$, which results in the formation of $BaO(s)$ and $CO_2(g)$, it is found that the equilibrium pressure is a function of temperature only. (This applies whether one starts with $BaCO_3$ alone, or two of $BaCO_3$, BaO and/or CO_2, or all three, $BaCO_3$, BaO, and CO_2.) Does the equilibrium system consist of pure crystals of $BaCO_3$ and pure crystals of BaO or does it contain a solid solution of these compounds? Only one of these alternatives is allowed thermodynamically. Explain. *Hint:* Consider the Phase Rule.

8. The following information was obtained from a set of cooling curves for Sb–Cd:

X_{Cd}	Break/(°C)	Halt /(°C)
0	—	630
0.188	550	410
0.356	461	410
0.412	—	410
0.480	419	410
0.600	—	439
0.683	400	295
0.925	—	295
1	—	321

Sketch the T vs. X_{Cd} phase diagram on graph paper. Label all areas.

9. The binary phase diagram for SiO_2 (quartz) and Al_2O_3 shows several interesting features. Pure SiO_2 crystallizes from the melt at 1735°C as **cristobalite.** At 1475°C cristobalite transforms to another quartz polymorph characterized by its small (often triplet) crystals, **tridymite.** Aluminum oxide crystallizes at 2020°C to produce **corundum,** a very hard mineral often used as an abrasive. In the SiO_2/Al_2O_3 phase diagram at 1540°C there is a eutectic. There is one compound in the binary phase diagram, and this is **mullite,** a rare clay mineral of disputed empirical formula, most likely Al_6SiO_{11}.

[12] Rudolph Julius Emmanuel Clausius (1822–1888) was born in Prussia and was a Professor of Mathematical Physics in Switzerland and Germany. Clausius made many contributions to thermodynamics, kinetic theory (he introduced the concept of mean free path), and electricity and magnetism.

Mullite melts incongruently at 1805°C. Sketch and label the SiO_2/Al_2O_3 binary phase diagram (with the ordinate as mass % Al_2O_3). Al_6SiO_{11} is 83.5 mass% in Al_2O_3.

10. Cooling curves for liquid mixtures of magnesium and zinc give the following results:

Mass% Zn	Freezing start/(°C)	Freezing end/(°C)
0	651	651
10	623	346
20	580	346
30	520	346
40	443	346
50	360	346
60	437	346
70	520	346
80	577	346
84.3	595	595
90	557	368
95	456	368
97.5	379	368
100	419	419

a. Draw and label the phase diagram, temperature vs. composition (mass% Zn) as accurately as possible.

b. If 100. g of a liquid mixture containing 75.0 mass% Zn is cooled from 620 to 460°C, what phase(s) will be formed and in what amounts?

11. The following data for the ternary system $Li_2SO_4/(NH_4)_2SO_4/H_2O$ show the mass% composition of the liquid phase in equilibrium with solids as indicated, at 30°C.

Liquid phase		
% Li_2SO_4	% $(NH_4)_2SO_4$	Solid phase(s)
25.1	0.0	$Li_2SO_4 \cdot H_2O$
23.7	6.7	$Li_2SO_4 \cdot H_2O$
23.0	9.4	$Li_2SO_4 \cdot H_2O$
21.9	12.4	$Li_2SO_4 \cdot H_2O + NH_4LiSO_4$
16.6	19.3	NH_4LiSO_4
10.8	28.7	NH_4LiSO_4
7.8	35.6	NH_4LiSO_4
6.5	39.5	$NH_4LiSO_4 + (NH_4)_2SO_4$
2.9	41.4	$(NH_4)_2SO_4$
0.0	44.1	$(NH_4)_2SO_4$

a. Draw the phase diagram using triangle coordinate graph paper. Label each area.

b. Consider a binary mixture of Li_2SO_4 and $(NH_4)_2SO_4$ containing 80.0 mass% Li_2SO_4. If this mixture is dissolved in water and then the resulting solution is evaporated to yield a ternary mixture containing 25% water by mass, what phase(s) will be present in the remaining mixture? Indicate this mixture on your phase diagram with an asterisk. (*Hint:* As long as the $Li_2SO_4/(NH_4)_2SO_4$ ratio is the same when the water is added, at all proportions of water this solution will fall on a straight line going from the original mixture to the water vertex.)

12. In a publication [H, Steinfink, J.S. Swinnea, Z.T. Suí, H.M. Hsu, and J.B. Goodenough, (1987), *Journal of the American Chemical Society*, **109**, 3348] concerning work on the high-temperature superconductor $YBa_2Cu_3O_{6.5}$ (which is made from Y_2O_3, BaO and CuO), the authors state "[a]t most three phases can coexist in this system when the temperature and partial pressure of O_2 (0.21 atm) are fixed." Justify this statement. Keep in mind that this is a ternary system.

13. Sketch the Ag/Sn phase diagram (T vs. X_{Sn}) phase diagram based on the following information. Pure Sn melts at 350°C; pure Ag melts at 962°C; Ag_3Sn is the only intermetallic compound formed and it melts incongruently at 440°C to give liquid and S_1, where S_1 is a solid rich in Ag; there is one eutectic (X_{Sn} = 0.91) and it melts at 210°C; Sn and Ag_3Sn are insoluble in the solid state; Ag_3Sn and S_1 are insoluble in the solid state. Label the diagram.

14. The pressure–temperature coexistence line for solid–vapor equilibrium (sublimation) is always less steep in a P vs. T plot for a pure material than is the melting line. Explain why. A diagram may be helpful.

15. The quotation at the start of Part III of this book reads, "Passed through the fiery furnace, As gold without alloy, Triumphant over trials, Accounting it all joy." The implication is that "gold without alloy" can pass through a fiery furnace without damage, whereas gold "with alloy" (i.e., impure gold) would be damaged in a furnace. Provide an explanation. Include a phase diagram.

16. The system $Na_2SO_4/Na_2MoO_4/H_2O$ was studied at 25°C by W.F. Linke and J.A. Cooper (1956), *Journal of Physical Chemistry*, **60**, 1662. They determined that the following solid phases just precipitate out of a saturated liquid at these compositions (compositions given are mass% in the liquid):

Liquid phase		Solid phase(s)
Mass% Na_2SO_4	Mass% Na_2MoO_4	
32.90	0.0	Na_2SO_4
28.14	5.89	Na_2SO_4
21.66	14.15	Na_2SO_4
15.40	22.73	Na_2SO_4
10.50	30.00	Na_2SO_4
9.47	31.89	$Na_2SO_4 + Na_2MoO_4 \cdot 2H_2O$
9.42	31.92	$Na_2SO_4 + Na_2MoO_4 \cdot 2H_2O$
9.46	31.90	$Na_2SO_4 + Na_2MoO_4 \cdot 2H_2O$
7.28	33.73	$Na_2MoO_4 \cdot 2H_2O$
3.88	36.65	$Na_2MoO_4 \cdot 2H_2O$
0.0	39.90	$Na_2MoO_4 \cdot 2H_2O$

The hydrate $Na_2MoO_4 \cdot 2H_2O$ is 85.11 mass% Na_2MoO_4.
   a. Plot the points and then draw the phase diagram on triangular coordinate graph paper.
   b. Consider a dry sample that is 40% by mass Na_2SO_4 (i.e., 60 mass% Na_2MoO_4). Imagine this sample to be completely dissolved in water and describe what happens as the water present is allowed to evaporate.

17. Polymorphism can be a considerable problem when dealing with chromophores. For example, a compound may have different optical properties in different polymorphs. Explain how temperature and pressure can influence the stability of a phase and the resulting optical properties of a compound.

18. The silver–lanthanum isobaric binary phase diagram has three compounds. LaAg melts congruently at 886°C; $LaAg_2$ melts incongruently at 864°C; $LaAg_3$ melts congruently at 955°C. In the La–Ag phase diagram there are also three eutectics: 11 atomic% La and 778°C; 43.5 atomic% La and 741°C; 71 atomic% La and 518°C. Pure silver melts at 962°C and pure lanthanum melts at 812°C. Sketch and label the La–Ag binary phase diagram.

19. In the gold–lead phase diagram, there are two compounds, Au_2Pb, which melts incongruently at 418°C, and $AuPb_2$, which melts incongruently at 254°C. The peritectic compositions are 44 atomic% Pb and 72 atomic% Pb, respectively. In addition, there is a eutectic at 215°C, 84.4 atomic% Pb. Pure gold melts at 1063°C and pure lead melts at 327°C. Sketch and label the Au–Pb binary phase diagram.

20. A bar of a material has a small concentration of impurity (initial value C_0), shown as a function of distance along the bar, x, in the diagram below. The portion of the melting point diagram near the pure material, showing the affect on the melting point of the addition of a small amount of this impurity, is also shown in Figure 9.50. The bar will be heated with a small heater such that successive regions of the bar will be melted, as shown schematically in Figure 9.51.

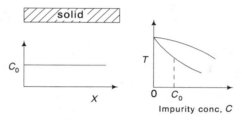

Figure 9.50. A bar of metal has an initial impurity concentration constant at C_0 along its length. The melting diagram of the material at low impurity concentration, C, also is shown.

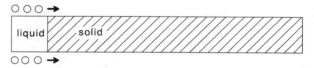

Figure 9.51. A heater passes along the bar melting material in the heater region.

a. At the region of the molten material, is the solid more or less pure than the liquid? Use the phase diagram in Figure 9.50 to answer this question.
b. As the heater moves along the sample from left to right, the purity profile (C vs. x) changes. Sketch the final C vs. x relationship, when the heater has traveled the full length of the bar, and then the bar has fully solidified.
c. This procedure is called **zone refinement** and it is used to purify materials such as silicon. How can the purity of the final bar (or regions of it) be improved further?

21. Compound A is **optically active**, that is it comes in two molecular forms that are chemically identical except that they are nonsuperimposable mirror images of each other. One form (*d*-A) rotates polarized light in one direction, and the other form (*l*-A) rotates polarized light in the other direction. In the solid state, there are (at least) two possible arrangements for the **racemic mixture** (this is equimolar *d*-A and *l*-A): the two isomers are totally insoluble in the solid state or the two isomers

form a racemic compound (d-A and l-A in a 1:1 mole ratio). Sketch binary phase diagrams (temperature vs. mole fraction of d-A in a mixture of d-A and l-A) for each of these possibilities.

22. In the phase diagram for H_2O/Fe_2Cl_6, six chemically distinct species are possible: Fe_2Cl_6, $Fe_2Cl_6 \cdot 4H_2O$, $Fe_2Cl_6 \cdot 5H_2O$, $Fe_2Cl_6 \cdot 7H_2O$, $Fe_2Cl_6 \cdot 12H_2O$, and H_2O. Use the definition of "component" to show why the Fe_2Cl_6/H_2O phase diagram is a binary (two-component) phase diagram.

23. For the mixing of two liquids,

$$\Delta_{mix}G = \Delta_{mix}H - T\Delta_{mix}S \tag{9.25}$$

and from Equation (6.71) it follows that

$$\frac{d\Delta_{mix}G}{dT} = -\Delta_{mix}S. \tag{9.26}$$

   a. For a binary system with an upper consolute temperature, what is the sign of $(d\Delta_{mix}G/dT)$? What does this imply about $\Delta_{mix}S$? Explain.
   b. For a binary system with a lower consolute temperature, what is the sign of $(d\Delta_{mix}G/dT)$? What does this imply about $\Delta_{mix}S$? Explain.

24. The slope of the solid-liquid line in the phase diagram of H_2O (Figure 9.5a) at atmospheric pressure is -135 atm K^{-1}. Consider an ice skater whose mass is 80 kg with a contact surface between the skates and the ice of 20 cm^2.
   a. Calculate the pressure exerted on the ice by the skater.
   b. Determine how much this pressure lowers the melting point of ice.
   c. From (b), do you think that the formation of a liquid layer under the skate is a plausible explanation for the ice skating phenomenon?

25. 4,4′-Dimethoxyazoxybenzene (DMAB) and 4,4′-diethoxyazoxybenzene (DEAB) both form liquid crystalline phases. Above 135°C, DMAB is an isotropic liquid, as is DEAB above 160°C. In the isotropic liquid phase, DEAB and DMAB mix with each other in all proportions. On cooling, at 135°C, pure DMAB transforms to a nematic liquid crystalline phase, as does pure DEAB at 160°C. The DMAB and DEAB nematic phases are miscible in each other in all proportions, and the isotropic liquid to nematic liquid crystal transition temperature changes smoothly as DEAB is added to DMAB. At 118°C, pure DMAB freezes to a crystalline solid. Similarly, pure DEAB freezes at 132°C. DMAB and DEAB are not soluble in the solid state, and they do not form compounds. However, they do form a eutectic at 97°C and $X_{DEAB} = 0.45$. (All data are for $P = 1$ atm.)
   a. Sketch and label the isobaric T vs. X_{DEAB} phase diagram.
   b. Deduce and draw the cooling curve (temperature vs. time) from 160 to 90°C for $X_{DEAB} = 0.45$.

26. The refractive index of a liquid is usually quite different from that of its vapor, mostly due to differences in density. Therefore, there is a discontinuity in refractive index on passing from liquid to vapor. At and near the critical point there are large fluctuations in refractive index, leading to a phenomenon known as **critical opalescence:** the material appears milky and opaque. Describe the light path to explain the origin of critical opalescence of a material in which both the liquid and vapor are transparent and colorless.

27. The following data are from Lightfoot and Prutton's 1947 study of the $CaCl_2$–KCl–H_2O system at 75°C.

Saturated solution (mass%)		
CaCl$_2$	KCl	Solid phase(s)
0	33.16	KCl
11.73	21.62	KCl
18.27	16.00	KCl
28.47	9.62	KCl
37.65	6.77	KCl
47.65	8.43	KCl
50.19	10.32	KCl, 2KCl·CaCl$_2$·2H$_2$O
50.92	9.36	2KCl·CaCl$_2$·2H$_2$O
53.85	6.21	2KCl·CaCl$_2$·2H$_2$O
56.33	4.51	2KCl·CaCl$_2$·2H$_2$O
57.62	3.60	2KCl·CaCl$_2$·2H$_2$O, CaCl$_2$·2H$_2$O
57.77	2.56	CaCl$_2$·2H$_2$O
58.58	0	CaCl$_2$·2H$_2$O

a. Sketch and label the isothermal isobaric phase diagram.

b. State what phases exist on evaporation of a solution that is initially 80 mass% H$_2$O, 10 mass% CaCl$_2$ and 10 mass% KCl, until no water remains.

28. Label the phases in the following phase diagrams: (a) Mg/Pb (Figure 9.52) and (b) Al/Zn (Figure 9.53).

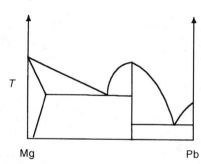

Figure 9.52. The Mg/Pb phase diagram.

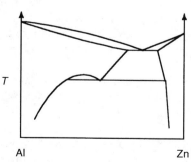

Figure 9.53. The Al/Zn phase diagram.

10

SURFACE AND
INTERFACIAL PHENOMENA

10.1 INTRODUCTION

Although we have concentrated until now on the bulk properties of materials, the surfaces of materials can have special and important properties that are distinct from those inside the bulk of the solid or liquid. As a simple example, consider that in the interior of an ionic crystal the charges balance out when all directions are averaged, whereas at the surface there can be a layer of ions all of the same charge.

One important property of a surface is its ability to interact with other atoms and molecules. These molecules may be **absorbed** *into* the interior of the material, as in the case of **graphite intercalates** (i.e., graphite with other types of atoms between the graphitic layers), an example of which is shown in Figure 10.1a.

The molecules could also be **adsorbed** *onto* the surface, as shown schematically in Figure 10.1b. The strength of the interaction between the **adsorbate** (adsorbed species) and the surface determines the type of adsorption. A relatively weak interaction ($\Delta H \approx -20$ kJ mol^{-1}, due to weak van der Waals interactions), is termed **physisorption** (or **physical adsorption**). A stronger interaction ($\Delta H \approx -200$ kJ mol^{-1}, due to covalent bonding) is termed **chemisorption** (or **chemical adsorption**).

Understanding the properties of surfaces can be very important in chemistry. For example, many reactions are catalyzed on surfaces that allow energetically favorable reaction pathways, primarily through chemisorption of reactants.

In the study of surfaces, it may be important to have a large surface area and/or to have very well defined surfaces. The large surface area may be particularly important in the case of an effect, such as the enthalpy change associated with adsorption, that scales with the amount of surface area. However, the surface must be regular for measurements to be meaningful.

For many years MgO was a popular surface to study, both for its homogeneity and its large surface area (several hundred square meters per gram); it is readily produced by combustion of Mg.

a.

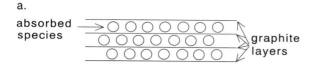

b.

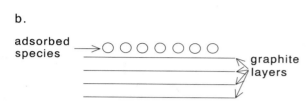

Figure 10.1. Cross-sectional views of absorption into and adsorption onto graphite. (a) Atoms or molecules can be absorbed into a graphite structure to form a graphite intercalate. (b) Atoms or molecules also could be adsorbed onto a graphite structure.

Another important surface, both for its homogeneity and its large surface area, is **exfoliated graphite.** This material, which is composed of graphitic particles, is produced from intercalated graphite that has been reacted to remove the intercalating species, blowing apart the structure and leaving behind islands of graphitic structure, with surface areas of several thousand square meters per gram.

10.2 SURFACE ENERGETICS

The Gibbs energy of a surface is always higher than that of the bulk material because the surface atoms are not able to achieve bonds in all directions as can the interior atoms. In this sense surfaces are unstable relative to the bulk material.

One consequence of this is that if there is flexibility within the surface, it will choose to have the minimum surface area. This is, for example, the reason why free-floating bubbles are spherically shaped; any other shape would require more surface area.

Another consequence is that it may be more favorable energetically for a material to be covered with another material. For a given mass, a group of small crystals is of higher energy (i.e., less favorable) than one large crystal, due to the decreased surface area of the large crystal. This fact can be used to grow large crystals from small ones, as follows. If the small crystals are placed in a saturated solution, and the temperature is increased, more material will dissolve (provided the temperature coefficient of solubility is positive, as is the usual situation). If the temperature is lowered, material will precipitate out due to the lowered solubility, and the most energetically favored way for the precipitation to occur is onto the largest crystal(s) present. Subsequent heating–cooling cycles can eventually transform many small crystals into a few larger crystals. This is shown schematically in Figure 10.2.

The energy at a surface is always higher than the bulk, but the **surface energy,** G_s, depends on the material in question. The data in Table 10.1 show one aspect of the driving force of the rusting process: metal oxide has a lower surface energy than pure metal. If the metal surface is coated with an organic material, this lowers the surface energy below that of the metal oxide, which is one reason why undercoating a car prevents rust formation, although the dominant effect is keeping oxygen away from the metal.

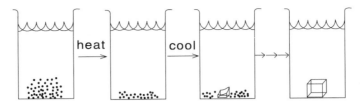

Figure 10.2. This diagram shows schematically how a group of small crystals (e.g., a powder) can be transformed to one large crystal, if the material is more soluble when the solution is heated (as is usual). Heat leads to greater solubility and cooling causes preferential precipitation on a seed crystal. After a number of cycles, powder can be transformed into a single crystal. This process is thermodynamically favored on the basis of lowered Gibbs energy for the sample with the lowest surface area (one large crystal).

TABLE 10.1.
Surface Energies, G_s, of Some Materials.[a]

Material	G_s / J m^{-2}
Metal	~ 1
Metal oxide	~ 0.3
Ice	~ 0.1
Wax	~ 0.07

[a] Data from R.G. Linford (1972), *Chemical Society Review,* **1,** 445.

10.3 SURFACE INVESTIGATIONS

Surfaces can be investigated by many means. For example, the enthalpies of adsorption of species can be important in gaining understanding of the surface-adsorbate interactions. Many microscopic techniques can be used to investigate surfaces, and one is highlighted here: the **scanning tunneling microscope.**

The scanning tunneling microscope (STM) was first described in 1986.[1] Its principle involves the overlap of the wavefunctions of a fine-tipped (atomic resolution) "wand" that scans the surface, with the wavefunctions of the surface atoms. When the overlap is sufficient, electrons can tunnel between the surface and the tip, completing an electrical circuit. This is shown schematically in Figure 10.3. The tunneling current is measured as a function of the position of the tip, giving images that represent an atomic resolution view of the surface. Some STM images are shown in Figures 10.4 and 10.5.

[1] Ch. Gerber, G. Binnig, H. Fuchs, O. Marti and H. Rohrer (1986). *Review of Scientific Instruments,* **57,** 221. Gerd Binnig (1947–) and Heinrich Rohrer (1933–) are researchers at IBM Zurich, Switzerland, and co-winners of the 1986 Nobel Prize in Physics, for their design of the scanning tunneling microscope; their Nobel Prize was shared with Ernst Ruska (1906–1988), a researcher from the Fritz-Haber Max Planck Institute in Berlin, who was cited for his design of the first electron microscope.

a.

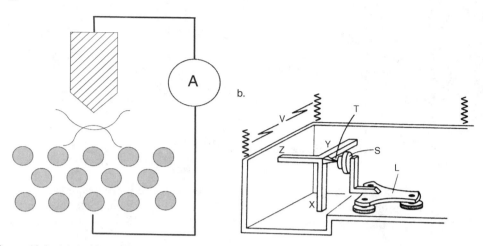

b.

Figure 10.3. (a) A schematic view of the scanning tunneling microscope. The overlap of the wavefunctions of the surface atoms with the scanning tip ("wand") completes the circuit and allows a measurable tunneling current. (b) The first scanning tunneling microscope, showing a piezotripod (X–Y–Z) for fine positioning of the tip T and a piezomotor L for rough positioning of the sample S. Reprinted, with permission, from Ch. Gerber, G. Binnig, H. Fuchs, O. Marti, and H. Rohrer (1986), *Review of Scientific Instruments,* **57,** 221, copyright American Institute of Physics.

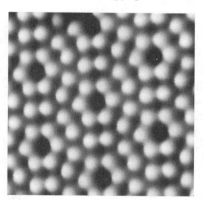

Figure 10.4. A scanning tunneling microscope image of Si, courtesy of M. Jericho.

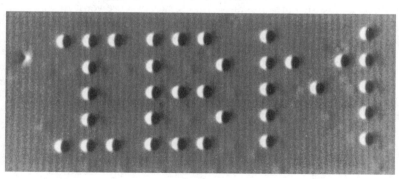

Figure 10.5. Individual Xe atoms on Ni have been manipulated to spell "IBM." Photo courtesy of D. Eigler.

10.4 SURFACE TENSION AND CAPILLARITY

Just as in the solid state, the molecules at the surface of a liquid are energetically different from those in the bulk of the liquid. This is shown schematically in Figure 10.6, where the molecules in the bulk are shown to be attracted equally in all directions, whereas the molecules at the surface have a net inward attraction to their neighbors. This again shows that the surface is high in energy with respect to the bulk, and the most favorable situation would be to have as little surface as possible. This is why very small drops of liquid are very nearly spherical (although larger ones may be flattened by their weight): spherical shape minimizes the surface area.

The energy required to increase the surface area leads to the definition of **surface tension, γ**,

$$\gamma = \frac{dG}{dA} \tag{10.1}$$

where G is the Gibbs energy and A is the surface area, so SI units of γ are J m^{-2}. (For a liquid, surface tension is the same as surface energy.)

The presence of surface tension leads to some interesting facts. For example, surface tension is a major factor in allowing insects to reside on the surface of a pond; although gravitational forces would dictate that the insect (which is denser than water) should be at the bottom of the pond, the cost in increasing the surface area, as the "skin" of the pond is ripped and the insect sinks, is too high. The energetic cost of increasing the surface area of the water in the pond is related to the contact angle of the insect with the water, and the degree of wetting of the solid phase (insect) by the liquid. The former can be seen by example: careful placement of a needle horizontally on a glass of water will allow it to float, but more vertical placement allows the needle to fall to the bottom while only making a tiny (low-energy cost) slit. The latter can be responsible for the old adage, "Don't touch a wet canvas tent or it will start a leak":

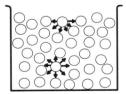

Figure 10.6. The attractive interaction of molecules with each other in liquids leads to no net force on an interior molecule, but a net inward force on surface molecules.

a.

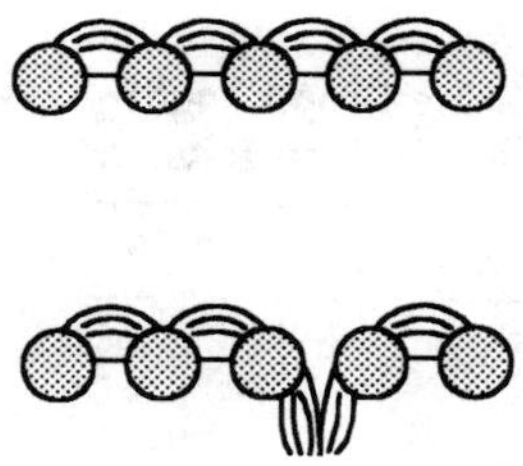

Figure 10.7. (a) Surface tension normally keeps water from flowing through a canvas tent roof. This is the basis of its water repellency. (b) Once a pathway has been established for a leak in a canvas tent roof by lowering the surface tension, the water will continue to pour through it.

b.

your finger (or more accurately the materials on your finger) will lower the surface tension and the water will then leak continuously where you touched. This is shown schematically in Figure 10.7.

For a spherical drop of radius R, the total surface Gibbs energy is $4\pi R^2 \gamma$; a decrease in radius by dR would lower the surface Gibbs energy by $8\pi R\gamma dR$. At equilibrium, the tendency to shrink is balanced by a pressure difference ΔP between the interior of the drop and its exterior such that the work against this pressure difference, $\Delta P 4\pi R^2 dR$ equals the difference in surface Gibbs energy:

$$\Delta P 4\pi R^2 dR = 8\pi R\gamma dR \tag{10.2}$$

which simplifies to

$$\Delta P = \frac{2\gamma}{R}. \tag{10.3}$$

Equation 10.3 shows the relationship between radius of curvature of a surface and surface tension; this can be used to determine surface tension experimentally.

Some typical values of surface tensions are given in Table 10.2. They can be seen to vary rather widely, and account for the large variation in sizes of free-falling drops of a material: the higher the surface tension, the larger the drops, as a large drop will have less surface area than many smaller drops with the same total volume. The size of the drop can be used to quantify surface tension, γ, using Tate's law:

$$m_{\text{ideal}} = 2\pi r\gamma/g \tag{10.4}$$

TABLE 10.2.
Surface Tension, γ, for Some Liquids at $T = 25°C$[a]

Liquid	$\gamma / 10^{-3}$ J m^{-2}
Br_2	40.95
H_2O	71.99
Hg	485.48
CS_2	31.58
CH_3OH	22.07
CH_3NH_2	19.15
CH_3CH_2OH	21.97

[a] Data from J.J. Jasper (1972), *Journal of Physical and Chemical Reference Data,* **1**, 841.

TABLE 10.3.
Correction Factors for the Drop Weight Method of Determining Surface Tension[a]

$r/V^{1/3}$	f
0.30	0.7256
0.40	0.6828
0.50	0.6515
0.60	0.6250
0.70	0.6093
0.80	0.6000
0.90	0.5998
1.00	0.6098
1.10	0.6280
1.20	0.6535

[a] r is the tube radius and V is the volume of the drop.

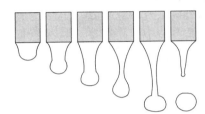

Figure 10.8. The sequence of shapes for a drop falling from an open tube. Note the residual liquid after the drop is detached.

where m_{ideal} is the ideal mass of a drop falling from an opening of a tube of radius r, and g is the gravitational constant. In this context, m_{ideal} is related to the observed mass, m_{obs}, by

$$m_{obs} = f m_{ideal} \qquad (10.5)$$

where f is a correction factor, given in Table 10.3,[2] that accounts for residual liquid left behind when the drop detaches, as shown in Figure 10.8.

Another way to determine surface tension is the **capillary effect.** If a liquid wets the walls of a capillary as shown in Figure 10.9, the radius of curvature, R, is related to the capillary radius, r,

$$\Delta P = \frac{2\gamma}{R} = \frac{2\gamma \cos \theta}{r} \qquad (10.6)$$

where now $\Delta P = P_1 - P_2$. P_1 and P_2 are pressures outside and inside the liquid, respectively, and θ quantifies the curvature of the meniscus (see Figure 10.9). $(P_1 - P_2)$ corresponds to the change in hydrostatic pressure of the liquid, $\rho g h$, where ρ is the density, g is the gravitational constant, and h is the capillary rise. Therefore,

$$\rho g h = \frac{2\gamma \cos \theta}{r} \qquad (10.7)$$

and surface tension can be determined by direct measurement of capillarity.

[2] Data from A.W. Adamson (1967), *Physical Chemistry of Surfaces,* 2nd Ed. John Wiley & Sons, Ltd.

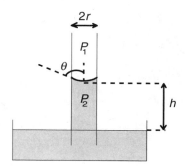

Figure 10.9. Rise of water in a glass capillary.

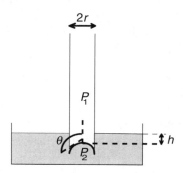

Figure 10.10. Mercury is depressed in a glass capillary.

Water wets a glass surface, making $h > 0$, whereas for mercury, $\theta \approx 140°$ (see Figure 10.10), which leads to $h < 0$, i.e., capillary depression results.

The **capillary rise** for water at room temperature would be 2.8 cm for a 1-mm diameter tube and 28 cm for a 0.1-mm-diameter tube. The rise of water in a capillary accounts for transport properties of fluids in plants and animals. It also partially accounts for the maximum possible height of a tree: it is limited by the molecular size, which dictates the minimum possible diameter of the capillaries involved.

Typical **capillary depression** for mercury is –0.79 cm for a 2-mm-diameter tube and –15.7 cm for a 0.1 mm diameter tube. Although today pressures are seldom measured using mercury **manometers,** when such manometers are used the effects of capillarity on the "true" height of the mercury need to be taken into account for accurate pressure determination, particularly in small capillaries.

COMMENT

NONSTICK COATING

A new coating made by copolymerization of esters of fluoroalkyl alcohols with functionalized vinyl monomers gives a nonstick material that cannot be wetted by solvents. Its crosslinked structure makes it very durable, and its nonstick properties arise from the orientation of the perfluoro groups on the surface. 3M plans to use this material for coatings on consumer products such as countertops and bathroom tile. Although the nonstick surface has some similarities to Teflon (also

(continued)

10.5 LIQUID FILMS ON SURFACES

When **amphiphilic** molecules[3] [molecules that are both **hydrophobic** (water-hating) and **hydrophilic** (water-loving)] are dispersed on a water surface, they tend to orient themselves with their hydrophobic ends (e.g., alkyl tails) away from the water and their hydrophilic ends (e.g., polar head groups) pointing into the water. This is shown schematically at low surface coverage in Figure 10.11a.

If the water surface is compressed, the surface layer becomes more ordered, until eventually the surface coverage is exactly one molecule thick, with each molecule occupying the smallest surface area that it can (Figure 10.11b); this is called a **monolayer.** As the water surface is compressed further, the structure will collapse (Figure 10.11c).

The pressure of the surface (which has units of N m^{-1}, compared with N m^{-2} for bulk pressure) can be measured by the force on a floating barrier, as shown in the trough in Figure 10.12. As the surface area is decreased, the pressure increases, until the monolayer is formed; with the collapse of the monolayer, the pressure falls, as shown in the phase diagram in Figure 10.13. The difference in surface tension above and below what we now know to be monolayer coverage was first described by Pockels.[4]

a.   b.   c.

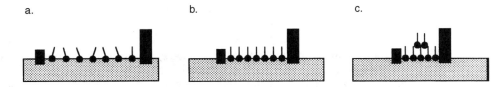

Figure 10.11. (a) A schematic view of surfactant molecules at low surface coverage on water. The polar head groups point into the water and the nonpolar tails point out of the water. (b) As the pressure is increased (i.e., the surface area is decreased), a one-molecule- thick film (monolayer) is formed. (c) At higher pressures, the film collapses. Here the pressure is defined as the force per unit length of the edge of the monolayer. The nonpolar tails of the surfactant molecules would be dynamically disordered.

[3] As we shall see in Chapter 11, amphiphilic molecules affect surface tension and therefore are usually called **surface-active agents,** often abbreviated as **surfactants.**

[4] Agnes Pockels (1862–1935) was born in Vienna and raised in Germany. Despite her lack of formal training (she had only high school education), she carried out seminal work related to surface films, starting with her first investigations at age 18. She carried out her work in her home, examining the effects of films on water. Ignored by most German scientists, she was encouraged by her brother, a professor of physics, to write to Lord Rayleigh about her results, and Lord Rayleigh aided her in publishing her work in *Nature* in 1891.

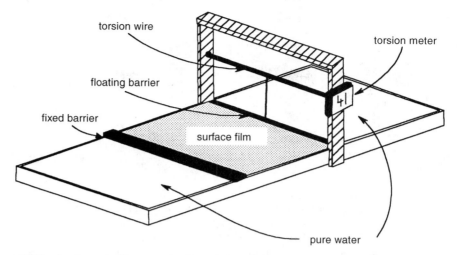

Figure 10.12. A schematic diagram of a floating-barrier trough for measuring the pressure of surface films. This is commonly referred to as a Langmuir trough.

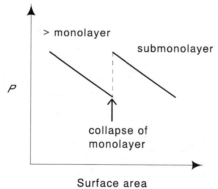

Figure 10.13. Two-dimensional pressure vs. area relationship for a Langmuir-Blodgett film showing the collapse of the film as the surface area is decreased.

Coherent (i.e., highly organized) films of this type are usually referred to as **Langmuir–Blodgett films.**[5] Langmuir–Blodgett films have many special properties due to the highly regular arrangement of molecules one molecular layer thick. Although these films themselves are rather fragile, developments in the past decade or so have led to transfer of these films to solid substrates, giving rise to many applications. The process of the transfer of one or more layers is shown schematically in Figure 10.14. This transfer allows stacking of as many as 1000 monolayers on a substrate, in either pure or doped forms.

---

[5] Irving Langmuir (1881–1957) was an American chemical physicist who spent most of his career as a research scientist at the General Electric Company in Schenectady, New York where he worked on gases, films, and electrical discharges. He was awarded the 1932 Nobel Prize in Chemistry for contributions to surface science. Katherine Burr Blodgett (1898–1979) also was a scientist working at General Electric. Blodgett was the inventor of antireflective coating for glass.

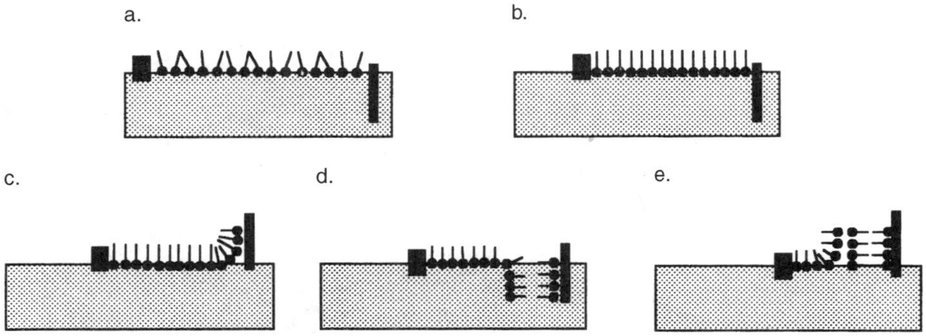

Figure 10.14. (a–e) Deposition of a Langmuir-Blodgett film on a solid substrate by successive passes of a movable barrier. In this case the layers are all head-to-head and tail-to-tail.

Langmuir–Blodgett films can lead to interesting electrical and optical properties, which can be tailor-made by changing the composition of the film. For example, insulating films can be placed on semiconducting substrates, enhancing their electrical properties for use as photovoltaic devices. The **microlithographic** (literally "small printing") possibilities of these films can lead to "wires" of 20–100 Å width. Special optical devices with second- or third-harmonic generation (recall that these take light of a certain frequency ν and produce light of frequency 2ν or 3ν, respectively; see Chapter 5) are possible by coating fibers with coherent films.

The highly organized Langmuir–Blodgett films also can be of use in adhesion, catalysis, and in the field of biosensors. Polymerization within layers after deposition is a recent development that can lead to increased mechanical stability.

The important factors that lead to the utility of the Langmuir–Blodgett films are nanometer control over the thickness of the film, the uniformity of the film, and the orientation and architecture of the molecules involved.

COMMENT

MORE USEFUL COATINGS

Special glass coatings made from tin or titanium oxides can change the reflection and transmission properties of glass. In certain conditions these coatings can allow greater amounts of solar energy into a building, with less energy loss. Due to the low emissivity, these are known as **low-E** coatings. In other circumstances the placement of the material can be reversed to prevent unwanted heating.

COMMENT

CAN MIRRORS REFLECT X-RAYS?

Ordinary mirrors do not efficiently reflect X-rays because they are not smooth enough. Since the roughness is on the same order of magnitude as the wavelength of light, the reflected beam is quite diffuse. However, X-ray mirrors being developed using Langmuir-Blodgett films can be so regular that they allow reflection of X-rays with a high degree of efficiency.

COMMENT

COMMAND SURFACES

When a light-sensitive material is coated on a surface, this layer can be used to modify the properties of the assembly. The azobenzene layer shown in Figure 10.15 changes orientation in the presence of light, and this in turn changes the alignment of the outer liquid–crystalline layer, as shown. The resultant change in optical and/or electrical properties can be controlled "on command" [K. Ichimura et al. (1993), *Langmuir,* **9,** 3298].

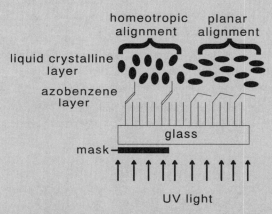

Figure 10.15 Light induces a change in the orientation of the azobenzene layer in this coating which, in turn, changes the orientation of the liquid crystalline layer and the optical and electrical properties of this assembly in the area in which the light is not masked.

Nanomaterials: A Tutorial

Many materials and devices have their properties associated with the extremely small dimensions of the repeat unit; these are often referred to as **nanomaterials** if their dimensions are of the order of a nanometer (10^{-9} m).

a. The band structure of a metal relies on the electrons being delocalized over ~10^{23} atoms. If a "nanostructure" is composed of atoms that in their bulk phase are metallic, do you expect the nanomaterial to be metallic? Explain.

b. Do you expect the band structure of a semiconductor material in a nanostructure to be the same as for the bulk material? Explain. Give an indication of the utility of a semi-conductor nanostructure.

c. There are several methods that can be used to produce nanostructures. Langmuir–Blodgett deposition is one; others include metal–organic **chemical vapor deposition** (CVD), **molecular beam epitaxy** (MBE; growth of a material from a molecular beam onto a substrate), and lithographic techniques. Explain how scanning tunneling micro-scopy (STM) could be used to characterize these structures.

d. In STM, the tunneling current through the tip is determined as a function of position on the surface. In another useful related method, called **atomic force microscopy** (AFM), the force on the tip is determined as a function of position. What causes this force? What further information can this give about a material?

e. What is the advantage of having such small-scale materials for the storage of informa-tion? Consider that, for example, in digital magnetic tape storage devices, the magnetic field orientation at each domain is oriented either "ON" or "OFF."

f. Can you imagine any use for nanostructure materials in optical devices? Consider in-terference effects, nonlinear optical effects, and any other pertinent effects.

COMMENT

SiC DOTS

A molecular beam of C_{60} can be reacted with silicon to produce regular dots of silicon carbide (SiC), about 10^{-6} m high. These new materials may have appli-cations in the semiconductor industry, as SiC is a hard material with high elec-tron mobility and thermal conductivity, combined with a wide band gap [R. Baum (1994), C_{60} Used to Produce Silicon Carbide, *Chemical & Engineering News,* November 7, 1994, 29].

FURTHER READING

General References

Many physical chemistry textbooks include information on surface science. In addition, the fol-lowing address many of the general points raised in this chapter:

Materials Research Bulletin, January 1996, Special Issue: "Polymer Surfaces and Interfaces."
V.K. Agarawal (1988). Langmuir–Blodgett Films. *Physics Today,* June 1988, 40.
F.J. Almgren, Jr. and J.E. Taylor (1976). The Geometry of Soap Films and Soap Bubbles. *Scientific American,* July 1976, 82.
A.E. Anwander, R.P.J.S. Grant, and T.M. Letcher (1988). Interfacial Phenomena. *Journal of Chem-ical Education,* **65,** 608.
D.J. Barber and R. Loudon (1989). *An Introduction to the Properties of Condensed Matter.* Cam-bridge University Press.

G. Cooke (1997). Laying It on Thin. *Chemistry in Britain,* April 1997, 54.

L.V. Interrante, L.A. Casper, and A.B. Ellis, Eds. (1995). *Materials Chemistry: An Emerging Discipline. Advances in Chemistry, Series 245,* American Chemical Society.

M. Lagally (1993). Atom Motion on Surfaces. *Physics Today,* November 1993, 24.

T. Richardson (1989). Langmuir–Blodgett Films. *Chemistry in Britain,* December 1989, 1218.

G. Somorjai (1998). From Surface Materials to Surface Technologies. *MRS Bulletin* **23**(5), 11.

G.A. Somorjai and G. Rupprechter (1998). The Flexible Surface: Molecular Studies Explain the Extraordinary Diversity of Surface Chemical Properties. *Journal of Chemical Education,* **75,** 162.

A. Swift (1995). Skimming the Surface. *Chemistry in Britain,* November 1995, 887.

Atomic Microscopy

R. Baum (1994). Chemical Force Microscopy. *Chemical & Engineering News,* October 3, 1994, 6.

C. Bustamante and D. Keller (1995). Scanning Force Microscopy in Biology. *Physics Today,* December 1995, 32.

H. Carmichael (1997). In Search of Atomic Resolution. *Chemistry in Britain,* November 1997, 28.

R. Dagani (1996). Putting the Nano Finger on Atoms. *Chemical & Engineering News,* December 2, 1996, 20.

D.M. Eigler and E.K. Schweizer (1990). Positioning Single Atoms with a Scanning Tunneling Microscope. *Nature,* **344,** 524.

M. Jacoby (1998). STM Tips Scientists on Atomic Details. *Chemical & Engineering News,* February 16, 1998.

J. Leckenby (1995). Probing Surfaces. *Chemistry in Britain,* March 1995, 212.

C. Lieber (1994). Scanning Tunneling Microscopy. *Chemical & Engineering News,* April 18, 1994, 28.

G. Meyer and K.H. Rieder (1998). Lateral Manipulation of Single Absorbates and Substrate Atoms with the Scanning Tunneling Microscope. *MRS Bulletin,* January 1998, 28.

D. Rugar and P. Hansma (1990). Atomic Force Microscopy. *Physics Today,* October 1990, 23.

Y.E. Strausser and M.G. Heaton (1994). Scanning Probe Microscopy. *American Laboratory,* April 1994, 20.

J.D. Todd and T.R.I. Cataldi (1991). STM—a New Image for Surfaces. *Chemistry in Britain,* June 1991, 525.

Nanomaterials

MRS Bulletin, August 1997, Special Issue: "Nanoscale Characterization of Materials."

Physics Today, June 1993, Special Issue: "Optics of Nanostructures."

R.M. Baum (1997). Nurturing Nanotubes. *Chemical & Engineering News,* June 30, 1997, 39.

R.M. Baum (1998). Nanotube Characterization. *Chemical & Engineering News,* January 5, 1998, 6.

R. Baum and M. Roubi (1998). Nantotube Functions as Molecular Transistor. *Chemical & Engineering News,* May 11, 1998, 6.

M. Brennan (1998). Empty Virus Packs Nanoscale Materials. *Chemical & Engineering News,* May 18, 1988, 8.

E. Corcoran (1990). Diminishing Dimensions. *Scientific American,* November 1990, 122.

R. Dagani (1996). Carbon Nanotube Ropes. *Chemical & Engineering News,* July 29, 1996, 5.

P. Day (1996). Room at the Bottom. *Chemistry in Britain,* July 1996, 29.

M. Freemantle (1996). Nanotubes shown to be flexible. *Chemical & Engineering News,* June 24, 1996, 10.

M. Freemantle (1996). Filled Carbon Nanotubes Could Lead to Improved Catalysts and Biosensors. *Chemical & Engineering News,* July 15, 1996, 62.

M. Freemantle (1998). Chemists and Their Nanotubes. *Chemical & Engineering News,* February 16, 1998, 4.

K.J. Gabriel (1995). Engineering Microscopic Machines. *Scientific American,* September 1995, 150.

J.R. Heath (1995). The Chemistry of Size and Order on the Nanometer Scale. *Science,* **270,** 1315.

A.M. Rouhi (1998). Nanotechnology. *Chemical & Engineering News,* April 20, 1998, 57.

R.W. Siegel (1993). Exploring Mesoscopia: the Bold New World of Nanostructures. *Physics Today,* October 1993, 64.

E. Wilson (1995). Microlithography Uses Metastable Argon Atoms. *Chemical & Engineering News,* September 4, 1995, 8.

E. Wilson (1998). Versatile Membranes from Carbon Nanotubes. *Chemical & Engineering News,* June 1, 1998, 12.

PROBLEMS

1. Calculate (a) the ideal mass and (b) the observed mass of a drop of water falling from an opening of radius (i) 1 mm, and (ii) 3 mm.

2. If a given surfactant molecule has a cross-sectional area of 6 Å^2, calculate
   a. the number of such molecules that will be dispersed in a monolayer film of dimensions 15 × 30 cm, and
   b. the volume of 0.1 M solution (where 1 M ≡ 1 mole of solute per liter of solution) of this surfactant in a nonpolar volatile solvent to produce the film in (a). (This can be dispersed drop-wise on the surface and the solvent will evaporate, leaving behind the surfactants on the surface of the water.)

3. STM and AFM are both very useful techniques for studying surfaces. Most researchers in this field agree that "the tip is everything." Explain this comment.

4. Often, after a shower, there is still some water in the shower head. This water may stay for some time but if some force is applied to the shower head (which can be as little as closing the door to the bathroom), the water will drain away. Explain this phenomenon. Include energetics in your explanation.

5. Nippon Telephone & Telegraph experienced trouble with snow sticking to wires and antennas in the Hida Mountains of Japan. As a result, NTT developed a coating to which snow would not stick. This coating, which can be painted on, is a fluorine-containing compound with a low surface Gibbs energy, which only weakly interacts with water. As a result, when applied to a roof with a slope of at least 45°, snow does not accumulate.
   a. Can you suggest other uses for a low water-surface interaction coating? Consider applications in which materials must move through water.
   b. The material does not absorb in the visible range and its fine particles on the surface disperse white light. What color is the coating? Does this lead to any practical applications that would be ruled out if it were another color?

6. Two bubbles, one smaller than the other, are each attached to ends of glass tubes, and the tubes are connected via a stopcock, as shown in Figure 10.16. When that stopcock is opened, will the bubbles change in size, and, if so, which will grow? Explain your answer.

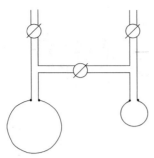

Figure 10.16. Two bubbles, one larger than the other, are attached to glass tubes as shown here.

7. Given the surface tension of water (7.2×10^{-2} J m^{-2}) at room temperature, calculate the minimum energy required to break a 1.0-cm-diameter sphere of water into 1×10^{-6} m-diameter droplets. (This is the minimum energy because breaking a sphere will produce droplets of the required size and also smaller droplets.)

8. A clever way to determine surface energies of solids has been devised [F.H. Buttner, E.R. Funk, and H. Udin (1952), *Journal of Physical Chemistry*, **56,** 657]. A number of wires of the material, all of the same initial diameter and length, have weights of various masses attached to them. The wires are heated to about 90% of the absolute temperature of their melting point to increase atomic mobility. The gravitational pull on the wire will tend to elongate the wires, but with increased atomic mobility surface tension will tend to decrease the length of the wire. The first of these will dominate with large weights and the second will dominate with small weights. At some intermediate weight, the two forces balance such that no elongation is observed. Derive a relationship between the surface energy (G_s), the wire radius (r), and the mass of the weight that causes no elongation (m).

9. a. A bubble has a volume of 10 cm^3. Calculate its surface area if its shape is a (i) cube; (ii) cylinder with its height equal to its radius; (iii) sphere.
   b. Which shape gives the smallest surface area?

10. AgNO$_3$ and NaCl are both quite soluble in water, whereas AgCl is virtually insoluble. When aqueous solutions of AgNO$_3$ and NaCl are mixed, a fine precipitate of AgCl forms. This precipitate has a much higher surface area than a large crystal of AgCl with an equivalent mass would have, and the large crystal would have a lower energy. Why is a large crystal of AgCl not formed?

11. a. A spherical bubble of volume V breaks into N smaller spherical bubbles, each of the same size. If A is the surface area of the original bubble, and A_N is the total surface area of the smaller bubbles, derive a relationship for A/A_N.
   b. What is A/A_N for (i) $N = 2$ (ii) $N = 10$ (iii) $N = 1000$ (iv) $N = 10^6$?

11

OTHER PHASES OF MATTER

11.1 INTRODUCTION

Many of the phases of matter discussed so far, such as gases, liquids, pure solids, solid solutions, liquid crystals, and orientationally disordered crystals, are **homogeneous** (i.e., uniform throughout). Both in nature and in the laboratory, materials can be combined (by alloying, coating, bonding, dispersion, lamination, etc.) to give new materials with different properties. There are important states of matter that are **heterogeneous** (not uniform throughout), and a Langmuir–Blodgett film is one example. The film is different from the substrate, and this is what makes it heterogeneous. Other examples of heterogeneous materials are described in this chapter.

11.2 COLLOIDS

A **colloid** is a system of particles too small to be visible in an optical microscope, typically 100 to 10,000 Å in size. A colloid can be stable (e.g., a macromolecular solution or a micellar system [see 11.3]) or metastable (e.g., a colloidal dispersion). The word colloid is derived from the Greek for "glue-like," as glue is a colloid. An important characteristic of colloids is that they do not pass through membranes that are permeable to (smaller) dissolved species.

A colloidal dispersion has its particles (**dispersed phase**) held in a **dispersing medium.** Gas, liquid, and solid can be either the dispersed phase or the dispersing medium; this leads to eight types of colloidal dispersion, listed with examples in Table 11.1.

In principle, colloidal dispersions are metastable with respect to the separated forms. This is apparent to anyone who has let whipped cream sit too long before eating it! However, they can be kinetically stabilized and can last for a long time; gold sols prepared by Michael Faraday[1] more than 150 years ago are still in their dispersed form today; these are on display in the

---

[1] Michael Faraday (1791–1867) was an English scientist who gained his scientific training when Sir Humphrey Davy hired him as an assistant during his temporary loss of sight due to a laboratory accident. Faraday made many major contributions to the theories of electricity and magnetism.

TABLE 11.1.
Colloidal Dispersion Compositions and Examples

Dispersed phase	Dispersing medium	System name	Example(s)
Solid	Solid	Solid dispersion	Colored glass
Solid	Liquid	Sol or suspension	Gold in water, ink
Solid	Gas	Aerosol of solid particles	Smoke
Liquid	Solid	Solid emulsion	Jelly
Liquid	Liquid	Emulsion	Milk, mayonnaise
Liquid	Gas	Aerosol of liquid droplets	Fog
Gas	Solid	Solid foam	Pumice, styrofoam
Gas	Liquid	Foam	Whipped cream

British Museum in London, England. Some colloidal dispersions may be kinetically stabilized by the presence of surface charges on the dispersed particles; they do not **flocculate** (coalesce) due to the mutual repulsion between particles.

Suspended charged particles can be made to move in electrical fields, and this is the basis for the separation process called **electrophoresis.** It can also be used preparatively, as an electrical field applied across a colloidal suspension of rubber latex can cause flocculation at one electrode. The flocculated latex particles can take on the shape of the electrode, and this is how rubber gloves, for example, are produced.

COMMENT

PROCESSING OF CERAMICS

Sol-gel processing can be used to prepare ceramics with considerable control, as schematically illustrated in Figure 11.1. The interactions between ceramic particles can be controlled by coating particles with surfactant (reducing the attraction) or with polymeric surface modifiers (reducing the repulsion). Sol-gel processing also is used to deposit thin films for chemical sensors and other sensors such as pH electrodes.

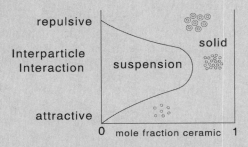

Figure 11.1. The balance between attractive and repulsive forces can determine the packing of ceramic particles in a gel phase, with packing fractions ranging from very low values to about 70%, depending on the interparticle interactions.

COMMENT

COLLOIDAL SPONGES

Colloidal microgels are polymeric materials with cells of about 10^{-9} to 10^{-6} m, swollen with solvent molecules. One example is poly(N-isopropylacrylamide), which in aqueous solution shrinks on heating due to conformational changes in the polymer; this makes the polymer–solvent interaction less favourable at high temperature. This reversible transformation reduces the volume by a factor of 60 on heating from 25 to 50°C. The gels are also sensitive to pH and the presence of other ions and molecules. Colloidal microgels have applications in drug delivery, and oil recovery and as viscosity modifiers.

11.3 MICELLES

Micelles are another heterogeneous phase of matter involving aggregates. At low concentrations of surfactant molecules (e.g., soap), individual ions are dispersed in aqueous solution. Above a certain temperature (the Krafft temperature, T_K) and above a certain concentration (the **critical micellar concentration,** abbreviated cmc) the solution will consist of ions and aggregates called **micelles.** A micelle is a three-dimensional structure that, in a polar solvent, usually has the hydrophilic ends of the molecules pointing out toward the solvent and the hydrophobic ends pointing inward. Examples of micellar structures are shown schematically in Figure 11.2. For $T > T_K$, the hydrophobic part (hydrocarbon tail) of the surfactant molecule is mobile and disordered in *trans* and *gauche* configurations, entropically stabilizing the micellar phase. For $T < T_K$, the stabilizing entropy effect is diminished and the molecules are more rigid, making the crystalline form more stable than the micelles.

A micelle, typically consisting of 50 to a few hundred monomer ions,[2] is a dynamic entity, with respect to both the motion of the hydrophobic ends of the constituent species and the exchange with the free ions in solution. The phase diagram of a micelle-forming solution is shown schematically in Figure 11.3.

The physical proof of the existence of micelles can come from many different types of experiments. Because the particle sizes are of the same order of magnitude as the wavelength of visible light, light scattering can be used to determine the average particle size. A typical light-scattering set-up is shown in Figure 11.4.

---

[2] Typical cmc values for aqueous solutions at 20°C are $CH_3(CH_2)_{11}N(CH_3)_3Br$, 14.4 mol L^{-1}, $\approx$ 50 monomers per micelle; $CH_3(CH_2)_{11}SO_4Na$, 8.1 mol L^{-1}, $\approx$ 62 monomers per micelle; $CH_3(CH_2)_{11}(OCH_2)_6OH$, 0.1 mol L^{-1}, $\approx$ 400 monomers per micelle.

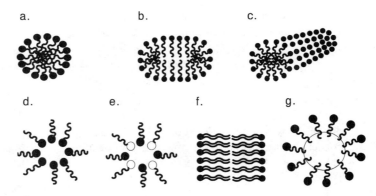

Figure 11.2. Schematic view of organized surfactant structures: (a) spherical micelle; (b) oblong micelle; (c) tubular micelle; (d) reverse micelle; (e) mixed reverse micelle; (f) lamellar micelle; (g) oil droplet solubilized by micelle. The nonpolar tails of the surfactant molecules would be dynamically disordered.

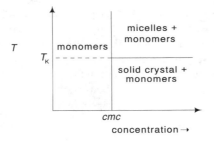

Figure 11.3. A schematic temperature-composition phase diagram for an aqueous solution of surfactant molecules indicating the region of micelle formation. T_K is the Krafft temperature, and for $T > T_K$ and concentrations $> cmc$, micelles form.

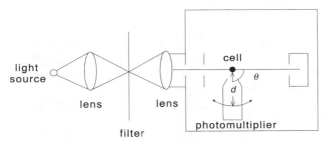

Figure 11.4. A schematic diagram of a light-scattering apparatus for determination of particle size.

From Chapter 4, we know that the intensity of scattered light of wavelength λ at angle θ, I_θ, at distance d from the sample is related to the incident light intensity, I_i, by

$$\frac{I_\theta}{I_i} = \frac{8\pi^4\alpha^2(1 + \cos^2\theta)}{d^2\lambda^4} \tag{11.1}$$

where α is the polarizability. The scattered light depends on the interaction of light with the sample, and this depends on the refractive index of the pure solvent (n_0) and the refractive index of the solution (n). Equation (11.1) can be rewritten[3] in terms of these quantities and the concentration (c, expressed in mass per unit volume) and the **mass-averaged molecular mass**[4] of the solute, $\bar{M}_m$:

$$\frac{I_\theta}{I_i} = \frac{2\pi^2 n_0^2(1 + \cos^2\theta)(n - n_0)^2 \bar{M}_m}{cN_A d^2\lambda^4} \tag{11.2}$$

where N_A is Avogadro's number. From Equation 11.2, the mass-averaged molar mass of the solute (micelle) can be determined.

The presence of the micelle can be determined by comparison with the calculated properties of a solution with the same concentration of surfactant molecules that are not aggregated: an ionic micellar solution will have lower electrical conductivity (it drops dramatically at the *cmc* due to aggregation of ions and some counterions) and less effective colligative properties such as freezing point depression, boiling point elevation, vapor pressure lowering, and osmotic pressure, due to the aggregation.

These properties can be used to determine the **average molecular mass**[5] of the aggregate in the micelle. For example, the **osmotic pressure**, Π, is the pressure that is developed inside a solution to bring it into equilibrium with pure solvent that is separated from it by a membrane (see Figure 11.5). The membrane is **semipermeable;** it allows passage of solvent but not solute. As the solvent passes into the solution, the pressure in the solution increases by Π, until equilibrium is attained. The increase in pressure may be reflected in the solution rising up a tube as shown in Figure 11.5, and the rise, *h,* will be proportional to Π:

$$\Pi = \rho g h \tag{11.3}$$

where g is the gravitational constant and ρ is the density of the solution. The osmotic pressure, Π, is directly related to the **number-averaged molecular mass,**[6] $\bar{M}_n$:

$$\Pi = \frac{\bar{M}_n RT}{V_m} \tag{11.4}$$

where V_m is the molar volume of the solvent.

---

[3] See R.A. Alberty (1987), *Physical Chemistry,* 7th Ed. John Wiley & Sons Ltd., p. 475, for a derivation.

[4] It is important to note that this is an **average** molecular mass, as not all micelles in a solution will have exactly the same composition. There will be a distribution of molecular masses, not just one molecular mass. This particular average is weighted by the mass of the solute particles.

[5] This average can be by mass, i.e., $\bar{M}_m$, or by number, i.e., $\bar{M}_n$, where the former is determined by light scattering and the latter is determined by osmotic pressure.

[6] That is, the average weighted in proportion to the probability of having a certain molecular mass.

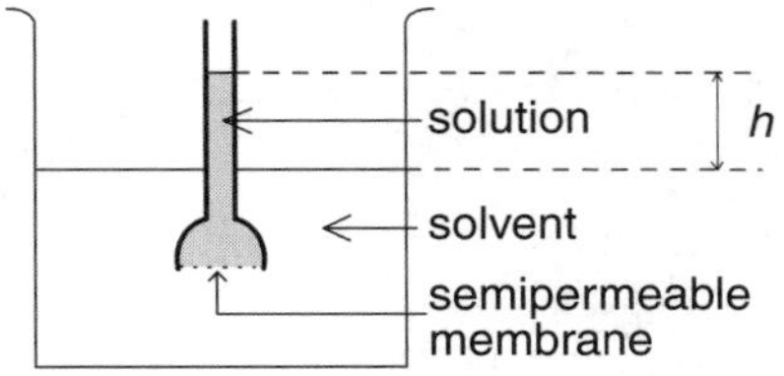

Figure 11.5. A schematic representation of an **osmometer,** which is a device used to measure osmotic pressure. The solvent flows through the semipermeable membrane into the solution to increase the pressure of the solution by Π, the osmotic pressure. This is reflected in h, the difference between the levels of the solution and the surrounding solvent.

11.4 SURFACTANTS

Knowledge of the properties of surfactant molecules can lead to the stabilization of emulsions. As can be seen from Table 11.1, an **emulsion** results from a colloidal suspension of one liquid in another. Oil and water will form two layers if poured gently together, but if shaken vigorously (with the mechanical work going to increase the surface area of the droplets) an emulsion can be formed.

Although an emulsion will tend to separate out again over time, an amphiphilic material added to the emulsion will act as a **surfactant** (which is an abbreviation for **surface active agent**) by lowering the interfacial tension between the oil and water. The surfactant material is able to do this by solubilizing one end in the oil and the other end in the water; this stabilizes the emulsion. Because surfactants lower the surface tension, they can lead to spreading, wetting, solubilization, emulsion and dispersion formation, and frothing. For example, foams are metastable, but can be made to last longer by the addition of a surfactant. For these reasons, surfactants are common additions to household products such as foods (e.g., as emulsifiers in salad dressing) and cleansing agents.

COMMENT

SURFACE TENSION IN LUNGS

It has now been shown that low surface tension is important in our lungs to keep them from collapsing. The absence of the surfactant that reduces the surface tension in the lungs of premature babies has been related to their respiratory difficulties.

COMMENT

BIOMIMETIC GELS

Polymer gels that mimic some biological activities have recently been developed. For example, a moving gel loop can be made to move by simple manipulations: in an electric field, surfactant molecules collect on the strand's top surface, and electrostatic forces cause the gel to shrink; with a change in field polarity, the surfactant enters the aqueous phase again, and the strand extends to a new position [Y. Osada and S.B. Ross-Murphy (1993). Intelligent Gels. *Scientific American,* May 1993, 82].

WAITING FOR THE PAINT TO DRY

Latex paints and Elmer's Glue® contain a mixture of water, small polymer particles, and a surfactant. The surfactant stabilizes the polymer particles and prevents them from aggregating in the aqueous solution; surfactant molecules surround the polymer particles, in a micellar form, with their nonpolar parts facing inward to solubilize the polymers and their polar head-groups facing outward into the aqueous solution. Over time the water evaporates, leaving the latex particles behind.

11.5 INCLUSION COMPOUNDS

Inclusion compounds are multicomponent materials in which one type of species forms a host in which other species reside. They can exist in solution or in the solid state. In the latter, the range of topologies of the host lattice allows several different types of structures, as shown in Figure 11.6. Inclusion compounds belong to the larger family of materials known as **supramolecular materials;** these are defined by properties that stem from *assemblies* of molecules. In the case of inclusion compounds, many physical and chemical properties are different in the supramolecular assembly than for the pure host or guest species. Supramolecular assemblies are generally held together by noncovalent forces such as hydrogen bonding, van der Waals interactions, and Coulombic forces. Industrial applications of inclusion compounds include fixation of volatile fragrances and drugs, and inclusion of pesticides to make them safer to handle.

A **clathrate** is a material in which the host lattice totally traps the guest species. The term was coined in 1948 by H.M. Powell,[7] and a number of materials are now known to form clathrates. Examples are **clathrate hydrates** (H_2O molecules form the host lattice; see Figure 11.7) and organic compounds such as hexakis(phenylthio)benzene (see Figure 11.8).

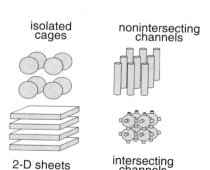

isolated
cages

nonintersecting
channels

2-D sheets

intersecting
channels

Figure 11.6. Schematic representations of possible inclusion compound topologies in the solid state. Isolated cages (e.g., clathrates), nonintersecting channels (e.g., urea compounds), sheets (e.g., graphite intercalates), and intersecting channels (e.g., zeolites) give extended structure in zero, one, two, and three dimensions, respectively.

[7] H.M. Powell (1948). The Structure of Molecular Compounds. Part IV. Clathrate Compounds. *Journal of the Chemical Society,* 61.

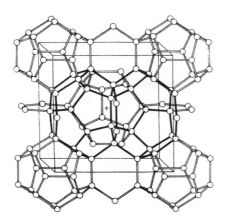

Figure 11.7. The structure of a clathrate hydrate. Only the oxygen atoms of the H_2O molecules are shown. Small molecular guests can reside in the cages. Reproduced, with permission, from G.A. Jeffrey (1994). *Inclusion Compounds,* Vol. 1. Academic Press.

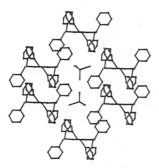

Figure 11.8. A schematic representation of the structure of the CBr_4 inclusion compound of hexakis(phenylthio)benzene. The CBr_4 molecules reside in cages formed by the substituted-benzene host molecules. The central phenyl ring in each host molecule is perpendicular to the page.

In an **intercalate** structure the host lattice forms layers with guests between the layers. For example, alkali metals can be intercalated into graphite.

Some host lattices form channels in which the guest species can reside. An example is urea's inclusion of *n*-alkanes, as shown in Figure 11.9. In these **channel compounds** the guests often can undergo translational motion within the channels.

a.   b.

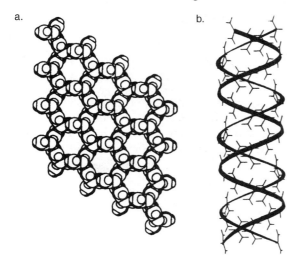

Figure 11.9. Schematic views of the structure of urea inclusion compounds. (a) Looking down into the channels of hydrogen-bonded urea molecules. Reproduced from K.D.M. Harris and J.M. Thomas (1990), *JCS Faraday Transactions,* **86,** 2985, by permission of The Royal Society of Chemistry. (b) A lateral look at a urea channel, showing the orientation of the hydrogen bonds in urea. Reproduced from M.D. Hollingsworth and K.D.M. Harris, in Chapter 7 in Volume 6 of *Comprehensive Supramolecular Chemistry,* J.-M. Lehn, J.L. Atwood, D.D. MacNicol, J.E.D. Davies, and F. Vögtle, Eds. Copyright 1996, Pergamon Press, p. 179, with kind permission from Elsevier Science Ltd.

Other host lattices can form interconnecting structures as shown in Figure 11.6; **zeolites** (which are aluminosilicates) are one example.

The possibility of including guest molecules in a host matrix can lead to many interesting chemical and physical properties. Most chemists have used molecular sieves to dry solvents; water is included in the cavities of these materials. As another example, consider that the cavities in zeolites can allow reactions that preferentially produce only certain geometric isomers. Inclusion compounds also can have interesting optical, electrical, and magnetic properties; in addition, they can be used to produce materials with unusual thermal properties.

COMMENT

MOLECULAR SELF-ASSEMBLY

In the self-assembly shown schematically in Figure 11.10, the π-electron-rich 1,5-dioxynaphthalene unit combines with the π-electron-deficient tetracationic cyclophane; this can be likened to threading a bead on a wire. The assembly can be monitored optically, as the resulting material has a charge transfer band in the visible region [R. Ballardini et al. (1993). A Photochemically Driven Molecular Machine. *Angewandte Chemie* (*International Edition in English*), **32**, 1301].

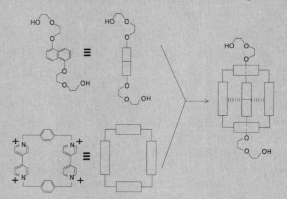

Figure 11.10. This self-assembly can be likened to threading a bead on a wire.

Hair Care Products: A Tutorial

A large portion of the economy is devoted to hair care products. In many ways these are related to materials science; the material involved is the biomaterial we know as "hair."

Several factors are considered when assessing hair quality: thickness of fiber (i.e., diameter), fiber density (number of hairs per square centimeter), stiffness, luster, configuration (curly or straight), and static. Some of these properties can be controlled, whereas others (e.g., density) cannot.

Hair is a protein fiber, with an outer scaly cuticle and an inner cortex. Most of the properties of the hair are concerned with the state of the cuticle.

a. Proteins have both acidic and basic groups, and so does hair. Because of this, hair is amphoteric. The acidic groups are –COOH, and the basic groups are –NH_2. The whole fiber appears neutral at pH 5.5, and the cuticle is neutral at pH 3.8. Why is the pH of a shampoo important? What happens to the functional groups of hair at low pH? What happens at high pH?

b. Hair conditioners are **cationic surfactants** (i.e., the surfactant ions carry a positive charge), usually with long chain fatty (hydrocarbon) portions. Would these be expected to attach to hair better at low pH or at high pH? Explain.

c. The fatty portion of the hair conditioner can act as a lubricant. Is this good or bad? How might the molecular mass of the surfactant play a role?

d. Hair conditioner can leave a film on hair. Is this a desirable feature or not? Consider the reflective properties of films.

e. At high pH, hair will swell and high-molecular-weight molecules can be absorbed into the cortex. Can you see an application for this?

f. One of the most annoying things in winter, as anyone with long hair can tell you, is the build-up of static electricity in hair. Why is this not a problem in summer? How could hair conditioner help prevent static, fly-away hair?

g. If hair is curled when wet and then allowed to dry, it retains its curl. Why?

h. How is curl in hair made "permanent"?

Applications of Inclusion Compounds: A Tutorial

Inclusion compounds can be synthesized in the laboratory, although many occur naturally.

a. **Clathrate hydrates** (clathrate compounds in which the last lattice is composed of water molecules) occur in nature, especially in polar regions and under high pressures in the oceans. In fact, **gas hydrates** (i.e., clathrate hydrates with gaseous guests) are thought to contain more than 50% of the earth's reserves of organic carbon, about twice the organic carbon reserves of fossil fuels. Although difficult to recover, the fuel potential in gas hydrates is substantial. Clathrate hydrates can melt congruently or incongruently, depending on the guest species, the composition, and the overpressure of gas. Tetrahydrofuran (THF) forms a clathrate hydrate of composition $THF \cdot 17H_2O$; it melts congruently at 5°C. Sketch the THF/H_2O phase diagram. Pure THF melts at −108.5 °C.

b. Decomposition of naturally occurring clathrate hydrates has been proposed as an explanation for the Bermuda Triangle (i.e., the mysterious disappearance of ships and aircraft in that region). Explain.

c. Explain how an inclusion compound could be used to produce a nonlinear optical material. Discuss advantages and disadvantages.

d. Explain how an inclusion compound could be used to produce a material with special electronic properties. In particular consider nanomaterials.

e. The few inclusion compounds that have been investigated to date appear to have thermal conductivities that increase with temperature at all investigated temperatures.[8] This characteristic is usually associated with glassy materials, but the inclusion compounds are crystalline. Can you explain this apparent contradiction?

FURTHER READING

General References

P. Ball (1994). *Designing the Molecular World.* Princeton University Press.

R. Dagani (1994). New Opportunities in Materials Science Draw Eager Organometallic Chemists. *Chemical & Engineering News,* October 3, 1994, 31.

R. Dagani (1996). Chemists Explore Potential of Macromolecules as Functional Materials. *Chemical & Engineering News,* June 3, 1996, 30.

M.F. Hawthorne and M.D. Mortimer (1996). Building to Order. *Chemistry in Britain,* April 1996, 32.

R.J. Hunter (1993). *Introduction to Modern Colloid Science.* Oxford University Press.

T.E. Mallouk and H. Lee (1990). Designer Solids and Surfaces. *Journal of Chemical Education,* **67,** 829.

Colloids and Surfactants

C.P. Ballard and A.J. Fanelli (1993). Sol-Gel Route for Materials Synthesis. Chapter 1 in *Chemistry of Advanced Materials,* C.N.R. Rao, Ed. Blackwell Scientific Publications.

J.A. Clements (1962). Surface Tension in the Lungs. *Scientific American,* December 1962, 121.

R. Dagani (1997). Intelligent Gels. *Chemical & Engineering News,* June 9, 1997, 26.

M. Hair and M.D. Croucher, Eds. (1982). *Colloids and Surfaces in Reprographic Technology. ACS Symposium Series 200.* American Chemical Society.

C.A. Murray and D.G. Grier (1995). Colloidal Crystals. *American Scientist,* **83,** 23.

Y. Osada and S.B. Ross-Murphy (1993). Intelligent Gels. *Scientific American,* May 1993, 82.

E. Pefferkorn and R. Varoqui (1989). Dynamics of Latex Aggregation. Modes of Cluster Growth. *Journal of Chemical Physics,* **91,** 5679.

L.L. Schramm (1993). *The Language of Colloid and Interface Science.* American Chemical Society.

D.R. Ulrich (1990). Chemical Processing of Ceramics. *Chemical & Engineering News,* January 1, 1990, 28.

H. B. Weiser (1949). *A Textbook of Colloid Chemistry.* John Wiley & Sons, Ltd.

K.I. Zamaraev and V.L. Kuznetsov (1993). Catalysts and Adsorbents. Chapter 15 of *Chemistry of Advanced Materials,* C.N.R. Rao, Ed. Blackwell Scientific Publications.

Composites

H.R. Clauser (1973). Advanced Composite Materials. *Scientific American,* July 1973, 36.

A. Kelly (1967). The Nature of Composite Materials. *Scientific American,* September 1967, 160.

---

[8] See, for example, D. Michalski and M.A. White (1997). Thermal conductivity of an Organic Clathrate. Possible generality of glass-like thermal conductivity in crystalline molecular solids. *Journal of Chemical Physics,* **106,** 6202.

Inclusion Compounds

Chemistry in Britain, special issue concerning supramolecular chemistry, January 1995.

Comprehensive Supramolecular Chemistry (1996). 11 Volume Series. Pergamon Press.

J.L. Atwood, J.E.D. Davies, and D.D. MacNicol, Eds. (1984). *Inclusion Compounds,* Vol. 1: *Structural Aspects of Inclusion Compounds Formed by Inorganic and Organometallic Host Lattices.* Academic Press.

J.L. Atwood, J.E.D. Davies, and D.D. MacNicol, Eds. (1984). *Inclusion Compounds,* Vol. 2: *Structural Aspects of Inclusion Compounds Formed by Organic Host Lattices.* Academic Press.

J.L. Atwood, J.E.D. Davies, and D.D. MacNicol, Eds. (1984). *Inclusion Compounds,* Vol. 3: *Physical Properties and Applications.* Academic Press.

J.L. Atwood, J.E.D. Davies, and D.D. MacNicol, Eds. (1991). *Inclusion Compounds,* Vol. 4: *Key Organic Host Systems.* Oxford University Press.

J.L. Atwood, J.E.D. Davies, and D.D. MacNicol, Eds. (1991). *Inclusion Compounds,* Vol. 5: *Inorganic and Physical Aspects of Inclusion.* Oxford University Press.

S. Borman (1995). Organic 'Tectons' Used to Make Networks with Inorganic Properties. *Chemical & Engineering News,* January 2, 1995, 21.

S. Borman (1997). Nanoporous Sandwiches Served to Order. *Chemical & Engineering News,* April 28, 1997.

J.F. Brown, Jr. (1962). Inclusion Compounds. *Scientific American,* July 1962, 82.

S. Carlino (1997). Chemistry between the Sheets. *Chemistry in Britain,* September 1997, 59.

R. Dagani (1998). The Shape of Things to Come. *Chemical & Engineering News,* June 8, 1998, 35.

J.E.D. Davies (1977). Species in Layers, Cavities and Channels (or Trapped Species). *Journal of Chemical Education,* **54,** 536.

J.M. Drake and J. Klafter (1990). Dynamics of Confined Molecular Systems. *Physics Today,* May 1990, 46.

M. Jacoby (1998). Hydrogen Supersponges. *Chemical & Engineering News,* May 25, 1998, 6.

J. Haggin (1996). New Large-pore Silica Zeolite Prepared. *Chemical & Engineering News,* May 27, 1996, 5.

K.D.M. Harris (1993). Molecular Confinement. *Chemistry in Britain,* February 1993, 132.

H. Kamimura (1987). Graphite Intercalation Compounds. *Physics Today,* December 1987, 64.

G.T. Kerr (1989). Synthetic Zeolites. *Scientific American,* July 1989, 100.

D. Michalski, M.A. White, P. Bakshi, T.S. Cameron, and I. Swainson (1995). Crystal Structure and Thermal Expansion of Hexakis(phenylthio)benzene and its CBr_4 Clathrate. *Canadian Journal of Chemistry,* **73,** 513.

M. Ogawa and K. Kuroda (1995). Photofunctions of Intercalation Compounds. *Chemical Reviews,* **95,** 399.

D. O'Hare (1992). Inorganic Intercalation Compounds. Chapter 4 in *Inorganic Materials,* D.W. Bruce and D. O'Hare, Eds. John Wiley & Sons Ltd.

M. Rouhi (1996). Container Molecules. *Chemical & Engineering News,* August 5, 1996, 4.

F. Vögtle, Ed. (1991). *Supramolecular Chemistry: An Introduction.* John Wiley & Sons, Ltd.

P. Zurer (1998). Polymer Self-Assembles into Large Hollow Spheres. *Chemical & Engineering News,* March 23, 1988, 8.

Molecular Engineering

R. Baum (1995). Chemists Create Family of 'Molecular Squares' Based on Iodine or Metals. *Chemical & Engineering News,* February 13, 1995, 37.

S. Borman (1994). Molecular 'Brake': Side Chain Reversibly Slows Rotating 'Wheel'. *Chemical & Engineering News,* April 25, 1994, 6.

R. Dagani (1991). Building Complex Multimolecule Assemblies Poses Big Challenges. *Chemical & Engineering News,* May 27, 1991, 24.

M.J. Snowden and B.Z. Chowdhry (1995). Small Sponges with Big Appetites. *Chemistry in Britain,* December 1995, 943.

F. Stoddart (1991). Making Molecules to Order. *Chemistry in Britain,* August 1991, 714.

G.M. Whitesides (1995). Self-Assembling Materials. *Scientific American,* September 1995, 146.

Other Materials

M. Bloom (1992). The Physics of Soft, Natural Materials. *Physics in Canada,* January 1992, 7.

P.-G. de Gennes and J. Badoz (1996). *Fragile Objects.* Springer-Verlag.

M. Jacoby (1998). Durable Organic Gels. *Chemical & Engineering News,* January 26, 1998, 34.

B. Kahr, J.K. Chow, and M.L. Petterson (1994). Organic Hourglass Inclusions. *Journal of Chemical Education,* **71**, 584.

PROBLEMS

1. Ethanol and water form a water-rich clathrate hydrate that melts incongruently at about –75°C. Sketch and label the H_2O/CH_3CH_2OH phase diagram.

2. Choosing two phases each from a list of three phases would lead to nine combinations, but only eight types of colloids are listed in Table 11.1. Why is there no entry for a gas dispersed in a gas?

3. **Aerogels** are an interesting phase of matter. They are a subset of aerosols, composed of a suspension of fine particles in air. They were first formed in the 1930s by the preparation of a gel phase (from sodium silicate and hydrochloric acid), followed by the removal of the solvent under supercritical conditions. (Supercritical conditions are required because if the solvent is removed as a liquid the gel collapses due to surface tension considerations.) Recent syntheses have produced aerogels with more than 99% air and densities as low as 3×10^{-3} g cm^{-3}. (The density of air is 1.2×10^{-3} g cm^{-3}.)
   a. One of the main uses of aerogels is in thermal insulation. Explain why an aerogel would conduct heat much less efficiently than the corresponding bulk material (e.g., silica aerogel compared with pure silica).
   b. How could the particle size within the aerogel be determined?
   c. Suggest other uses for aerogels.

4. When potassium atoms are intercalated between the layers of carbon atoms in graphite, the extent of intercalation can be followed by color changes. For example, stage 2 intercalates are blue, whereas stage 1 intercalates are gold colored. (The designation "stage n" indicates that there are n layers of graphite between each pair of intercalant layers.) On the basis of the colors, do you expect potassium-graphite intercalates to be semiconductors or metals? Explain.

5. An advertisement for NonScents® claims that it is a "100% natural odor magnet," "a zeolite" that, by virtue of its "negative molecular ion charge" attracts airborne odor molecules that have a "free ride on dust particles, which have a positive molecular ion charge." The advertisement further claims that "millions of tiny micropores . . . . give the material a great adsorbent surface area." Is NonScents® more likely to be useful to trap large or small organic molecules? Explain.

6. An emulsion of two transparent liquids (e.g., oil and water) leads to a milky-white liquid if the refractive indices of the two phases differ. Why does the emulsion not appear transparent?

P A R T

ELECTRICAL AND MAGNETIC
PROPERTIES OF MATTER

The missing link between electricity and magnetism was found in 1820, by Hans Christian Ørsted, who noticed that a magnetic compass needle is influenced by current in a nearby conductor. His discovery literally set the wheels of modern industry in motion.

—Rodney Cotterill
The Cambridge Guide to the Material World

12

ELECTRICAL

PROPERTIES

12.1 INTRODUCTION

Although all the properties of materials—optical, thermal, electrical, magnetic, and mechanical—are related, it is perhaps the electrical properties that most distinguish one material from another. This can be as simple a distinction as metal vs. nonmetal, or it can involve more exotic properties such as superconductivity. In this chapter the principles that determine the electrical properties of matter are discussed.

12.2 METALS, INSULATORS, SEMICONDUCTORS: BAND THEORY

The **resistance** to flow of electric current in a material, designated R, is determined by the dimensions of the material (length L and cross-sectional area A) and the **intrinsic resistivity** (also known as **resistivity,** and represented by ρ) of the material:

$$R = \rho\left(\frac{L}{A}\right)$$

(12.1)

where R is in units of ohms[1] (abbreviated Ω) and ρ is typically in units Ω m. The intrinsic resistivity depends not just on the specific material, but also on the temperature. (We will see later how this can be used to produce electronic thermometers.) Some typical resistivities are given in Table 12.1.

---

[1] Georg Simon Ohm (1789–1854) was a German physicist. Ohm published his law ($V = iR$, where V is voltage, i is current, and R is resistance) in 1827, but it received little attention for nearly 20 years. It is now known to be one of the most fundamental principles of electronics, and is called Ohm's Law.

TABLE 12.1.
Electrical Resistivities, ρ, and Conductivities,
$\sigma(=\rho^{-1})$, of Selected Materials at 25°C

Material	ρ / $(\Omega\ m)$	$\sigma/(\Omega^{-1}\ m^{-1})$
Metals		
Ag	1.61×10^{-8}	6.21×10^{7}
Cu	1.69×10^{-8}	5.92×10^{7}
Au	2.26×10^{-8}	4.44×10^{7}
Al	2.83×10^{-8}	3.53×10^{7}
Ni	7.24×10^{-8}	1.38×10^{7}
Hg	9.58×10^{-7}	1.04×10^{6}
Semiconductors		
Ge	0.47	2.1
Si	3×10^{3}	3×10^{-4}
Insulators		
Diamond	1×10^{14}	1×10^{-14}
Quartz	3×10^{14}	3×10^{-15}
Mica	9×10^{14}	1×10^{-15}

The **electrical conductivity, σ,** of a material is the reciprocal of its resistivity, ρ:

$$\sigma = \frac{1}{\rho} \tag{12.2}$$

so typical units of σ are $\Omega^{-1}\ m^{-1}$. Electrical conductivity also can be equivalently expressed in terms of **current density J** (units $A\ m^{-2}$, where A is amperes[2]) and **electric field ε** (units $V\ m^{-1}$, where V is volts[3]):

$$\sigma = \frac{J}{\varepsilon} \tag{12.3}$$

so equivalently units of σ can be expressed as $A\ m^{-1}\ V^{-1}$ ($\equiv \Omega^{-1}\ m^{-1}$).

The distinguishing feature separating metals, semiconductors, and insulators is the range of σ: $> 10^4\ \Omega^{-1}\ m^{-1}$ for metals, 10^{-3} to $10^4\ \Omega^{-1}\ m^{-1}$ for semiconductors, and $< 10^{-3}\ \Omega^{-1}\ m^{-1}$ for insulators (see Figure 12.1 and also Table 12.1).

In contrast with thermal conductivities, which span a range of about 6 orders of magnitude from the poorest thermal conductor to the best (see Chapter 8), electrical conductivities (see Figure 12.1) span more than 22 orders of magnitude, from the poorest electrical conductor to the best. The difference in ranges of thermal and electrical conductivity reflect the fact that thermal conductivity requires collective action of phonons, whereas electrical conductivity is more related to the motion of relatively independent electrons along a path of least resistance.

[2] André Marie Ampère (1775-1836) was a French mathematician and physicist who taught at École Polytechnique in Paris. Ampère carried out many experiments in electromagnetism and provided a mathematical theory for electrodynamics.

[3] Count Alessandro Volta (1745-1827) was a member of the Italian nobility who carried out many important electrical experiments. Volta was the inventor of the electrical battery and the discoverer of electrolysis of water. His home town, Como, Italy, has a museum that houses many Volta artifacts, including experimental apparatus and laboratory notes.

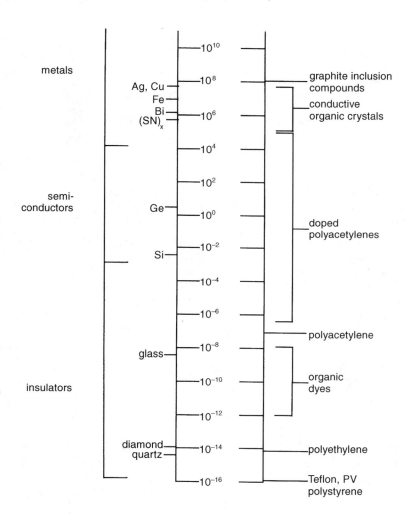

Figure 12.1. This diagram shows the wide range of electrical conductivity of common materials at room temperature.

The wide variety of electrical conductivities in different materials is associated with the availability of energy levels directly above the filled energy levels in the material. The presence (and width) of a gap between the filled electronic states and the next available electronic states leads to very different electrical properties for insulators and noninsulators.

Metals

As we have seen in Chapter 3 (particularly Figures 3.2, 3.6, and 3.7), there is a continuum of electronic energy levels (called **bands**), and for a metal there are unoccupied electronic energy levels (empty bands) immediately above the highest occupied levels. This is shown in Figure 12.2

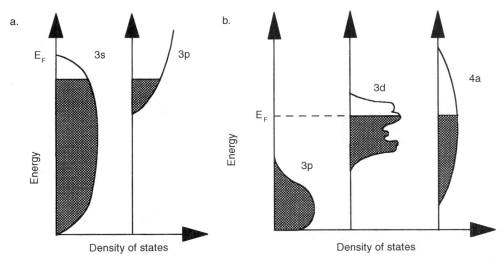

Figure 12.2. Electronic energy levels for metals. E_F is the Fermi energy. The shaded areas represent the filled energy levels, and bands below the valence bands are omitted. In both cases, energy levels are available immediately above the filled bands, and this is what gives rise to metallic properties. (a) Energy bands of magnesium showing the overlap in energy between the 3s band and the 3p band. E_F represents the energy at which the probability of occupation is one-half, i.e., the highest energy level that would be occupied if the 3p band were empty. (b) Energy bands of cobalt showing the overlap of the energy range of the 3d and 4s bands.

for two specific metals where realistic density of states is shown; these figures can be contrasted with the simpler representations of Figures 3.3, 3.7 and 3.8.

When there are available electronic energy levels immediately above the filled levels (i.e., there is no gap), the material is a metal, and excitation of the electrons takes place at all temperatures above 0 K. The presence of excited electrons allows facile conduction of electric current and gives rise to the high electrical conductivity (low electrical resistivity) of a metal. The conduction electrons also are responsible for the shiny appearance of metals (see Chapter 3) and their high thermal conductivity (see Chapter 8).

The **work function, Φ**, for a metal is defined as the minimum energy required to remove an electron into vacuum far from the atom (see Figure 12.3a). Different metals would have different values of work functions (Figure 12.3b), such that when dissimilar metals are in contact an electrical potential results (Figure 12.3c). This **contact potential** can be used in devices such as thermometers (see the Tutorial at the end of this chapter).

Semiconductors

A semiconductor has a gap (forbidden region) between the valence band and the conduction band (Figure 12.4). At a temperature of 0 K, all the electrons of a semiconductor are in the valence band. However, for $T > 0$ K, there will be some thermally excited population of the conduction band, as shown in Figure 12.4. It is this small population of the conduction band that allows slight electrical conductivity in semiconductors.

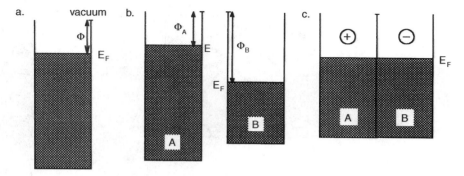

Figure 12.3. (a) The work function, Φ, for a metal is defined as the minimum energy required to remove an electron to vacuum far from the atom. (b) Different metals, shown here as A and B, would have different work functions. (c) Dissimilar metals, when placed in contact with each other, equalize their Fermi energies, giving a contact potential, shown here as an electrical field.

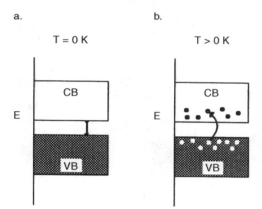

Figure 12.4. The energy bands of a semiconductor at (a) $T = 0$ K and (b) $T > 0$ K. Thermal energy can cause some slight population of the conduction band of a semiconductor, and this gives it a nonzero (but low) electrical conductivity.

The structure of pure germanium, which is a semiconductor, is shown in Figure 12.5, along with a schematic representation of its electrical conductivity mechanism.

The element tin occurs as two allotropes, white and gray tin. White tin is the more familiar form, as it is stable above 13°C, and we know this form to have the typical characteristics of a metal: it is shiny, ductile, and a good conductor of heat and electricity. White tin is denser than gray tin (their densities are 7.2 and 5.7 g cm^{-3}, respectively), and this allows its electronic energy bands to overlap more, giving it a metallic character. In contrast, gray tin has a band gap of 0.1 eV, an electrical resistivity of 10^{-6} Ω m, and acts as a semiconductor. Folklore has it that Napoleon lost many of his soldiers during his winter invasion of Russia due to the white tin to gray tin transformation of the tin buttons holding their uniforms closed. Since the gray form is not metallic, it is crumbly, and this is said to have led to the soldiers perishing in the cold.[4]

---

[4] Although the transformation from white to gray tin is at 13°C, like many first-order transitions (e.g., freezing), the high-temperature phase can be supercooled. Therefore the soldiers could have experienced temperatures slightly below 13°C without loss of their jacket closures, but in the extreme cold of the Russian winter the tin was cooled sufficiently to ensure complete transformation of the buttons to the crumbly gray form.

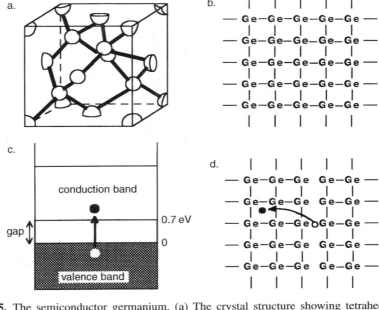

Figure 12.5. The semiconductor germanium. (a) The crystal structure showing tetrahedral bonding among Ge atoms. (b) A schematic two-dimensional representation of the bonding in Ge. (c) A look at the energy level diagram of Ge, showing the formation of a "hole" (absence of electron) in the valence band and an electron in the conduction band. This electron has been promoted from the valence band to the conduction band by thermal energy. (d) An electron-hole pair in Ge.

Insulators

An insulator has a very large gap between its valence band and the conduction band. Although thermal energy could promote electrons across this gap, in practice the required thermal energy is sufficient to melt the solid! This large gap essentially prevents significant electrical conductivity in insulators.

COMMENT

CONDUCTING POLYMERS

Some conjugated organic polymer materials, which are normally semiconductors with a substantial gap between the highest occupied molecular orbital (HOMO) and the lowest unoccupied molecular orbital (LUMO), can be made metallic by oxidation or reduction. The induced charge provides new energy states within the band gap. For example, polyacetylene can be converted to an electrically conducting material by oxidation or reduction. One of the most important properties of polymers is that they are quite robust, and this undoubtedly will lead to new materials and processes using conducting polymers. Materials scientists are trying to control the impurities and defects in these materials to make the properties more reproducible.

COMMENT

ONE-DIMENSIONAL CONDUCTORS

Although most molecular systems are insulators, some have high electrical conductivity; one such system is the charge-transfer complex of tetrathiafulvalene (TTF) with tetracyanoquinodimethane (TCNQ). Molecular structures for TTF and TCNQ are shown in Figure 12.6.

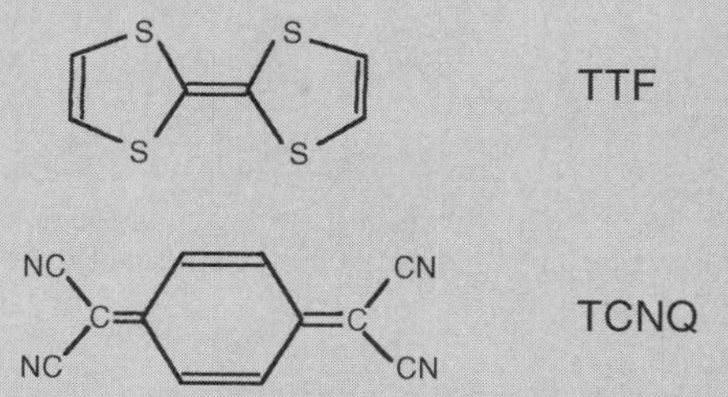

Figure 12.6. The molecular structures of tetrathiafulvalene (TTF) and tetracyanoquinodimethane (TCNQ).

In the crystalline complex, TTF and TCNQ form segregated regular stacks, and there is considerable delocalization of electrons (i.e., overlap of wave functions) along these columns. TTF–TCNQ forms a charge-transfer complex; on average, each TTF molecule donates 0.59 electrons to a TCNQ. This allows appreciable conductivity; σ is $5 \times 10^4\ \Omega^{-1}\ m^{-1}$ along the chains, but 100 times less in directions perpendicular to the chains, i.e., this material is highly anisotropic. The unusually high metal-like electrical conductivity of TTF–TCNQ persists down to 60 K.

COMMENT

ELECTRICAL ANISOTROPY

TTF–TCNQ is not the only material that is electrically anisotropic. Other examples include graphite, asbestos, and $K_2Pt(CN)_4Br_{0.3}\cdot3H_2O$. The latter is composed of $Pt(CN)_4$ units stacked to give Pt–Pt distances of 2.89 Å, i.e., nearly the same distance along these chains as in Pt metal (2.78 Å). However, the Pt–Pt distance is much greater between chains, and this results in anisotropic structure and properties—the crystals are highly reflective and electrically conducting in the direction of the chains, and transparent and electrically insulating in directions perpendicular to the chains.

12.3 TEMPERATURE DEPENDENCE OF ELECTRICAL CONDUCTIVITY

The temperature dependence of the electrical properties of a material can be used to provide important devices such as switches and thermometers. In this section we examine the principles

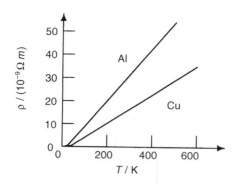

Figure 12.7. Resistivity as a function of temperature for two typical metals, aluminum and copper.

involved in temperature dependence of electrical conductivity (or resistivity) for metals and pure (intrinsic) semiconductors.

Metals

The resistivity of a metal increases with temperature (see Figure 12.7) and this increase is essentially linear, except at the lowest temperatures.

The increase in resistivity (decrease in electrical conductivity) with increased temperature can be directly associated with the increase in thermal motion of the atoms on their lattice sites at increased temperature. The increase in motion serves to scatter the conduction electrons and decrease their mean free path, hence decreasing their ability to carry electrical charge.

COMMENT

WIEDEMANN–FRANZ LAW: A RELATIONSHIP BETWEEN ELECTRICAL AND THERMAL CONDUCTIVITIES FOR METALS

The Debye expression for thermal conductivity (Equation 8.15) can be combined with the electronic contribution to the heat capacity of a metal (Equation 6.46) to yield

$$\kappa = \frac{1}{3}\gamma T v \lambda \qquad (12.4)$$

where γ is the electronic coefficient of the heat capacity, v is the electron's speed, and λ is the electron's mean free path. Since γ is related to the Fermi energy (E_F) for N electrons per unit volume as

$$\gamma = \pi^2 N k^2 / 2 E_F \qquad (12.5)$$

and the electrical conductivity is given by

$$\sigma = N e^2 \lambda / m v_F \qquad (12.6)$$

(continued)

COMMENT

WIEDEMANN–FRANZ LAW: (continued)

where v_F is the electron speed at the Fermi energy and m and e are the mass and charge of the electron, respectively, assuming $v \approx v_F$, this leads to a ratio of

$$\frac{\kappa}{\sigma T} = \frac{\pi^2 k^2}{3e^2} \tag{12.7}$$

which has a value of 2.45×10^{-8} W Ω K^{-2} ($\equiv$ L, the **Lorenz constant**) for all metals. Equation 12.7, which shows that the ratio of thermal conductivity to electrical conductivity at a given temperature is constant for metals, is known as the **Wiedemann–Franz Law** and it holds quite well. However, in very pure samples in the temperature range where phonon scattering is important, electron–phonon interactions (which have been neglected in this derivation) lead to a breakdown of the Wiedemann–Franz Law. The Wiedemann–Franz Law was important in the history of our understanding of metals, as it shows the electrical conductivity to be inversely proportional to temperature, in support of modeling metals as gases of electrons.

Intrinsic Semiconductors

In contrast with metals, the electrical resistivity of intrinsic (pure) semiconductors decreases (electrical conductivity *increases*) with increasing temperature. Germanium is shown as an example in Figure 12.8.

This can be understood as follows. At low temperatures, few electrons are able to jump the gap from the valence band to the conduction band; in fact, none has jumped at $T = 0$ K. As the temperature is increased, the number of electrons that have sufficient thermal energy to jump from the valence band to the conduction band is increased. From Fermi's expression for $P(E)$, the probability of having a free electron at energy E (Chapter 3):

$$P(E) = \frac{1}{e^{(E-E_F)/kT} + 1} \tag{12.8}$$

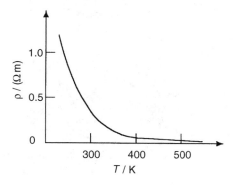

Figure 12.8. The electrical resistivity of pure germanium as a function of temperature.

can be rewritten as

$$P(E) = \frac{e^{-(E-E_F)/kT}}{1 + e^{-(E-E_F)/kT}} \qquad (12.9)$$

and for $|E-E_F|/kT \ll 1$ (i.e., near the Fermi energy), this can be approximated by

$$P(E) = \frac{1}{2} e^{-(E-E_F)/kT} \qquad (12.10)$$

which has the same form as the Boltzmann distribution:

$$n(E) \propto e^{-(E-\bar{E})/kT} \qquad (12.11)$$

where $n(E)$ is the number of species in energy state E and $\bar{E}$ is the average energy of the system. For an intrinsic semiconductor, $\bar{E}$ is E_F [since $P(E) = \frac{1}{2}$ at the Fermi energy, by definition, and the distribution is symmetric[5] about E_F] and the most probable energy of an excited (conduction) electron will be $E_F + \frac{1}{2} E_g$ (i.e., the bottom of the conduction band), so Equation 12.11 becomes

$$n(E) \propto e^{-E_g/2kT} \qquad (12.12)$$

where $n(E)$ is the number of conduction electrons. Since the electrical conductivity, σ, is proportional to the number of conduction electrons, this gives

$$\sigma = \sigma_0 e^{-E_g/2kT} \qquad (12.13)$$

where σ_0 is a proportionality constant.

Equation 12.13 shows an interesting feature: from measurements of electrical conductivity as a function of temperature, the energy of the band gap, E_g, can be determined for a semiconductor (see Problem 12.3).

COMMENT

SILICON

Silicon-based semiconductors are at the heart of most of the computer devices that we use daily. Silicon's properties make it very useful: it is one of the most common elements on earth; it is hard and chemically rather inactive (mainly reacts with oxygen); it has a high melting point (1685 K), which allows for doping by solid-state diffusion at high temperatures; and it has a low band gap (1.1 eV), which allows for low-power applications (Figure 12.9).

Figure 12.9. A photograph of silicon, a very useful element.

[5] Strictly speaking, the distribution is symmetric only if the density of states in the valence band and conduction band are equal, but even if they are not, the correction for asymmetry is usually small.

12.4 PROPERTIES OF EXTRINSIC (DOPED) SEMICONDUCTORS

Defects in a semiconductor can give rise to localized electronic states, as seen previously in the discussion of optical properties of semiconductors (Chapter 3).

For example, if Si (which has four valence electrons) is doped with P (which has five valence electrons), there is an extra electron in the lattice and a localized energy level within the gap corresponding to this extra electron. This extra level (Figure 12.10) is called a **donor level,** since P donates an extra electron. Electrons can be promoted from this level to the conduction band. Since the resultant current carriers are electrons, which are negatively charged, these are called **n-type extrinsic semiconductors.**

Holes also could carry charges, and, since holes are effectively positive charges (they represent the absence of negative charge), these are termed **p-type extrinsic semiconductors.** An example is Si (four valence electrons) doped with Al (three valence electrons); the extra localized level in the gap due to electron deficiency in Al is an **acceptor level,** since Al accepts electrons from Si. This is depicted in Figure 12.11.

The importance of p-type and n-type semiconductors is realized when they are sandwiched together.

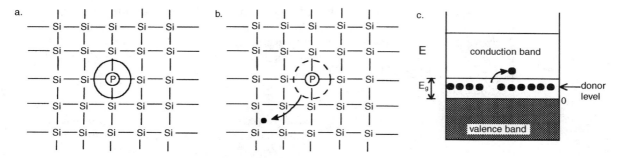

Figure 12.10. An n-type extrinsic semiconductor. (a) P substitutes for Si in a lattice. (b) The extra electron of P (compared with Si) can be donated to the conduction band to carry electrical charge. (c) The band model showing the donor level in the band gap.

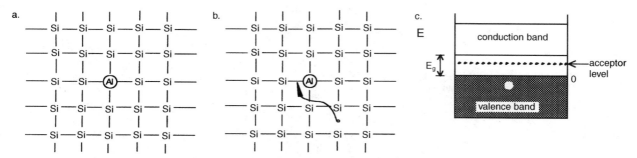

Figure 12.11. A p-type extrinsic semiconductor. (a) Al substitutes for Si in a lattice. (b) The missing electron of Al (compared with Si) can accept an electron from the valence band, thus leaving an electron hole in the valence band that can act as an electrical charge carrier. (c) The band model showing the acceptor level in the band gap and a hole in the valence band.

12.5 ELECTRICAL DEVICES USING EXTRINSIC SEMICONDUCTORS

Several important electronic devices can be designed using the combined properties of n-type doped semiconductors and p-type doped semiconductors. Some of these are described here.

p,n-Junction

The junction of a p-type semiconductor with an n-type semiconductor is referred to as a **p,n-junction** and it has a special property called rectification. When an n-type semiconductor is placed next to a p-type semiconductor, the small number of conducting electrons in the n-type are drawn toward the boundary, as are the few holes in the p-type. This electron-hole attraction is a simple consequence of unlike charges attracting. Since, for the configuration shown in Figure 12.12a, electrons are flowing left to right in the n-type semiconductor, and holes (which are positive charges) are flowing right to left (implying flow of effective negative charge from left to right) in the p-type semiconductor, there is net (small) flow of electrons from left to right across the device. Because this exists even in the absence of an applied electric field, it is referred to as **intrinsic electrical conductivity.**

Now, if an electrical field is placed across the p,n-junction such that the negative terminal is next to the n-type semiconductor and the positive terminal is next to the p-type semiconductor (this arrangement is referred to as **forward bias**), the interactions of the moving charges (electrons or holes) with the field will enhance the flow of charge, and there will be a larger flow of electrons left to right, as shown in Figure 12.12b. The field-induced current is called **extrinsic electrical conductivity.**

If the field is reversed (**reverse bias**), the electrons in the n-type semiconductor will be pulled in both directions—toward the holes on the p side and toward the positive terminal. Similarly, the holes in the p-type semiconductor will be attracted in both directions, toward the negative charge of the n-type semiconductor, and toward the negative terminal (see Figure 12.12c. Therefore, reversing the potential field across a p,n-junction has the effect of turning a large current into a small current; this property is called **rectification.** This is contrary to what is observed in normal conductors (i.e., metals), where a reversal of the field causes only the direction of the current flow to change, not its magnitude.

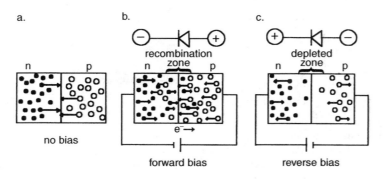

a. no bias

b. recombination zone forward bias e⁻→

c. depleted zone reverse bias

Figure 12.12. A p,n-junction showing rectification. Electrons are shown as ● and holes as ○. (a) With no field (no bias) there is a small net flow of negative charge due to electron-hole annihilation at the interface. (b) With a forward bias, there is net electron flow from left to right. (c) With a reverse bias, extrinsic conductivity as in (b) disappears and only a small amount of intrinsic electrical conductivity as in (a) remains.

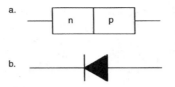

Figure 12.13. (a) A diode and (b) its electrical schematic symbol.

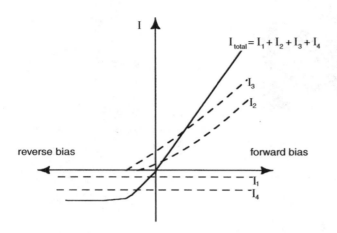

Figure 12.14. The total current, I_{total}, in a p,n-junction diode as a function of bias. Also shown are the contributions to the current: I_1, due to mobile electrons in the p $\rightarrow$ n direction; I_2, due to mobile electrons in the n $\rightarrow$ p direction; I_3, due to mobile holes in the p $\rightarrow$ n direction; I_4, due to mobile holes in the n $\rightarrow$ p direction.

The p,n-junction is often referred to as a **diode,** which is an electrical device that makes use of the electrical properties of the p,n-junction. The symbol for a diode is shown in Figure 12.13; the arrow has been chosen to show the motion of current (opposite to electron motion) under forward bias.

The principle of rectification in a diode can be seen in Figure 12.14, where the total current is shown as a function of bias (forward and reverse), and the contributions to the current are explicitly shown.

The properties of a p,n-junction also can be understood in terms of the electronic energy distributions. Figure 12.15 shows the distributions for an n-type semiconductor and a p-type

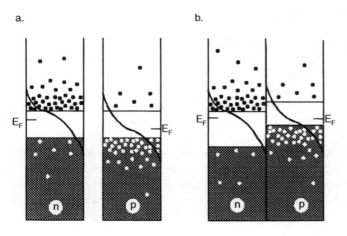

Figure 12.15. Electronic distributions for n-type semiconductors and p-type semiconductors (a) when they are separated and (b) when they are adjacent [showing the shift to equal Fermi energies, defined as the energy at which $P(E) = \frac{1}{2}$].

a.   b.   c.

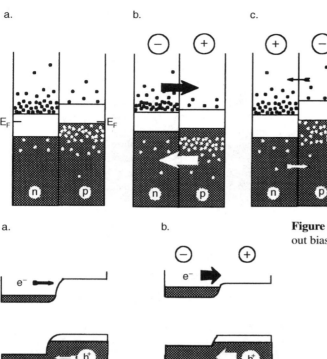

Figure 12.16. Electronic distributions for a p,n-junction (a) with no bias; (b) with a forward bias (giving carriers in both the valence band and the conduction band), and (c) with a reverse bias (giving only a few carriers).

a.   b.

Figure 12.17. Electronic distributions of a p,n-junction (a) without bias and (b) with forward bias.

n   p   n   p

semiconductor when they are separated and when they are placed adjacent to each other. When the two are in contact, the gaps shift so as to equalize the Fermi energies. This shift causes a contact potential (due to the difference in work function, see Section 12.2), as electrons are moved from the valence band of the n-type semiconductor to the conduction band of the p-type semiconductor.

In an electric field, the effect of bias on a p,n-junction can be seen in the electronic distributions shown in Figures 12.16 and 12.17.

With rapid change in bias of an ac source, and the rectification property of a p,n- junction, a device constructed of a p,n-junction can be used to turn alternating current into direct current, for example in a wall adapter for use with a lap-top computer.

A **Zener diode** is a special type of diode that has little current flow at low reverse bias as usual but more flow at large reverse bias (see Figure 12.18). This is because large reverse biases can be sufficient to cause electrons to tunnel through the junction. That is, the overlap of the wavefunctions of the electrons on both sides of the junctions increases the probability of them getting across the junction. This is shown schematically in Figure 12.18. As the reverse bias in a Zener diode is increased, the electric field across the depletion region becomes larger. When a charge carrier gets into the depletion region, it is efficiently carried to the appropriate side of the junction. This can be so efficient that the motion of the charge causes ionization (creation of a free electron and a hole), which means there are more charge carriers, and an amplification of current referred to as an **avalanche effect.** A particularly useful feature of

a.

b.

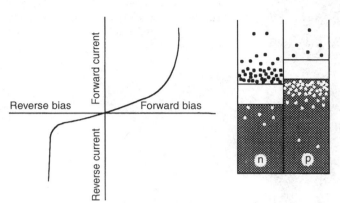

Figure 12.18. Zener diode showing (a) an increase in current at large reverse bias and (b) the electronic bands that allow tunneling through the junction (from the conduction band of the n-type semiconductor to the valence band of the p-type semiconductor).

Zener diodes is that slight changes in the composition of the doped semiconductors can be used to tailor-make the threshold voltage for the dramatic reverse-bias current.

CARBON MONOXIDE SENSORS

The electrical properties of semiconductors can be very sensitive to the state of purity. For example, the introduction of band gap impurities can greatly change the resistivity of a semiconductor.

In addition, electrical properties can be used to sense slight changes in the semiconductor due to reactivity. This property can be used to design chemical sensors.

Semiconducting metal oxides, such as tin oxide and zirconium oxide, have been used in sensors to detect the presence of carbon monoxide (Figure 12.19). CO is a very good reducing agent, as CO_2 is more stable than CO. Metal-oxide semiconductors will be reduced in the presence of CO, and the change in chemical composition will lead to a change in electrical properties, which can be used to sound an alarm in the event of harmful levels of CO.

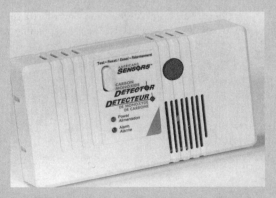

Figure 12.19. A carbon monoxide sensor makes use of the sensitivity of the electrical properties of a semiconductor to the presence of a reducing agent.

Figure 12.20. The original contact transistor made December 23, 1947. It amplified electrical signals by passing them through a semiconductor sandwich. Photo is property of the AT&T Archives. Reprinted with permission of AT&T.

Transistors

Two back-to-back p,n-junctions can work together to amplify an electrical signal. This can be either a p,n,p-junction or an n,p,n-junction; either of these form a **transistor.** A photo of the original point-contact transistor is shown in Figure 12.20.

In the p,n,p-junction shown in Figure 12.21, the applied electric field causes the left p,n-junction to be biased in the forward direction and the right p,n-junction to be biased in the reverse direction. Holes move from the left p-type semiconductor (called the **emitter**) toward the n-type semiconductor (called the **base**); if the n-layer is sufficiently thin, some of the holes will pass through the n-semiconductor to the reverse-biased junction and into the far p-layer (called the **collector**). Because holes are attracted to the negative electrode at the far side of this junction, these holes greatly increase the reverse current and cause amplification. Furthermore, the number of holes reaching the reverse-bias junction increases with an increase in the potential difference between the emitter and the base, so the **gain** of the amplification (i.e., the amplification factor) can be easily controlled.

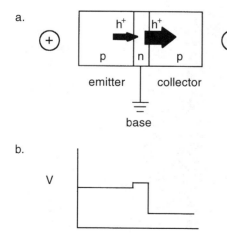

Figure 12.21. (a) A schematic view of a transistor. The p,n,p-device acts to amplify the current. (b) The corresponding voltage, V, across the transistor.

COMMENT

FIELD-EFFECT TRANSISTOR (MOSFET)

A **metal-oxide semiconductor field-effect transistor,** abbreviated **MOSFET,** allows current control with very little power consumption. The general structure of a MOSFET, shown in Figure 12.22, consists of two n-type regions in a p-type semiconductor (e.g., a silicon wafer); one n-region is the **source** and the other is the **drain.** Between the source and the drain, the wafer is covered with an insulating layer (e.g., a 10^{-7}-m-thick layer of silicon dioxide), which is covered by a metal that acts as the **gate.** At a positive gate voltage, electrons from the n-type semiconductor are attracted to the surface and holes are repelled, allowing a current to pass from the source to the drain in the region near the surface. Therefore, it is the field at the gate that controls the current, and since the gate does not consume much power, this low-power device can be very small (less than 10^{-3} mm^2). **Integrated circuits** consist of several such transistors made on one wafer (or chip), with the transistors connected to each other through layers of metal deposited on the chip's surface.

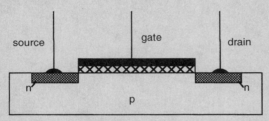

Figure 12.22. A field-effect transistor consists of an n-type source and n-type drain in a p-type wafer that is coated with an insulator and a metal. When a field is applied to the metal a gate controls the current between the source and the drain.

COMMENT

PELTIER COOLING DEVICE

The **Peltier**[6] **effect** (conversion of electrical energy into a temperature gradient) is a useful **thermoelectric property.** If current is passed through a bank of n- and p-type semiconductors, as shown in Figure 12.23, heat is absorbed at one side and released at the other. (These heat effects are in addition to the Joule heating, where the energy released is expressed as i^2R, where i is the current and R is the resistance.) The Peltier effect can be used to make cooling devices, using only electrical means and no moving parts.

(continued)

[6] Jean Charles Athanase Peltier (1785–1845) was a French clockmaker. When his mother-in-law died in 1815, the inheritance allowed him to retire from his trade and take up scientific pursuits. These included studies of phrenology, electricity, microscopy, and meteorology. A field trip to collect data concerning electrical charges in the atmosphere and cloud formation led to a cold from which he did not recover.

PELTIER COOLING DEVICE (continued)

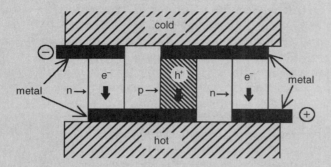

Figure 12.23. A schematic view of a Peltier cooling device. As current flows through the device, electrons in the n-semiconductor and holes in the p-semiconductor carry heat away from the upper surface, making it cooler than the lower surface. If the direction of the current is reversed, the flow of the electrons and holes is reversed, and so is the flow of heat. In a working Peltier cooler there would be a number of banks of n- and p-type semiconductors similar to that shown here.

12.6 DIELECTRICS

A **dielectric** is an insulating material that reduces the Coulombic force between two charges. When placed between the plates of an electrical capacitor, a dielectric increases its capacitance. If C and C_0 are the capacitances with and without the substance present, respectively, then ε, the **dielectric constant** (also called the **relative permittivity**) of the material, is defined as

$$\varepsilon = \frac{C}{C_0}.$$ (12.14)

Dielectric constants of some materials are listed in Table 12.2.

Dielectric molecules are polarized when they are placed in the electric field of the capacitor. That is, their center of negative charge is displaced from the center of their positive charge, and there is an induced dipole moment (i.e., nonzero net vector of charge distribution in space). The dipole moment per unit volume is called the **polarization,** and is given the symbol P. The dipoles align themselves in the electric field, and the aligned dipoles create an electric field that opposes the applied field, as shown schematically in Figure 12.24. Thus the presence of the dielectric has the effect of reducing the electric field, which gives a lower voltage, V, between the capacitor plates. Since

$$C \propto \frac{1}{V}$$ (12.15)

capacitance increases when the dielectric is present.

TABLE 12.2.
Dielectric Constants for Selected Materials
at $T = 300$ K (unless specified otherwise)

Material	ε^a
n-Hexane	1.89
CCl_4	2.23
Benzene	2.28
Diamond	5.7
$AgNO_3$	9.0
Ge	16
Ethanol	24.3
Methanol	32.6
BaO	34
Water	78.54
Ice Ih	99 (at 243 K)
Ice VI	193 (at 243 K)
$PbZrO_3$	200 (at 400 K)

^{a} Note that ε is dimensionless.

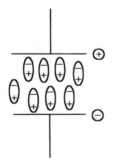

Figure 12.24. The dipoles of a dielectric material tend to align themselves between the plates of a capacitor. This opposes the applied electric field, so that the capacitance increases with the presence if the dielectric relative to the absence of a dielectric.

An "ordinary" dielectric material will be unpolarized in the absence of an external electric field and can be polarized by the application of an external electric field. The sign of the polarization can be reversed by reversing the field, as is shown in Figure 12.25a.

A **pyroelectric** crystal can have a nonzero electric polarization even in the absence of an applied electric field. The source of this behavior is the arrangement of the ions in the internal

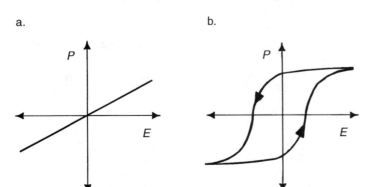

a.

b.

Figure 12.25. Polarization, P, as a function of electric field, E, for two types of materials: (a) a normal dielectric, showing zero polarization in zero field; (b) a ferroelectric material at $T < T_c$ showing spontaneous polarization even in the absence of an electric field, and also showing hysteresis.

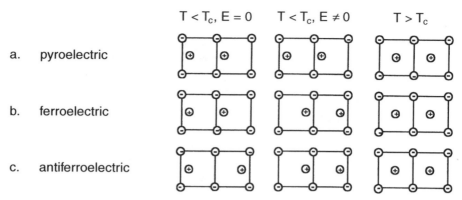

Figure 12.26. Schematic representations of structures of solids, all of which are paraelectric at temperatures in excess of T_c: (a) pyroelectric, showing a net spontaneous polarization even in the absence of an electric field; (b) ferroelectric, showing a net spontaneous polarization that can be reversed by application of an electric field; (c) antiferroelectric, showing microscopic ferroelectric regions that are exactly balanced by regions with the opposite polarity, so that there is no net polarization in the absence of a field, but polarization can be induced by the application of an electric field.

structure of the crystal, as shown schematically in Figure 12.26a. A material is no longer pyroelectric above a phase transition temperature (also called the **critical temperature,** T_c) because the available thermal energy brings about a more symmetric structure. This state is called **paraelectric,** indicating that it is not polarized but can be polarized under other circumstances, namely lowered temperature.

A crystal is **ferroelectric** if it is pyroelectric and the spontaneous polarization can be reversed by application of a small electric field. This is shown schematically in Figure 12.26b. Ferroelectrics can be distinguished from ordinary dielectrics by their very large dielectric constants (>1000) and by their retention of polarization even after the field is turned off. This leads to a **hysteresis loop** (i.e., different polarizations on increasing and decreasing the electric field) as shown in Figure 12.25b. The term "ferroelectric" is derived by analogy with the term "ferromagnetic"; in the former, electric dipoles are aligned and in the latter magnetic moments are aligned. Again, a ferroelectric material becomes paraelectric at high temperatures.

There are other crystal structures in which there is no net polarization of a cell in the absence of an electric field, but the cell is polarized when a field is applied. These structures are **antiferroelectric** as shown schematically in Figure 12.26c. In this case, cells (or layers) have their own nonzero polarization, but they cancel out over the whole crystal, so there is no net polarization. Again, at higher temperatures, $T > T_c$, an antiferroelectric crystal will have no net polarization, and so it will be paraelectric.

Some ferroelectric and antiferroelectric materials and their phase transition temperatures are presented in Table 12.3.

12.7 SUPERCONDUCTIVITY

The **superconducting** state, in which a material has zero direct current resistivity, was first discovered in 1911 by Gilles Holst, a student of Professor Kamerlingh Onnes at the prominent

TABLE 12.3.
Some Ferroelectric and Antiferroelectric Materials and
Their Transition Temperatures

Material	T_c/K
Ferroelectric	
K_2SeO_4	93
$SrTiO_3$	110
KH_2PO_4	123
Thiourea	169
KD_2PO_4	213
$BaTiO_3$	408
$NaNO_2$	436
$BaMF_4$ (M = Mn, Ni, Zn. Mg, Co)	FE at all temperatures
Antiferroelectric	
$NH_4H_2PO_4$	148
NaOD	153
KOH	227
$ND_4D_2PO_4$	242
KOD	253
$PbZrO_3$	506
WO_3	1010

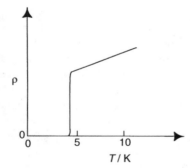

Figure 12.27. Schematic view of the transition to the superconducting state in mercury, as studied by immersion of mercury in
liquid helium (at its boiling point of 4.2 K).

low-temperature laboratory in Leiden. The first material found to be superconducting was mercury, and the first evidence came from measuring its resistivity at $T = 4.2$ K, the boiling point of helium.[7] The transition to the superconducting state, as seen in electrical resistivity, is shown schematically in Figure 12.27.

The property of superconductivity is particularly useful in that superconducting magnets, with their zero electrical resistivity, carry current indefinitely. For example, superconducting materials are used in nuclear magnetic resonance to achieve very high magnetic fields. Since they have no electrical losses (resistance along wires causes electrical losses), superconductors can have many useful applications. For this reason, much effort has been expended toward

[7] It is said that Kamerlingh Onnes strongly disputed his student's results and made the student repeat the measurement many times before accepting the fact that a material could have zero resistivity.

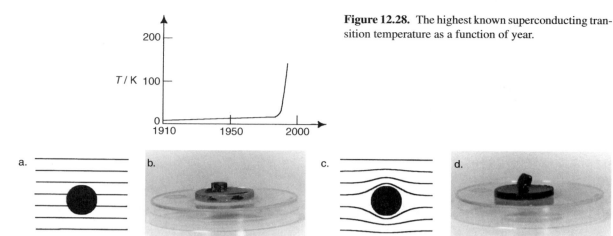

Figure 12.28. The highest known superconducting transition temperature as a function of year.

Figure 12.29. (a) A magnetic field can penetrate a material when it is not in its superconducting state. (b) In its normal state, a superconducting material (the lower black pellet is a 1–2–3 superconductor material in its normal state) is penetrated by the magnetic field of the magnet (the upper, smaller pellet), and the magnet rests on the 1–2–3 pellet. (c) In the superconducting state, the external magnetic field is repelled. This is an example of the Meissner effect. (d) The Meissner effect causes levitation of the magnet above a 1–2–3 superconductor at the boiling point of liquid nitrogen, $T = 77$ K, which is below the superconducting transition temperature for the pellet.

increasing the superconducting transition temperature, with the aim to produce materials that are superconductors at room temperature. Although the transition temperature has been rising over the years (the highest superconducting transition temperature in early 1998 was 135 K for $HgBa_2Ca_2Cu_3O_{8+\delta}$; see Figure 12.28 for a chronology), the progress of increasing the superconducting temperature is unpredictable. Advancement in theoretical understanding of the superconducting state may be necessary before a room-temperature superconductor is realized.

One of the unusual properties of a superconductor is its interaction with a magnetic field. In its normal (nonsuperconducting) state, the magnetic field can penetrate a material (see Figure 12.29a and b). Due to internal currents, the magnetic field is repelled by a material in its superconducting state (Figure 12.29c). This causes a material in its superconducting state to have a strong repulsive interaction with a magnetic field, giving **levitation** of a magnet above a superconductor (Figure 12.29d) and even the possibility of magnetic levitation trains (highly efficient due to low frictional losses) using superconductors. The expulsion of magnetic field by a superconductor is called the **Meissner effect.**

Since the late 1980s there has been considerable interest in producing superconductors with very high superconducting temperatures. The main goal is to achieve a useful superconducting material in a convenient temperature range. Prior to 1986, all known superconductors were metals or alloys with very low transition temperatures. Tremendous excitement in this field arose when, in 1986, Bednorz and Müller at IBM, Zürich found ternary perovskite-related structures with superconducting temperatures above 30 K.[8] This was followed in 1987 by the

---

[8] J. Georg Bednorz (1950–) and K. Alexander Müller (1927–) are researchers at IBM, Rüschlikan, Switzerland and co-recipients of the 1987 Nobel Prize in Physics for their important breakthrough in the discovery of superconductivity in ceramic materials.

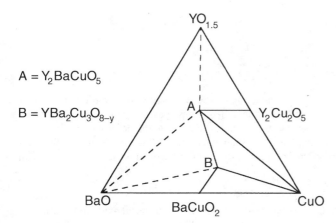

$A = Y_2BaCuO_5$

$B = YBa_2Cu_3O_{8-y}$

Figure 12.30. The $YO_{1.5}/BaO/CuO$ ternary phase diagram at 1000°C.

discovery, by Ching-Wu (Paul) Chu and his colleagues at the Universities of Houston and Alabama, that related compounds could be made to superconduct by cooling with liquid nitrogen (i.e., $T_c > 77$ K). The importance of this discovery was that liquid nitrogen (which boils at $T = 77$ K) is relatively easily and inexpensively prepared by condensing air. The other cryogenic (low-temperature) fluid that had been used to cool earlier superconductors ($T_c \ll 77$ K) is liquid helium (boiling point 4.2 K). The main disadvantages of liquid helium are its high cost and the fact that helium is a nonrenewable resource since helium gas easily escapes the earth's gravitational field.

One of the most important new high-temperature superconductors (also known as **high-T_c superconductors,** where T_c is the critical temperature below which the material is in the superconducting state) is $YBa_2Cu_3O_{8-y}$. This can be prepared from Y_2O_3, $BaCO_3$, and CuO in the mole ratios 1:4:6. The overall stoichiometry of Y:Ba:Cu in the final compound often leads to the abbreviation **1–2–3 superconductor.** The treatment required to produce $YBa_2Cu_3O_{8-y}$ is rather specific and complicated—it is a bit of a wonder that the material was discovered! The preparation method requires the following steps: a grind the initial reactants in stoichiometric amounts, heat to 960°C; hold for 10 hours; cool, grind, sieve, and press into pellets; heat to 960°C for 17 hours; cool; in flowing oxygen, heat to 980°C and hold for 8 hours, then cool slowly to 800°C, and then more quickly to room temperature. The Y_2O_3, BaO, CuO phase diagram, showing the 1–2–3 compound, is shown in Figure 12.30. The final result is a brittle black pellet (Figure 12.9) with a superconducting transition temperature of about 120 K.

One theory of superconductivity, commonly called **BCS theory** after its initiators Bardeen, Cooper, and Schrieffer,[9] attributes the lack of electrical resistivity in the superconducting state to interactions between pairs of electrons (**Cooper pairs**), particularly at low temperatures.

---

[9] John Bardeen (1908–) is an American physicist and the first person to win the Nobel Prize in Physics twice (in 1956 with William Shockley and W.H. Brattain for work leading to the invention of the transistor; in 1972 with L.N. Cooper and J.R. Schrieffer for superconductivity theory); Leon N. Cooper (1930–) is an American physicist who shared the 1972 Nobel Prize in Physics with Bardeen and Schrieffer for superconductivity theory work carried out while he was a research associate in Bardeen's laboratory; John Robert Schrieffer (1931–) is an American physicist who shared the 1972 Nobel Prize in Physics with Bardeen and Cooper for work on the theory of superconductivity carried out as a Ph.D. student under Bardeen's supervision.

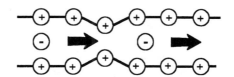

Figure 12.31. A schematic view of BCS superconducting theory. The electrons travel in pairs, one of each spin. As the first electron moves through the lattice, the cations (positively charged ions) are attracted to it, and this deforms the lattice somewhat. The electric field from the deformation attracts the second electron, which then moves with little effort. An analogue to the BCS motion is that the first person to push through a busy subway station makes a path along which a second person can saunter with little resistance.

Although electrons normally repel each other due to their like charge, at very low temperatures electrons can have a special affinity for one another. This is particularly so if they are paired with opposite spin, so that a pair (one of each spin) travels together in the lattice. In the ionic lattice of a superconductor, one electron in the pair interacts with the cationic (i.e., positively charged) portion of the lattice, by attracting it, causing a small deformation in the cationic sublattice (see Figure 12.31), corresponding to a lattice wave (phonon). This slight deviation from the equilibrium lattice leads to a potential gradient, which makes it very favorable for the second electron in the pair to move to where the first electron just was. In this simplified view, the motion of the second electron is so facilitated that electrical resistivity is zero. The fuller picture of the superconducting state is one of a single coherent state involving many pairs of electrons.

Although it explains many features of low-temperature superconductivity, BCS theory works only if the temperature is low enough that the available thermal energy is insufficient to break up the Cooper pairs. The 1–2–3 superconductors ($YBa_2Cu_3O_{8-y}$ and its analogues) exist in a superconducting state at temperatures that exceed the range of BCS theory—there would be too much thermal energy for the electrons to stay paired. Furthermore, there is experimental information concerning the high-temperature superconductors that does not fit BCS theory: isotopic substitution (e.g., replacement of ^{16}O with ^{17}O) in the lattice does not change the superconducting transition temperature as BCS theory predicts. The frequencies of the phonons would be affected by isotopic substitution, and in BCS theory this would change the electron–phonon interactions that give rise to the superconducting transition.

The present directions of research in superconductivity include semirational approaches to find new high-T_c materials. Another area of research involves increasing the current capacity and mechanical properties of 1–2–3 superconductors. Considerable effort is going into further experiments and theory to try to understand the mechanism(s) of superconductivity in these materials. In addition, new materials with interesting superconducting properties continue to be found; one recent example is the Fullerene derivative, K_3C_{60}.

Thermometry: A Tutorial

Temperature is one of the most fundamental properties required in physical measurement. Temperature can be measured many ways, from the rate of appearance of drops of sweat on a brow to the most sophisticated electronic instruments.

a. One of the most common methods of measurement of temperature in everyday application involves thermal expansion (e.g., of the fluid in a capillary tube). Can you suggest applications where this would not be a convenient method?

b. Suggest properties that could be desirable for a thermometer.

c. If the two ends of a piece of metal wire are at different temperatures, there is an electrical potential gradient (voltage difference, ΔV) along the wire. The relationship between ΔV and the temperature difference, ΔT, is given by

$$\Delta V = S(T)\Delta T \qquad (12.16)$$

where $S(T)$ is the thermoelectric power of the wire, which depends on the wire's composition. If two wires of different material are connected as shown in Figure 12.32, then the measurement of the voltage across them, ΔV_T

$$\Delta V_T = \Delta V_1 + \Delta V_2 \qquad (12.17)$$

(where ΔV_1 and ΔV_2 are given by Equation 12.16 for each material, 1 and 2) is a direct measure of ΔT, as defined in Figure 12.32. This device is a thermocouple.

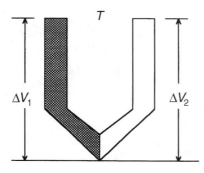

Figure 12.32. When dissimilar metals are joined together so that the junction is at a temperature different from the other ends, there is a voltage across them as shown. This voltage arises from the different work functions of the two materials.

  i. Could both arms of a thermocouple be made of the same metal? Why or why not?
 ii. A thermocouple measures temperature differences. How could this device be used to measure absolute temperatures?
iii. Would a better (more sensitive) thermocouple have a higher or lower value of thermoelectric power?
 iv. The thermoelectric power of a metal depends on both its temperature and its composition. A "typical" thermoelectric power is 1×10^{-5} V K^{-1}. If a sensitivity of 0.1 K is needed in a measurement, how accurately does the thermocouple voltage need to be determined?

d. Many other electrical principles can be used to measure temperature. We have seen that the resistance of a metal will vary with temperature. How could this be used to produce a resistance thermometer? What important properties should be considered? (*Hint:* One of the most useful resistance thermometers is the platinum resistance thermometer.) What determines the sensitivity of a resistance thermometer?

e. The resistance of intrinsic (pure) semiconductors changes with temperature. Could this be used to produce a semiconductor resistance thermometer? Do you foresee any problems at very low temperatures?

f. Semiconductors are notoriously difficult to purify, and many of their physical properties (including electrical resistivity) depend on their purity. One way to ensure their purity is to purposefully add impurities that overwhelm any small traces of unwanted impurities. Does this solve the low-temperature problem of using pure semiconductors? How?

g. Metal oxides that have been sintered and treated in oxidizing and/or reducing atmospheres to produce n-type semiconductors on the outside and p-type semiconductors on the inside (or vice versa) give useful thermometers called **thermistors.** A typical room-temperature resistance for a thermistor might be 10 kΩ.
   i. Would its resistance increase or decrease as the temperature is lowered? (Remember, they are semiconductors.)
   ii. Since thermistors have small particles, the electrical contact between the particles can be varied by the manufacturing process. How can this be an advantage for thermometry?
   iii. Thermistors can be made to be very small (less than 0.1 mm diameter). Why is this an advantage?

h. Figure 12.33 shows the temperature dependence of the forward-bias voltage of silicon and gallium arsenide diodes as functions of temperature.

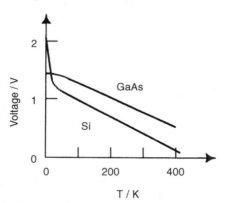

Figure 12.33. Forward-bias voltages of Si and GaAs diodes as functions of temperature.

Could these be used to measure temperature? What voltage accuracy would be needed for each thermometer to determine the temperature at about 200 K to within 1 K?

FURTHER READING

General References

M.Ya Axbel, I.M. Lifshitz, and M.I. Kganov (1973). Conduction Electrons in Metals. *Scientific American,* January 1973, 88.
P. Ball (1994). *Designing the Molecular World.* Princeton University Press.

R.J. Borg and G.J. Dienes (1992). *The Physical Chemistry of Solids.* Academic Press.

J.K. Burdett (1995). *Chemical Bonding in Solids.* Oxford University Press.

R.A. Butera and D.H. Waldeck (1997). The Dependence of Resistance on Temperature for Metals, Semiconductors, and Superconductors. *Journal of Chemical Education,* **74,** 1090.

W.D. Callister, Jr. (1994). *Materials Science and Engineering,* 3rd Ed. John Wiley & Sons.

B.S. Chandrasekhar (1998). *Why Things Are the Way They Are.* Cambridge University Press.

P. Chaudhari (1986). Electronic and Magnetic Materials. *Scientific American,* October 1986, 136.

A.K. Cheetham and P. Day, Eds. (1992). *Solid State Chemistry. Compounds.* Oxford University Press.

A. H. Cottrell (1967). The Nature of Metals. *Scientific American,* September 1967, 90.

P.A. Cox (1987). *The Electronic Structure and Chemistry of Solids.* Oxford University Press.

R.A. Dunlap (1988). *Experimental Physics.* Oxford University Press.

H. Ehrenreich (1967). The Electrical Properties of Materials. *Scientific American,* September 1967, 195.

A.B. Ellis, M.J. Geselbracht, B.J. Johnson, G.C. Lisensky, and W.R. Robinson (1993). *Teaching General Chemistry: A Materials Science Companion.* American Chemical Society.

A. Guinier and R. Jullien (1989). *The Solid State.* Oxford University Press.

T.S. Hutchison and D.C. Baird (1963). *The Physics of Engineering Solids.* John Wiley & Sons, Ltd.

C. Kittel (1996). *Introduction to Solid State Physics,* 7th Ed. John Wiley & Sons, Ltd.

P.G. Nelson (1997). Quantifying Electrical Character. *Journal of Chemical Education,* **74,** 1084.

C.N.R. Rao and J. Gopalakrishnan (1997). *New Directions in Solid State Chemistry.* Cambridge University Press.

H.M. Rosenberg (1988). *The Solid State,* 3rd ed. Oxford University Press.

J.F. Shackelford (1988). *Introduction to Materials Science for Engineers,* 2nd Ed. Macmillan.

L. Smart and E. Moore (1992). *Solid State Chemistry.* Chapman & Hall.

A.P. Sutton (1993). *Electronic Structure of Materials.* Oxford University Press.

L.H. Van Vlack (1989). *Elements of Materials Science and Engineering,* 6th Ed. Addison-Wesley,

A. Vijh (1989). Electrochemical Physics of Materials. *Transactions of the Royal Society of Canada,* V(IV), 283.

F. Vögtle, Ed. (1991). *Supramolecular Chemistry: An Introduction.* (Chapter 10 "Organic Semiconductors, Conductors and Superconductors" and Chapter 11 "Molecular Wires, Molecular Rectifiers and Molecular Transistors.") John Wiley & Sons, Ltd.

M.T. Weller (1994). *Inorganic Materials Chemistry.* Oxford University Press.

P.S. Zurer (1998). Plastics with an Electrical Bent. *Chemical & Engineering News,* April 13, 1998, 41.

Disordered Systems

B.L. Al'tshuler and P.A. Lee (1988). Disordered Electronic Systems. *Physics Today,* December 1988, 36.

R.A. Webb and S. Washburn (1988). Quantum Interference Fluctuations in Disordered Metals. *Physics Today,* December 1988, 46.

Electrochromic Materials

P.M.S. Monk, R.J. Mortimer, and D.R. Rosseinsky (1995). Through a Glass Darkly. *Chemistry in Britain,* May 1995, 380.

Electronic Devices

MRS Bulletin (1997). Special Issue: Polymeric and Organic Electronic Materials and Applications. June 1997.

MRS Bulletin (1997). Special Issue: GaN and Related Materials for Device Applications. February 1997.

MRS Bulletin (1996). Special Issue: Electroceramic Thin Films, Part II: Device Applications. July 1996.

Scientific American (1997). Special Issue: Solid-State Century.

R.T. Bate (1988). The Quantum-Effect Device: Tomorrow's Transistor? *Scientific American,* March 1988, 96.

R. Dagani (1994). New Transistor Class is Based on GaS, GaAs. *Chemical & Engineering News,* March 21, 1994, 8.

R. Dagani (1996). Two Photons Shine in 3-D Data Storage. *Chemical & Engineering News,* September 23, 1996.

R. Dagani (1997). Simpler Method for Organic Pixel Patterning. *Chemical & Engineering News,* May 5, 1997, 56.

R. Dagani (1998). The Dope on Quantum Dots. *Chemical & Engineering News,* January 5, 1998, 27.

M. Emmelius, G. Pawlowski, and H.W. Vollmann (1989). Materials for optical data storage. *Angewandte Chemie (International English Edition),* **28,** 1445.

N.G. Hingorani and K.E. Stahlkopf (1993). High-Power Electronics. *Scientific American,* November 1993, 78.

A.C. Jones (1995). The Chip Revolution Continues. *Chemistry in Britain,* May 1995, 389.

G.F. Neumark, R.M. Park, and J.M. DePuydt (1994). Blue-Green Diode Lasers. *Physics Today,* June 1994, 26.

R.E. Newnham (1983). Structure-Property Relations in Ceramic Capacitors. *Journal of Materials Education,* **5,** 941.

I.M. Ross (1997). The Foundations of the Silicon Age. *Physics Today,* December 1997, 34.

B.C. Sales (1998). Electron Crystals and Phonon Glasses: A New Path to Improved Thermoelectric Materials. *MRS Bulletin,* January 1998, 15.

R.J. Sullivan and R.E. Newnham (1993). Composite Thermistors. Chapter 16 of *Chemistry of Advanced Materials,* C.N.R. Rao, Ed. Blackwell Scientific Publications.

E.J. Winder, A.B. Ellis, and G. Lisensky (1996). Thermoelectric Devices: Solid-State Refrigerators and Electrical Generators in the Classroom. *Journal of Chemical Education,* **73,** 940.

G.M. Whitesides (1988). Materials for Advanced Electronic Devices. Chapter 9 in *Biotechnology and Materials Science,* M.L. Good, Ed. American Chemical Society.

Ice

B. Vonnegut (1965). Orientation of Ice Crystals in the Electric Field of a Thunderstorm. *Weather,* **20,** 310.

Metals

J. Emsley (1996). By Jove, Metallic Hydrogen! *Chemistry in Britain,* June 1996, 14.

W.A. Harrison (1969). Electrons in Metals. *Physics Today,* October 1969, 23.

A.R. Mackintosh (1963). The Fermi Surface of Metals. *Scientific American,* July 1963, 110.

Molecular Materials

Molecular Electronics. *Chemistry in Britain,* August 1991 (special issue with several articles on this topic).

D. Bloor (1995). Good Conduct. *Chemistry in Britain,* May 1995, 385.

S. Borman (1994). 'Wires' of Nanometer Dimensions Developed. *Chemical & Engineering News,* June 20, 1994.

D.O. Cowan and F.M. Wiygul (1986). The Organic Solid State. *Chemical & Engineering News,* July 21, 1986, 28.

R.B. Kaner and A.G. MacDiamid (1988). Plastics That Conduct Electricity. *Scientific American,* February 1988, 105.

P. Yam (1995). Plastics Get Wired. *Scientific American,* July 1995, 82.

Semiconductors

G.H. Döhler (1983). Solid-State Superlattices. *Scientific American,* November 1983, 144.

M.W. Geis and J.C. Angus (1992). Diamond Film Semiconductors. *Scientific American,* October 1992, 84.

G.C. Lisensky, R. Penn, M.L. Geselbracht, and A.B. Ellis (1992). Periodic Properties in a Family of Common Semiconductors. *Journal of Chemical Education,* **69,** 151.

S. O'Brien (1996). The Chemistry of the Semiconductor Industry. *Chemical Society Reviews,* **25,** 393.

A. Pisanty (1991). The Electronic Structure of Graphite. *Journal of Chemical Education,* **68,** 804.

E. Sandre, A. LeBlanc, and M. Danot (1991). Giant Molecules in Solid State Chemistry. *Journal of Chemical Education,* **68,** 809.

H.I. Smith and H.G. Craighead (1990). Nanofabrication. *Physics Today,* February 1990, 24.

Superconductors

Superconductors. *Chemistry in Britain,* September 1994 (special issue with several papers on this topic).

High-Temperature Superconductivity. *Physics Today,* June 1991 (special issue with several articles on this topic).

F.J. Adrain and D.O. Cowan (1992). The New Superconductors. *Chemical & Engineering News,* December 21, 1992, 24.

J. Bardeen (1990). Superconductivity and Other Macroscopic Quantum Phenomena. *Physics Today,* December 1990, 25.

R.J. Cava (1990). Superconductors beyond 1-2-3. *Scientific American,* August 1990, 42.

L.L. Chang and L. Esaki (1992). Superconductor Quantum Heterostructures. *Physics Today,* October 1992, 36.

P.C.W. Chu (1995). High-Temperature Superconductors. *Scientific American,* September 1995, 162.

C.D. Cogdell, D.G. Wayment, and D.J. Casadonte, Jr. (1995). A Convenient, One-Step Synthesis of $YBa_2Cu_3O_{7-x}$ Superconductors. *Journal of Chemical Education ,* **72,** 840.

D.L. Cox and M.B. Maple (1995). Electronic Pairing in Exotic Superconductors. *Physics Today,* February 1995, 32.

R. Dagani (1995). Superconductor Milestone: Flexible Tape Carries Record-High Current. *Chemical & Engineering News,* April 24, 1995, 6.

R. Dagani (1996). New Entry in the Superconducting Tape Derby. *Chemical & Engineering News,* November 11, 1996, 38.

J. de Nobel (1996). The Discovery of Superconductivity. *Physics Today,* September 1996, 40.

P.I. Djurovich and R.J. Watts (1993). A Simple and Reliable Chemical Preparation of $YBa_2Cu_3O_{7-x}$ Superconductors. *Journal of Chemical Education,* **70,** 497.

P.P. Edwards and C.N.R. Rao (1994). A New Era for Chemical Superconductors. *Chemistry in Britain,* September 1994, 722.

T.H. Geballe (1993). Superconductivity: From Physics to Technology. *Physics Today,* October 1993, 52.

R.M. Hazen (1988). Perovskites. *Scientific American,* June 1988, 74.

H. Hogan (1997). Superconductor Products Poised for Market. *The Industrial Physicist,* March 1997, 7.

J.R. Kirtley and C.C. Tsuei (1996). Probing High-Temperature Superconductivity. *Scientific American,* August 1996, 68.

G.B. Lubkin (1995). Applications of High-Temperature Superconductors Approach Marketplace. *Physics Today,* March 1995, 20.

G.B. Lubkin (1996). Power Applications of High-Temperature Superconductors. *Physics Today,* March 1996, 48.

B. Raveau (1992). Defects and Superconductivity in Layered Cuprates. *Physics Today,* October 1992, 53.

A.M. Wolsky, R.F. Giese, and E.J. Daniels (1989). The New Superconductors: Prospects for Applications. *Scientific American,* February 1989, 60.

PROBLEMS

1. Consider the following electrical resistivities, at 0°C: Cu, 16 Ω nm; Ni, 69 Ω nm; Zn, 53Ω nm; brass (Cu–Zn), ~60 Ω nm; and constantan (Cu–Ni), ~500 Ω nm.
   a. From these data, how does the resistivity of a solid solution (such as brass or constantan) compare with its pure components?
   b. Suggest a microscopic (atomic scale) explanation for the findings of (a).

2. On the basis of the information in Figure 12.34, arrange the materials noted in order of decreasing electrical conductivity. Comment on how this correlates with position in the periodic table and atomic size.

3. From Figure 12.35, which shows the intrinsic conductivity of pure germanium as a function of reciprocal temperature, estimate the band gap, E_g, for Ge.

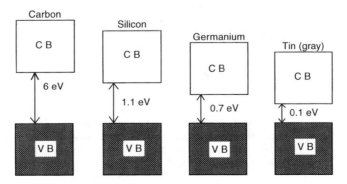

Figure 12.34. Various semiconductor materials, schematically showing their band gaps.

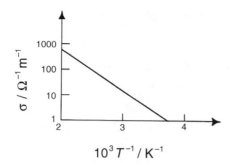

Figure 12.35. Intrinsic electrical conductivity of Ge as a function of reciprocal temperature.

4. The black-body radiation of a material can be used as a form of thermometer.
   a. Would this be more accurate at low or high temperatures? Explain.
   b. What would have to be measured to determine the temperature of a black-body radiator?

5. Semiconductors are known to have low packing fractions (i.e., low densities), which could mean that they could be easily compressed. We saw with tin that a change in density could be associated with a phase transformation that could make a semiconductor into a metal. In general, compression of a semiconductor will increase E_g, even if there is no phase change. Can you suggest an application for this principle?

6. An electronic security device involves a beam of light passing across the entryway to a home. When the beam is interrupted (e.g., as an intruder blocks it), a security bell sounds. What sort of material is likely used as the beam detector? Explain your reasoning.

7. Boron-doped diamonds are blue (see Chapter 3). Would you expect these diamonds to be electrically conducting? Explain.

8. A cholesteric (also called "twisted nematic") liquid crystal has a structure as shown in Figure 5.2.
   a. If the liquid crystal molecules are polar, explain how this material can be made ferroelectric.
   b. Can the degree of polarity of the material be controlled? If so, how?
   c. Suggest an application that makes use of controlled polarity of the material.

9. The ice crystals in storm clouds are ferroelectric.
   a. Will the local charge of the earth just prior to an electrical storm influence the orientation of the ice crystals in the clouds? If so, how?
   b. How might this affect the optical properties of the cloud?
   c. If clouds were made of nonpolar molecules would thunderstorms be more or less dramatic?

10. In Figure 12.15, the Fermi energy of an n-type semiconductor is shown as being higher than that of a p-type semiconductor. Explain why. (Assume that the n- and p-semiconductors have the same dominant material, to which either n- or p-impurities have been added.)

11. The intralayer electronic band structure of graphite is shown schematically in Figure 12.36; the bonding σ and π states are completely full at $T = 0$ K, and, since the band gap to the antibonding π^* molecular orbital is essentially zero, the slightest thermal energy can promote electrons. The electronic structure is quite anisotropic in graphite, because the interatomic distance between the layers is 3.35 Å, compared with 1.415 Å within the layers. Use this information to explain the following:

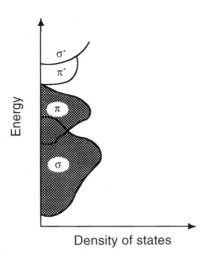

Figure 12.36. Band structure of graphite within the layers.

a. Graphite is a poor electrical conductor in the direction perpendicular to the layers and much better within the layers;

b. Graphite is an excellent thermal conductor in the direction of the layers and a poor thermal conductor in the direction perpendicular to the layers.

12. In most metals, the thermal conductivity is enhanced (over insulators) due to the transport of heat by free electrons. However, for this to happen, the electrons must be able to interact with lattice vibrations (phonons) that have the thermal energy. In most superconductors, the superconducting state is due to electrons that travel together as "Cooper pairs" and these electrons, because of their long wavelengths, cannot interact with phonons. On this basis, predict whether a material will have diminished or enhanced thermal conductivity in its superconducting state.

13. Hydrogen, which is an insulator under 1 atm of pressure, could become metallic at very high pressures. Explain why an insulator could become metallic at high pressures.

14. The difference between energy levels in a system can be expressed in many different units. The band gap of a typical semiconductor is of the order of 1 eV. Calculate the equivalent to 1 eV for the following units or parameters: a. J mol^{-1}; b. K (thermal energy); c. cm^{-1} (wavenumber); d. s^{-1} (frequency); e. nm (wavelength).

15. Ionic solids such as NaCl are normally considered to be insulators. However, such crystals always contain defects, e.g. where a Na$^+$ or Cl$^-$ ion at a particular lattice site is missing.

a. Propose a mechanism for electrical conductivity due to such defects.

b. The number of defects in an ionic crystal is normally very small. What does this indicate about the electrical resistivity of such a crystal?

c. The electrical conductivity of NaCl can be enhanced by doping with Ca^{2+} or Mg^{2+} ions. Explain why.

16. The field at which dielectric breakdown occurs is called the **dielectric strength.** Air is a dielectric with a relatively small dielectric strength. Use this information to explain how lightning occurs.

17. Calculate the thermal energy, kT, at room temperature in comparison with the band gap energy of a semiconductor with a band gap that corresponds to a wavelength of 1000 nm. Use this information to explain why infrared detectors using semiconductors are usually cooled to low temperatures (e.g., 77 K, using liquid nitrogen) to reduce the **dark current** (i.e., electrical signal in the absence of an IR source).

18. A **bolometer** is a cooled semiconductor used to detect far-infrared radiation (wavelengths as long as 5 mm). Use a density of states diagram to indicate why the bolometer's resistance changes in the presence of infrared radiation.

19. In Chapter 3 we saw that amorphous selenium (used in photocopiers) has a band gap of 1.8 eV. Another form of selenium, which is crystalline, has a band gap of 2.6 eV. Explain why the crystalline form has a higher band gap.

20. Phase diagram points with zero degrees of freedom are often used as standard points to calibrate electronic thermometers. Why?

21. Electrical conductivity in a metal can be changed by changing temperature, mechanical deformation, impurity concentration, or the size of crystallites. Discuss.

22. From Equation 12.6 and consideration of the temperature dependence of the mean free path of conduction electrons in a metal, explain the sign of the temperature dependence of the electrical conductivity of a metal.

C H A P T E R

13

MAGNETIC
PROPERTIES

13.1 INTRODUCTION

Magnetic materials have revolutionized our lives, from the recording and playing of our favorite music on magnetic tapes, to readable magnetic strips on our charge cards, to magnetic materials storing information in our computers. The great advantage of magnetic materials is that they can store so much information in such a small space.[1] In this chapter we explore the principles behind magnetic materials.

13.2 ORIGINS OF MAGNETIC BEHAVIOR

The magnetic properties of a material can be understood in microscopic terms when we take into account some of the special properties of moving electrons. It is for this reason that electrical and magnetic properties are related, and some of the terms are parallel in their use. It is the motion of a charge (i.e., electrons; the motion of the nuclei contributes only about 0.1% as much as the electrons to magnetic properties) that creates a magnetic field. Before we can proceed to consider this in detail, it is useful to define some terms.

A **magnetic field** is a region surrounding a magnetic body. This region can induce magnetic fields in other bodies. Magnetic materials, such as iron filings, can align themselves to represent these **field lines** (also called **flux lines**), as shown in Figure 13.1.

A magnet can be considered to be formed from two poles—a north pole and a south pole. We are familiar with this in a bar magnet. However, no matter how finely a bar magnet is divided, both poles are always present. Although one has never been observed, the unit pole con-

---

[1] Although magnetic materials can contain considerable information, it is interesting to compare this means of storage with other information storage. The information stored in an "average" brain corresponds to 10^{10} bytes, i.e., the information on 5000 2Mbyte computer disks, or 10^{10} pages of printed material.

Figure 13.1. Magnetic field patterns, as indicated by aligned iron filings on a plexiglass surface above a bar magnet.

cept is a useful theoretical idea to describe quantitative aspects of magnetism. A **unit pole** (or **magnetic monopole**) is defined as a body that will repel an equal unit pole placed 1 cm away in a vacuum, with a force of 10^{-5} N.

The **magnetic field strength**, $\bar{H}$, is a vector that measures the force acting on a unit pole placed at a fixed point in a magnetic field (in vacuum). $\bar{H}$ also measures the magnitude and direction of the magnetic field. The SI units of the magnitude of the magnetic field strength, H, are A m^{-1}.

The lines of force of a magnet describe the free path traced by an imaginary magnetic monopole in a magnetic field. For the space surrounding the source of the magnetic field, there is an **induction** whose magnitude, B, is the **flux density.** The induction and the magnetic field strength are related in vacuum by:

$$\bar{B} = \mu_0 \bar{H}$$

(13.1)

where μ_0 is the **permeability of vacuum.** If there is a material in the magnetic field the relation is similar,

$$\bar{B} = \mu \bar{H}$$

(13.2)

where μ is the **permeability** of that material.

Equation 13.2 is analogous to the electrical relation

$$J = \sigma \frac{V}{l}$$

(13.3)

where J is the electrical current density, σ is the electrical conductivity, and V/l is the voltage gradient. Comparison of Equations 13.2 and 13.3 shows that magnetic induction (B) is analogous to electric current density, magnetic field strength (H) is analogous to voltage gradient, and permeability (μ) is analogous to electrical conductivity.

Equation 13.2 can be rewritten to show the contribution of the material to the induction:

$$\bar{B} = \mu \bar{H} = \mu_0 (\bar{H} + \bar{M})$$

(13.4)

where $\bar{M}$ is the **magnetization** of the material and the term $\mu_0 \bar{M}$ represents the additional magnetic induction field associated with the material. The SI units for induction (B) are tesla[2]

[2] Nikola Tesla (1856–1943) was born in Croatia and made his major contributions as an electrical engineer in the United States. He is best known as the inventor of the ac induction motor, an idea that he sold to George Westinghouse for commercialization. Tesla was an eccentric reclusive inventor; he carried out many dangerous electrical experiments in his private laboratory, often putting out the power in his region.

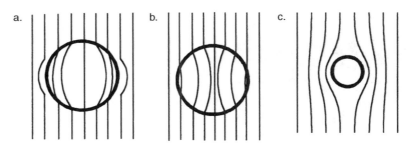

Figure 13.2. Magnetic flux density contours in (a) a diamagnetic material; (b) a paramagnetic material, and (c) a superconducting material.

(abbreviated as T) and for magnetization, M, are A m^{-1}. The **relative permeability**, μ_r, of a material is defined as

$$\mu_r = \frac{\mu}{\mu_0} \tag{13.5}$$

and the **magnetic susceptibility**, χ, is defined as

$$\chi = \frac{\bar{M}}{\bar{H}} \tag{13.6}$$

where both μ_r and χ are unitless.

The sign and magnitude of χ indicate the magnetic class of a material. If χ is negative and independent of H, it means that the material is expelled from a magnetic field, and this type of material is **diamagnetic.** If χ is small and positive (10^{-5} to 10^{-2}) and independent of H, it means that the material is slightly attracted to a magnetic field; this sort of material is **paramagnetic.** If χ is large and positive (10^{-2} to 10^6) and dependent on H, it means that the material is very attracted to a magnetic field; this sort of material is **ferromagnetic.** The effects of the material on the magnetic field lines for a diamagnet (flux density decreases), a paramagnet (flux density increases), and a superconductor (magnetic field repelled by the Meissner effect) are shown in Figure 13.2.

COMMENT

SUPERCONDUCTORS AS THERMAL SWITCHES

One of the properties of the superconducting state is that electrical resistivity is exactly zero. Another property is that its normal resistivity is restored in the presence of magnetic fields in excess of a **critical magnetic field,** B_c. The critical field depends on the critical temperature of the material, T_c, approximately as

$$B_c = B_0[1 - (T/T_c)^2] \tag{13.7}$$

where B_0 is the value of B_c as $T \rightarrow 0$ K.

(continued)

COMMENT

SUPERCONDUCTORS AS THERMAL SWITCHES (continued)

Although conduction electrons allow metals to conduct heat more efficiently than in insulators, the paired electrons (**Cooper pairs**) in a superconductor do not interact with thermal phonons, and therefore a superconductor does not conduct heat as well as a metal. This can allow a superconductor to be used as a thermal switch.

For example, a link of tin or lead can provide a connection between a low-temperature sample and its surroundings. The sample is in its metallic state allowing it to be an efficient thermal conductor when this is required (e.g., during cooling processes). A magnetic field in excess of B_c puts the link in its superconducting state to thermally isolate the system (e.g., to carry out thermal measurements). Such a thermal link avoids problems such as vibration and heating associated with mechanical heat switches.

The origins of diamagnetism, paramagnetism, and ferromagnetism are all related to the electrons in the materials, and understanding this relationship draws heavily on quantum mechanics. Briefly, there are two important ways in which electrons contribute to magnetism: **orbital magnetism** (due to the motion and resulting orbital angular momentum of the electrons in "orbits" about the nucleus) and **spin magnetism** (due to the spin-up and spin-down properties of an electron).

Diamagnetic behavior, which is characterized experimentally by the (usually weak) repulsion of the material from a magnetic field, arises due to the ("orbital") motion of the electrons being affected by the presence of a magnetic field. In a diamagnetic material, all the spins are paired (represented as $\uparrow\downarrow$, i.e., spin up and spin down) so the atoms have no resultant magnetic moment, and an applied magnetic field has little effect except induction associated with the motion of the electrons (current flows in a direction to produce a magnetic field that opposes the inducing field and repulsion results). All materials have this diamagnetic effect, but in materials with unpaired spins it can be overwhelmed by more powerful effects that attract the magnetic field, such as paramagnetism and ferromagnetism.

If there are unpaired electrons, as in many transition metal species and also some other materials such as O_2, the total electronic spin must be nonzero and paramagnetism results. We have already seen that spin can contribute to the magnetic properties of the system. In **ordinary paramagnets** (often referred to as **Pauli**[3] **paramagnets**) the magnetic properties are defined by the properties of the individual atoms, and spin is one of two important considerations. The other is orbital magnetism, which can arise for certain electronic states of some atoms and can give additional attraction of this system to a magnetic field, similar to the magnetic effect associated with an electric current flowing in a closed loop of wire.

In contrast with diamagnets and ordinary paramagnets, a ferromagnet relies on more than

[3] Wolfgang Pauli (1900–1958) was an Austrian theoretical physicist and winner of the 1945 Nobel Prize in Physics for the enunciation of what we now call the Pauli Exclusion Principle. Although a brilliant theoretician, his experimental contemporaries considered Pauli to be "jinxed" and he was considered to be bad luck in a laboratory. One story has it that an important experiment went disastrously wrong and this was blamed on Pauli passing through that city on the night train.

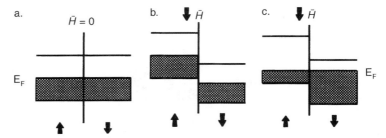

Figure 13.3. Electronic energy levels of a ferromagnetic material. (a) In the absence of a field there are equal numbers of electrons with both spins. (b) At the instant when a field is applied as shown, the aligned electrons have lower energies and the electrons with opposing alignment have increased energy. (c) In the magnetic field, some electrons realign their spins into lower energy levels, producing net magnetism.

the single-atom properties to derive its interaction with a magnetic field. The origin of this strong, attractive force is cooperative interaction between the magnetic moments of individual atoms arranged on a lattice. Iron is the archetypal example from which this class of materials derives its name. There are two competing interactions involving magnetic spins to keep in mind: **exchange interactions** tend to want to keep spins aligned ($\uparrow\uparrow$, due to electron–electron and electron–nuclear interactions) and **magnetic dipole interactions** tend to align spins antiparallel ($\uparrow\downarrow$, due to long-range interactions that favor this configuration).

In a ferromagnetic metal, the importance of many-atom effects in determining magnetism is directly related to the electronic band structure. A high density of states in the bands makes it possible to have reduced electron repulsion (i.e., reduced energy) by having electrons with parallel spins singly occupying energy levels near the Fermi energy. For this to be possible, this requires a high density of unoccupied states near the Fermi level. In the absence of a magnetic field, averaged over the whole sample, there would be approximately equal numbers of spins "up" and "down," as shown in Figure 13.3a. An external magnetic field lowers the energy of electrons aligned with the field, and raises the energy of those electrons that are aligned opposing the applied magnetic field (Figure 13.3b). To compensate for this, some electrons realign their spins into the lower energy levels, as shown in Figure 13.3c, leading to net magnetization. For metals with nearly full, narrow 3d bands (e.g., iron, nickel, cobalt), the density of states near the Fermi level is particularly high, and the cost of promoting electrons to higher levels is so small that it is very favorable energetically to have large numbers of unpaired electrons. As a result, these materials are ferromagnetic. There is a particularly favorable situation for ferromagnetism in the middle of the first row of the transition elements; the electronic band structures of some metals are shown in Figure 13.4.

In the absence of a magnetic field, a piece of iron (or other ferromagnet) might not be magnetized, depending on the sample's history. The reason is that the spins are aligned in a given direction only within small volumes within the crystal (about 10^{-14} m^3 each), called **domains.** Different domains have different magnetic orientations, such that the overall crystal is not magnetized. With only a few spins, the preferred orientation is aligned, but as more spins are added, magnetic dipolar interactions tend to favor antiparallel arrangements. The result is domains of different spin orientations, as shown in Figure 13.5.

Diamagnetic, paramagnetic, and ferromagnetic effects are summarized in Table 13.1.

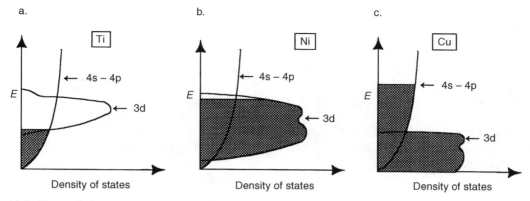

Figure 13.4. Electronic band structure diagrams for (a) titanium (diamagnetic as most valence electrons are in the diffuse 4s–4p band), (b) nickel (ferromagnetic since the Fermi level is in the region of high density of states, which makes exchange interactions energetically feasible), and (c) copper (diamagnetic since the Fermi level is in the diffuse 4s–4p band). The occupied levels are shaded.

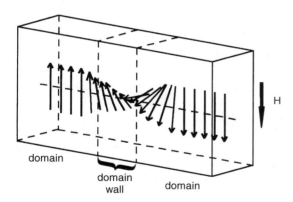

Figure 13.5. Domains in a ferromagnetic material. The region between domains is called the **domain wall.**

TABLE 13.1.
Diamagnetic, Paramagnetic, and Ferromagnetic Effects

Effect	Field	χ	Dominant origin	Magnitude
Diamagnetism	Weak repulsion	<0, independent of field strength	Induction associated with orbital motion; due to paired electrons; always present	Small
Pauli paramagnetism	Weak attraction	10^{-5} to 10^{-2}, independent of field strength	Unpaired spins of individual atoms + possibly orbital magnetism	Small
Ferromagnetism	Strong attraction	10^{-2} to 10^{6}, depends on field strength	Cooperative interactions; exchange ($\uparrow\uparrow$) and magnetic dipole ($\uparrow\downarrow$) compete	Large

MOLECULAR MAGNETS

Chemists have been attempting to design new magnetic materials with molecular building blocks to have control over their properties. The first-discovered molecular magnet, $[Fe(C_5(CH_3)_5)_2]^{\bullet+}[(NC)_2C{=}C(CN)_2]^{\bullet-}$ is ferromagnetic below 4.8 K. Studies of molecular magnets have shown that spin coupling is not sufficient to cause ferromagnetism; intermolecular interactions also are required. This can be investigated by making small changes to the molecular structure. Depending on the spin interaction, molecular magnets can be either ferromagnetic or antiferromagnetic.

13.3 MAGNETIC INDUCTION AS A FUNCTION OF FIELD STRENGTH

Equations 13.1 and 13.2 show the relations between induction, B, and magnetic field strength, *H,* for free space and a material, respectively. For a paramagnetic material, $\mu > \mu_0$ (typically $\mu = 1.01\mu_0$), and this leads to the induction–field strength relation shown in Figure 13.6. For a diamagnetic material, $\mu < \mu_0$ (typically $\mu = 0.99995\mu_0$); this induction–field strength relation is also shown in Figure 13.6.

Ferromagnetic materials show much more dramatic induction with increasing field strength, as shown in Figure 13.7 (see also Table 13.2). This is associated with their very large relative permeabilities that arise, as mentioned already, from cooperative effects among the magnetic moments within the material.

From Figure 13.7, the induction in a ferromagnetic has some further interesting properties that are not observed for diamagnetic and paramagnetic materials. When the magnetic field strength is increased, the induction reaches a maximum value, called the **saturation induction** (B_s). Furthermore, when the field is reduced to zero, there is still some induction, known as the **remanent induction,** B_r. (This is a well-known phenomenon; when a previously "unmagnetized" ferromagnetic material is brought up to a magnet, magnetism can be induced and re-

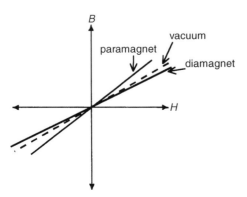

Figure 13.6. Induction (B) as a function of magnetic field strength (H) for free space (vacuum), a typical paramagnetic material, and a typical diamagnetic material. The differences between the slopes in vacuum (slope $= \mu_0$) and the paramagnetic material (slope $\approx 1.01\ \mu_0$) and diamagnetic material (slope $\approx 0.99995\ \mu_0$) are exaggerated here.

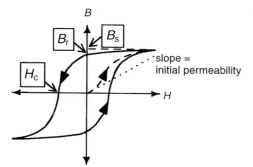

Figure 13.7. Induction (B)–magnetic field strength (H) relations for a ferromagnet. The broken line is the initial magnetization and the solid line is after the first magnetization. B_s is the saturation induction and B_r is the remanent induction (induction remaining when the field is removed). H_c is the coercive field required to reach zero induction. The hysteresis loop shows that the induction on increasing field is not the same as the induction on decreasing field.

main even after the first magnet has been removed.) If the field is reversed, at a certain value (known as the **coercive field,** H_c) the induction finally becomes zero. The induction–field strength relation on increasing field is not the same on increasing field as on decreasing field, and this gives rise to a **hysteresis loop** (Figure 13.7).

It is the cooperative effects of the magnetic moments on adjacent atoms that give such large magnetic effects in the case of a ferromagnet. This is shown schematically in Figure 13.8, where a trip along the hysteresis loop, starting at zero induction, first shows saturation to be a result of maximum spin alignment with the applied field. When the field is removed, there are still some residual aligned domains. As the field is reversed, more domains align in the reverse direction until finally they are fully aligned (saturation in the reverse direction).

The hysteresis loop of a ferromagnet can give rise to applications of these materials. The hysteresis loop can be very broad (these are called **hard magnets** because the domain walls are difficult to move) or narrow (these are called **soft magnets** because the domain walls are easier to move); this is illustrated in Figure 13.9.

One major use of metallic ferromagnets is in the core of power transformers. Soft magnets are used there because of the small width of their hysteresis loop and consequent higher **energy efficiency** (the area of the loop represents the energy loss in one pass through the loop). The ease of magnetization and demagnetization makes soft magnets particularly useful in

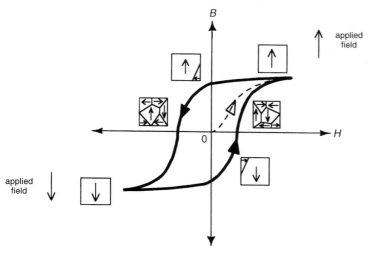

Figure 13.8. The induction–magnetic field hysteresis loop of a ferromagnetic, showing schematic domain structures at various points in relation to the applied magnetic field.

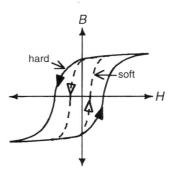

Figure 13.9. Comparison of typical hysteresis loops for a hard and a soft ferromagnet.

TABLE 13.2.
Properties of Selected Soft Magnets

Material	Initial relative permeability (μ_r at $B \approx 0$)	Hysteresis loss/ (J m^{-3} per cycle)	Saturation induction, B_s/(T)
Commercial Iron	250	500	2.16
Fe–4% Si, random	500	50–150	1.95
Fe–3%Si, oriented	15,000	35–140	2.0
45 Permalloy (Ni$_{0.45}$Fe$_{0.55}$)	2,700	120	1.6
Mumetal (Ni$_{0.75}$Cu$_{0.05}$Cr$_{0.02}$Fe$_{0.18}$)	30,000	20	0.8
Supermalloy (Ni$_{0.79}$Fe$_{0.15}$Mo$_{0.05}$)	100,000	2	0.79

alternating current applications. It is also important that a magnet used in this application have sufficiently high saturation induction to minimize the size of the transformer core. Properties of some soft magnetic materials are given in Table 13.2.

Hard magnets are useful as so-called **permanent magnets** because their large hysteresis leads to large residual magnetization values. Alloys such as samarium–cobalt, platinum–cobalt, and alnico (an alloy of aluminium, nickel, and cobalt) are useful hard magnets.

Experimentally, a soft magnet can be distinguished from a hard magnet by the value of the coercive field, H_c, necessary to return the induction to zero following magnetization, with $|H_c| > 1000$ A m^{-1} as the barrier beyond which a magnet is considered to be hard (permanent magnet). Another criterion is $|BH|_{max}$, as defined in Figure 13.10. This value is indicative of the energy required for demagnetization, and is a measure of the magnetic strength a material. Typical values of $|BH|_{max}$, B_r, and H_c are given in Table 13.3.

TABLE 13.3.
Properties of Selected Hard Magnets

Material	Remanence B_r/ (V s m^{-2})	Coercive field, H_c/ (kA m^{-1})	Maximum demagnetizing product, $|BH|_{max}$/ (kJ m^{-3})
Carbon steel	1.0	−4	1
Alnico V	1.2	−55	34
Ferroxdur (BaFe$_{12}$O$_{19}$)	0.4	−150	20
Rare earth-cobalt	1.0	−700	200
Nd$_2$Fe$_{14}$B	—	−1600	—

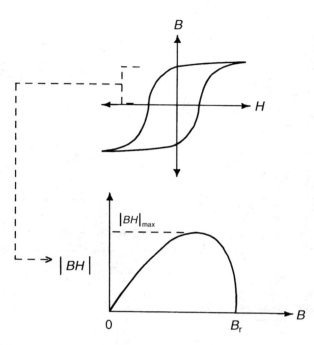

Figure 13.10. Plotting of the product |*BH*| from the demagnetization quadrant defines a maximum, designated |*BH*|$_{max}$, for a magnet, where |*BH*|$_{max}$ is a measure of the power of a permanent magnet.

13.4 TEMPERATURE DEPENDENCE OF MAGNETIZATION

For a normal paramagnetic material, where the paramagnetic species act independently, the susceptibility obeys the **Curie[4] Law** as the temperature, *T* (in kelvin), changes:

$$\chi = \frac{C}{T} \tag{13.8}$$

where *C* is the **Curie constant.** This is shown schematically in Figure 13.11a, and the temperature dependence reflects the increase in thermal randomization with greater thermal energy.

When there is a possibility of cooperative magnetic behavior, the temperature dependence is different. At very high temperatures there is too much thermal energy to allow cooperative magnetism; below a threshold temperature such cooperation is possible. For a ferromagnet, this temperature is the **Curie temperature,** T_C, and the Curie Law becomes

$$\chi = \frac{C}{T - T_C}, \qquad T > T_C. \tag{13.9}$$

---

[4] Pierre Curie (1859–1906) was a French physicist and co-discoverer (with his brother Jacques-Paul Curie) of piezoelectricity in 1877. His famous Ph.D. thesis concerned important work on magnetism; in 1903 he shared the Nobel Prize in Physics with his wife, Marie Sklodowska Curie, and Henri Becquerel for their work on radioactivity. His life ended prematurely when he was hit by a vehicle while crossing rue Dauphine in Paris.

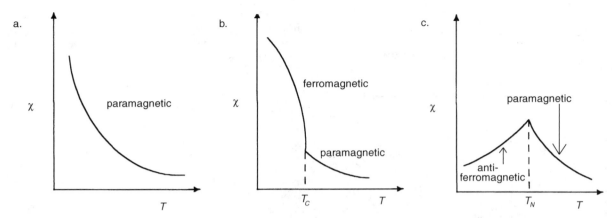

Figure 13.11. The temperature dependence of magnetic susceptibility for (a) a paramagnet, (b) a ferromagnet (showing the ferromagnetic–paramagnetic transition), and (c) an antiferromagnet (showing the antiferromagnetic–paramagnetic transition).

TABLE 13.4.
Curie Temperatures (T_C, Ferromagnet) and Néel
Temperatures (T_N, Antiferromagnet) for Selected Materials

Ferromagnets	T_C/K	Antiferromagnets	T_N/K
Co	1400	NiO	530
Fe	1043	CoO	292
Ni	631	FeO	198
CrO_2	392	MnO	122
Gd	292	NiF_2	83
EuO	69		

Below T_C, a ferromagnet has aligned magnetic moments that give rise to spontaneous magnetization even in the absence of a field. The moments (aside from domain effects) would be totally aligned at $T = 0$ K. The Curie temperatures of selected ferromagnets are given in Table 13.4.

For some materials there is a different temperature dependence to χ, expressed as

$$\chi = \frac{C}{T + T_N}, \qquad T > T_N \qquad (13.10)$$

where T_N is the **Néel**[5] **temperature.** These materials are **antiferromagnetic:** their magnetic spins are aligned in layers but alternate layers are of opposite magnetic orientation, such that there is no net alignment of spins in the system. An example of an antiferromagnetic structure is NiO; its magnetic structure is shown in Figure 13.12a. Antiferromagnetism arises in this case due to the **superexchange interaction,** driven by the covalent Ni–O–Ni bonding, as follows. The interaction of the d orbitals of Ni with the oxygen 2p orbital gives some covalent bonding, with partial transfer of an electron from the oxygen to the metal orbitals. As shown in Fig-

---

[5] Louis Eugène Néel (1904–) is a French physicist and winner of the 1970 Nobel Prize in Physics for work on antiferromagnetism and ferrimagnetism.

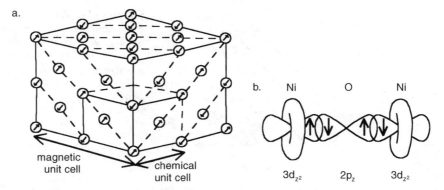

Figure 13.12. Antiferromagnetism arises from Ni atoms in NiO. (a) The antiferromagnetic structure of this material. Alternate layers of aligned Ni spins are oriented in opposing directions, such that there is no net magnetic moment. (b) Antiferromagnetism in NiO arises from the Ni–O–Ni bonding arrangement that dictates that alternate nickel atoms have opposing spins; this is an example of superexchange.

Figure 13.13. A schematic view of a ferrimagnetic structure, as for Fe_3O_4. Unbalanced magnetic moments arise from different numbers of unpaired electrons on Fe^{3+} and Fe^{2+}.

ure 13.12b, if one Ni has spin up, only a spin-down electron from oxygen can be transferred to this orbital (Pauli exclusion principle). Therefore bonding to an adjacent Ni must involve the spin-up electron of oxygen, which can be transferred to Ni only if its unpaired electron is in the spin-down state; this leads to an antiferromagnetic structure.

Another type of magnetism, **ferrimagnetism,** sometimes known as **unbalanced antiferromagnetism,** is illustrated in Figure 13.13; in ferrimagnetism there are two spin alignments but they are not balanced out. **Ferrites** are ferrimagnets with the general formula $Fe_2O_3 \cdot MO$ where M is a metal cation (e.g., Zn, Cd, Fe, Ni, Cu, Co, Mg). Magnetite, Fe_3O_4, is an example of a ferrite that has been known and used (e.g., as a compass) since ancient times. More recently ferrites have been used in recording tapes and transformer cores. Ferrites are among the most important materials in magnetic applications.

In practice, paramagnets, ferromagnets, and antiferromagnets can be distinguished by plots of χ^{-1} as a function of temperature, as shown in Figure 13.14. Equations 13.8, 13.9, and 13.10 can be generalized:

$$\chi = \frac{C}{T - \Theta} \tag{13.11}$$

where Θ, called the **Weiss**[6] **constant,** is zero for a paramagnet, positive for a ferromagnet, and negative for an antiferromagnet.

---

[6] Pierre Weiss (1865–1940) was a French physicist who developed a phenomenological theory of ferromagnetism.

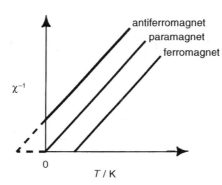

Figure 13.14. Variation of χ^{-1} as a function of temperature (extrapolated from high temperature) for a paramagnet, ferromagnet, and antiferromagnet. In each case the $\chi^{-1} = 0$ intercept value allows the calculation of the Weiss constant, Θ, from Equation 13.11.

MAGNETIC RESONANCE

Magnetic fields can affect the energy states of atoms and ions, and this can be used to probe the structure and dynamics of materials.

For a paramagnetic ion of spin $\frac{1}{2}$, with a doubly degenerate ground state in the absence of a magnetic field, the energy levels will split as a function of magnetic field B (see Figure 13.15a), with a gap of $g\beta B$, where g is the **Landé splitting factor** (related to the spin of the system in question) and β is the **Bohr magneton** (a constant related to the magnetic moment of an electron). If, at a given magnetic field B, the system is subject to radio frequency ν (energy $h\nu$) such that $h\nu = g\beta B$, the radio frequency will be absorbed, and g can be determined. The value of g (and its anisotropy) can indicate the magnetic environment of a given nucleus. In the slightly more complex case of a zero-field splitting of the ground state (Figure 13.15b), measurements at more than one field can be used to determine g.

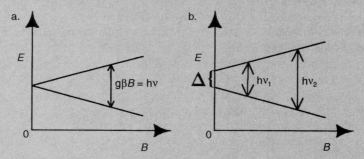

Figure 13.15. (a) A paramagnetic ion with a doubly degenerate ground state would have the ground state degeneracy lifted in a magnetic field. (b) The zero-field splitting of a ground state can be determined by energy absorptions at two different magnetic fields: $h\nu_1 = \Delta + g\beta B_1$ and $h\nu_2 = \Delta + g\beta B_2$.

Resonance experiments at microwave frequencies ($\approx 10^{10}$ s^{-1}) can be used to investigate electronic states of paramagnetic ions; this is **electron spin resonance** (esr), also called **electron paramagnetic resonance** (epr).

(continued)

COMMENT

MAGNETIC RESONANCE (continued)

Similar principles can be used to investigate states associated with the nucleus of an atom with nonzero spin; this is **nuclear magnetic resonance** (nmr). This is likely the most important structural technique in all of chemistry today, as it allows the atomic bonding arrangement (i.e., structure of molecules) to be determined. Nmr is a very sensitive technique; it is said that nmr imaging of brain processes during exposure to music can distinguish persons who have had musical training from those who have not.

Magnetic Devices: A Tutorial

a. Although much of the discussion in this chapter has concerned pure materials, one of the beauties of materials science is that it does not limit itself only to pure materials, and in purposefully added impurities, new properties can arise. Magnetic materials are no exception: steel (which can be considered to be impure iron) is a much harder magnet than very pure iron. Why? Consider domain growth in a magnetic field. Will domains generally affect hard or soft magnetic materials more?

b. Many of the first row transition metals can have very useful ferromagnetic properties; however, ferromagnetism is not observed in the second and third row transition metals. Given that the 4d and 5d bands are more diffuse than the 3d bands, can you reconcile this finding? Consider the cost of promotion of electrons (exchange interaction).

c. The first row transition elements can be combined with lanthanides to create some of the most powerful magnets known. Examples are $SmCo_5$ and $Nd_2Fe_{14}B$. Although they could potentially contribute directly to magnetism, the f electrons associated with the lanthanide atoms are too localized to exhibit band structure. In pure ferromagnetic lanthanide metals, the source of the ferromagnetism is delocalized d electrons; these d electrons interact with the localized f electrons to cause alignment of the d and f electrons (via exchange interactions) to reduce electron repulsion. Suggest qualitatively how such powerful magnetism arises in the transition metal–lanthanide alloys.

d. Chromium dioxide, CrO_2, is probably most familiar to you from audio cassette technology, where it is chosen for its magnetic properties. CrO_2 is a ferromagnet with a Curie temperature of 392 K. Its 3d orbitals form a very narrow band, allowing ferromagnetism. Later transition metal oxides (e.g., MnO_2) have localized 3d electrons and are insulators or semiconductors. The earlier transition metal oxide, VO_2, is a semiconductor at room temperature, but becomes metallic above 340 K; although it is paramagnetic for $T > 340$ K the spins are localized and not ordered cooperatively. This shows that CrO_2 has a special place among the first row transition metal oxides.

   i. The position of CrO_2 among the first row transition metal oxides is somewhat similar to that of Fe, Co, and Ni in the pure transition metals: it is at the balance point between wide bands of delocalized electrons (earlier elements) and localized electrons (later elements). Why does this balance point occur at atomic number 24 for the oxides and somewhat later (atomic numbers 26, 27, and 28) for the pure elements? Consider the metal–metal separation for the oxides compared to the pure metals, and how the metal–metal separation affects the overlap of the 3d orbitals and the width of the band.

ii. CrO_2 is used in audiotapes. A recording tape consists of a polyester tape impregnated with needle-like crystals of CrO_2. The sound to be recorded activates a diaphragm in a microphone and the vibrations cause fluctuating electric current in a coil of wire wrapped around an iron core in the recorder head, over which the tape passes. The varying current causes a varying magnetic field in the iron core, which in turn magnetizes the CrO_2 particles on the tape. The direction and magnetization of the CrO_2 particles are a record of the electrical impulses from the microphone. To play back a tape, the process is reversed. Sketch diagrams of the recording and playback mechanisms showing the magnet and the tape. Should the iron core be a hard ferromagnet or a soft ferromagnet? What about CrO_2? Are there any other considerations that are important for a good magnetic tape material? (You might consider the value of the Curie temperature for CrO_2 among other factors.)

e. For a computer memory device, it is most useful to have a magnetic material that switches cleanly between two different states (full induction in one direction or the other direction) as the magnetic field changes. Sketch the ideal B–H hysteresis loop for such a material. Examples of nearly ideal behavior are $(CoFe)O$ and Fe_2O_3.

FURTHER READING

General References

Physics Today (1995). April. Special Issue: Magnetoelectronics.

B.S. Chandrasekhar (1998). *Why Things Are the Way They Are.* Cambridge University Press.

A.K. Cheetham and P. Day, Eds. (1992). *Solid State Chemistry. Compounds.* Oxford University Press.

P.A. Cox (1987). *The Electronic Structure and Chemistry of Solids.* Oxford University Press.

J.H.W. de Wit, A. Demaid, and M. Onillon, Eds. (1992). *Case Studies in Manufacturing with Advanced Materials,* Vol. 1. North-Holland.

A. Guinier and R. Jullien (1989). *The Solid State.* Oxford University Press.

D. Jiles (1991). *Introduction to Magnetism and Magnetic Materials.* Chapman and Hall.

C.N.R. Rao and J. Gopalakrishnan (1997). *New Directions in Solid State Chemistry.* Cambridge University Press.

H.M. Rosenberg (1988). *The Solid State,* 3rd Ed. Oxford University Press.

J.F. Shackelford (1985). *Introduction to Materials Science for Engineers,* 2nd Ed. Macmillan.

L. Smart and E. Moore (1992). *Solid State Chemistry.* Chapman and Hall.

L. Van Vlack (1989). *Elements of Materials Science and Engineering, 6th Ed.* Addison-Wesley.

Disordered Systems

D.S. Fisher, G.M. Grinstein, and A. Khurana (1988). Theory of Random Magnets. *Physics Today,* December 1988, 56.

D.L. Stein (1989). Spin Glasses. *Scientific American,* July 1989, 52.

Magnetic Devices and Processes

M. Emmelius, G. Pawlowski, and H.W. Vollmann (1989). Materials for Optical Data Storage. *Angewandte Chemie (International Edition in English)* **28,** 1445.

M.H. Kryder (1987). Data-Storage Technologies for Advanced Computing. *Scientific American,* October 1987, 117.

S. Langenbach and A. Hiller (1991). Adiabatic Demagnetization of Antiferromagnetic Systems: A Computer Simulation of a Plane-Rotor Model. *Physical Review B,* **44,** 4431.

R.M. White (1980). Disk-Storage Technology. *Scientific American,* August 1980, 138.

Molecular Materials

D.O. Cwan and F.M. Wiygul (1986). The Organic Solid State. *Chemical & Engineering News,* July 21, 1986, 28.

J.S. Miller and A.J. Epstein (1994). Forthcoming Attractions. *Chemistry in Britain,* June 1994, 477.

J.S. Miller and A.J. Epstein (1995). Designer Magnets. *Chemical & Engineering News,* October 2, 1995, p. 30.

Other Magnetic Systems

R. Dagani (1992). New Material Is Optically Transparent, Magnetic at Room Temperature. *Chemical & Engineering News,* July 20, 1992, 20.

L.M. Falicov (1992). Metallic Magnetic Superlattices. *Physics Today,* October 1992, 46.

PROBLEMS

1. Although manganese is diamagnetic, some alloys containing manganese, such as Cu_2MnAl, are ferromagnetic. The Mn–Mn distance in these alloys is greater than in pure manganese metal. Suggest why this could lead to ferromagnetism.

2. The sizes and orientations of magnetic domains in a material can be altered by mechanical means. For example, domain orientations for two magnetic materials are shown in Figure 13.16 where the "textured" (oriented) structure is a result of cold rolling. Sketch the initial portion (starting at the origin) of the magnetization (B–H) plot for the random and for the textured materials.

3. Is there such a thing as a "non–magnetic" material? Explain.

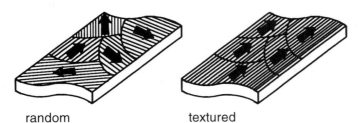

random   textured

Figure 13.16. The initial magnetic structure of this material is random, but cold rolling leads to a textured structure with magnetic alignment in the direction of rolling.

4. Suggest possible uses for the Meissner effect exhibited by superconducting magnetic materials.

5. Are magnetic effects in crystalline noncubic solids inherently isotropic or anisotropic? Explain.

6. From Equation 13.11 derive equations for the three lines in Figure 13.14.

7. So-called permanent magnets maintain their magnetization even in the absence of magnetic fields. This is achieved by using fine-grained alloys with "nonmagnetic" inclusions that inhibit domain wall movement. In addition, rod-shaped particles aligned along the direction of magnetization will hinder demagnetization. Which would make a better permanent magnet: Fe with a coercive field of 50 A m^{-1} and remanence of 1.2 T or SmCo$_5$ with a coercive field of 6×10^5 A m^{-1} and remanence of 0.9 T? Explain.

8. Arrange the antiferromagnetic transition metal oxides listed in Table 13.4 in order of increasing covalency on the basis of trends in their Néel temperatures.

9. Ferroelectric materials also can be described as "hard" or "soft" by analogy with ferromagnetic materials. Sketch the diagram (analogous to Figure 13.7) that distinguishes a hard ferroelectric material from a soft ferroelectric material.

MECHANICAL PROPERTIES
OF MATERIALS

[C]onsider a giant man sixty feet high—about the height of Giant Pope and Giant Pagan in the illustrated Pilgrim's Progress of my childhood. These monsters were not only ten times as high as Christian, but ten times as wide and ten times as thick, so that their total weight was a thousand times his, or about eighty to ninety tons. Unfortunately the cross-sections of their bones were only a hundred times those of Christian, so that every square inch of giant bone had to support ten times the weight borne by a square inch of human bone. As the human thigh-bone breaks under about ten times the human weight, Pope and Pagan would have broken their thighs every time they took a step. This is doubtless why they were sitting down in the picture I remember.

—J.B.S. Haldane
On Being the Right Size

14

MECHANICAL
PROPERTIES

14.1 INTRODUCTION

In many cases, mechanical properties are the most important factor in determining potential applications of a material. Stiffness, tensile strength, and elastic properties are important in material applications as seemingly diverse as the sound production from piano strings to the strength of dental porcelain to the protective use of a bulletproof vest. Some of the new high-temperature superconductors have very useful electrical and magnetic properties, but are limited in their applications due to their mechanical properties (they are brittle and it has been difficult to form them into wires). This chapter provides a basis for consideration of mechanical properties, based on a microscopic picture that has developed from experimental observation. We begin with relevant definitions.

Many of the mechanical properties of a material take into account how it behaves under certain forces. If a force, F, is applied to a material of cross section A, it develops a **stress, σ**,

$$\sigma = \frac{F}{A} \tag{14.1}$$

where the units of σ (e.g., N m^{-2} ≡ Pa) show that stress and pressure are similar concepts.

Under this force, there also will be a **strain, ε**, due to a change from the original length, l_0, by an amount Δl,

$$\varepsilon = \frac{\Delta l}{l_0} . \tag{14.2}$$

It is apparent from Equation 14.2 that strain is unitless.

At low stress and low strain, stress is proportional to strain, i.e.

$$\sigma \propto \varepsilon \tag{14.3}$$

so, at low σ and low ε,

$$\frac{\sigma}{\varepsilon} = E \tag{14.4}$$

where E, which is constant for a material at low ε, is called **Young's modulus** (also known as **elastic modulus**) for the material. Equation 14.4, which gives a linear relationship between force and extension, is a form of Hooke's law (see also Chapter 7). The proportionality constant, E, is a measure of the **stiffness** (i.e., resistance to strain) of a material, but it is quite distinct from the strength. A small value of E means that a small stress gives a large extension (as in rubber); a large value of E indicates that the material is very stiff. For example, E for diamond is 1.2×10^6 MPa whereas E for rubber is about 7 MPa. A related concept is **Poisson's ratio,** ν, defined as

$$\nu = -\frac{\Delta l_2 / l_2}{\Delta l_1 / l_1} \tag{14.5}$$

where l_1 and l_2 are the original dimensions, respectively, in the direction of and perpendicular to the extension force, and Δl_1 and Δl_2 are the respective changes in length.

Equation 14.4 describes the mechanical behavior of a material at low stress; a more complete picture is given in Figure 14.1. From this figure, the proportionality of stress to strain at low stress is shown, as well as the deviation from linearity at higher stress. The stress at which elongation is no longer reversible is called the **elastic limit.** Below the elastic limit, i.e., in the region of elastic behavior, strain is reversible: it disappears after the stress is removed. For metallic and ceramic materials, the elastic region shows a linear relationship between stress and strain, but for polymers such as rubber the relationship can be nonlinear.

The **yield strength** is defined as the value of the stress when the strain is 0.2% more than the elastic region would allow, i.e., the strain is 1.002 times the linearly extrapolated (elastic) strain.

Plastic deformation (plastic strain) is permanent strain and it occurs beyond the yield strength, as removal of the stress leaves the material deformed.

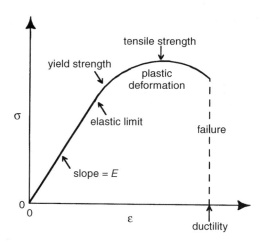

Figure 14.1. The stress (σ)–strain (ε) relationship for increasing strain for a ductile material. At low stress, the material is elastic and obeys Hooke's law with a proportionality of E, Young's modulus. The yield strength is defined as the stress at which the strain is 0.2% greater than predicted by Hooke's law. Plastic deformation, in which the sample remains deformed even after the stress is removed, occurs beyond the yield strength. The tensile strength is the maximum value of the stress. Ductility is defined as percent elongation at failure = $100\% \times \varepsilon_{\text{failure}}$.

TABLE 14.1.
Typical Values of Tensile Strength and Young's Modulus (*E*)
for Various Materials at Room Temperature

Material	Tensile strength/MPa	E/MPa
Diamond	$1.05×10^6$	$1.2×10^6$
Kevlar	4000	$1.8×10^5$
High-strength carbon fiber	4500	$2×10^5$
High-modulus carbon fiber	2500	$2×10^5$
High-tensile steel	2000	$2×10^5$
Superalloy	1300	$2×10^5$
Titanium	1200	$1.2×10^5$
Spider webs (drag line)	1000	$1×10^4$
Aluminium	570	$7×10^4$
Bone	200	$2×10^4$
Nylon	100	$3×10^3$
Rubber	100	$≈ 7$
Lexan	86	$2.4×10^3$
Wood	80	$3×10^4$

TABLE 14.2.
Poisson's Ratio for Various Materials
at Room Temperature

Material	ν
Fe	0.17
TiC	0.19
borosilicate glass	0.2
Al_2O_3	0.26
W	0.28
carbon steel	0.3
Al	0.33
Cu	0.36
Pb	0.4
Nylon 66	0.41

The **tensile strength** is the maximum stress experienced by a material during a test in which it is being pulled in tension, and **ductility** is the strain at failure (usually expressed as percent).

Toughness is a measure of the energy required to break a material. It is different from strength (a measure of the stress required to break or deform a material): tensile strength is the maximum stress value on a stress–strain curve, and toughness is related to the area under the stress–strain curve at failure. Toughness varies with the material: it is 0.75 MPa $m^{1/2}$ for silica glass, 4 MPa $m^{1/2}$ for diamond, and up to 100 MPa $m^{1/2}$ for steel.

Hardness is another mechanical property. It is the resistance of a material to penetration of its surface. For example, the **Brinell**[1] **hardness number (BHN)** is a hardness index based on the area of penetration of a very hard ball under standardized load.

Typical values of mechanical properties for some materials are given in Tables 14.1 and 14.2 and Figure 14.2.

---

[1] Johan August Brinell (1849–1925) was a Swedish metallurgist. His apparatus for testing hardness was first shown at the Paris Exhibition of 1900.

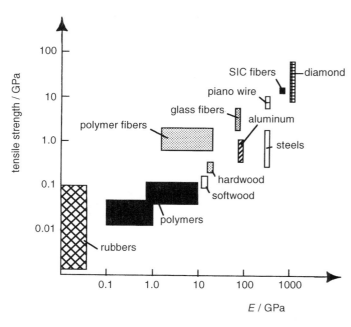

Figure 14.2. Tensile strength and Young's modulus ranges for a number of types of materials.

Figure 14.1 is a summary of experimental findings for a **ductile** material (i.e., one that undergoes plastic deformation). The low stress data (i.e., below the elastic limit), can be understood in terms of the interatomic potential energy that is responsible for holding the solid together. This potential, V, is shown in Figure 14.3 as a function of atomic separation, r. The corresponding force, F, written for the simplified one-dimensional case (see Equation 5.2), is given by

$$F = -\frac{dV}{dr} \tag{14.6}$$

and $F(r)$ is also shown in Figure 14.3. The corresponding stress–strain relationship, as shown in Figure 14.1, is similar to the force–separation relationship, $F(r)$, of Figure 14.3 [with stress comparable to force (but with opposite sign)] in the elastic region. Young's modulus, E, is related to the interatomic potential as

$$E = \frac{1}{r_0}\left(\frac{\partial^2 V}{\partial r^2}\right)_{r=r_0} \tag{14.7}$$

which shows that E depends on the curvature of the potential well at $r = r_0$. However, the pairwise intermolecular potential does not directly explain stress–strain curves beyond the elastic limit because after this point the shapes of these curves are determined by progressive destruction of the crystalline lattice, which cannot be understood in terms of two-atom interactions.

14.2 ELASTICITY AND RELATED PROPERTIES

Many materials with low values of Young's modulus are also very elastic, i.e., reversibly deformed. As for many other properties of materials, this property can be strongly temperature dependent.

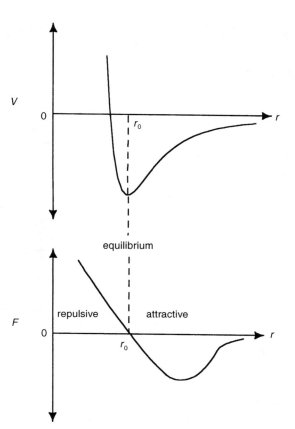

Figure 14.3. An intermolecular potential, $V(r)$, as a function of interatomic distance, r, leads to a force–separation relationship, $F(r)$, which can be used to deduce the elastic portion of the stress–strain relationship for a material (i.e., below its elastic limit). The intermolecular separation marked r_0 corresponds to the equilibrium position ($F = 0$).

An example of a highly elastic material is a polymer. Its elasticity stems from the coiling of long molecular chains; extension can take place by uncoiling without much change in interatomic distance, so it takes relatively little energy (see Figure 14.4). However, polymers with a high degree of structural organization (either a high degree of crystallinity or extensive cross-linking) are not very elastic, so it takes more energy to deform their structures, and the deformation is not reversible. It is found that polymer properties such as the glass transition temperature (T_g; whether the temperature is less than or greater than T_g reflects the rigidity of the polymer; see Chapter 6) can be related to mechanical properties such as Young's modulus, strength, and toughness. For example, **Lexan** (a very strong polycarbonate with such diverse uses as bubbles on space helmets and impact-resistant bumpers) has a very high glass transition temperature ($T_g = 149°C$, compared with 45°C for nylon and −45°C for polyethylene), which means that Lexan is in its rigid glassy state at room temperature. This correlates with Lexan's good strength and toughness characteristics at room temperature; polymers with lower T_g values would require lower temperatures for similar mechanical properties.

Figure 14.4. Uncoiling of an elastic polymer takes relatively little energy because the interatomic distances are not affected much.

coiled   stretched

COMMENT

STRETCHING A POLYMER

As a polymer is stretched, the polymer itself becomes more ordered due to the alignment of the previously disorganized chains. Thus, if the polymer is the thermodynamic system, entropy considerations indicate

$$\Delta S_{\text{system}} < 0. \tag{14.8}$$

However, from the second law of thermodynamics, any process that can happen must have

$$\Delta S_{\text{total}} > 0 \tag{14.9}$$

and since

$$\Delta S_{\text{total}} = \Delta S_{\text{system}} + \Delta S_{\text{surroundings}} \tag{14.10}$$

it follows that

$$\Delta S_{\text{surroundings}} > 0. \tag{14.11}$$

From the definition of reversible entropy change for the surroundings,

$$\Delta S_{\text{surroundings}} = \frac{q}{T} \tag{14.12}$$

and since $T > 0$, it follows that, on extension of a piece of elastic polymer,

$$q > 0. \tag{14.13}$$

In other words, extension of the elastic polymer is an ordering process that must therefore be exothermic, as can be verified by extending an elastic band while holding it next to your cheek or lip (very sensitive thermometers!). You will find that the polymer heats up when stretched.

Young's modulus, E, is related to the **velocity of sound** in a material, v, by

$$v = \left(\frac{E}{\rho}\right)^{1/2} \tag{14.14}$$

where ρ is the density of the material. The velocity of sound can be important in determining the desired properties for a material. For example, low-density (0.4–0.5 g cm^{-3}) pine with high Young's modulus (11–18 GPa) can be used to produce materials with high sound velocities (4800–6000 m s^{-1}) used for construction of violins.

Young's modulus is a measure of the elasticity of a material as it is extended or compressed. Considering compression further, by analogy with Equation 14.4, we can write

$$P = Kc \tag{14.15}$$

where P is the hydrostatic pressure, c is the compression of the material (change in dimension ÷ original dimension), and the proportionality constant in this version of Hooke's law, K, is the **isothermal bulk modulus,** defined as

$$K = -V\left(\frac{\partial P}{\partial V}\right)_T \tag{14.16}$$

i.e., K is the reciprocal of the isothermal compressibility (β_T) met in Chapter 6.

COMMENT

SEEING STRESS WITH POLARIZED LIGHT

When an isotropic material is under a nonuniform mechanical stress, it can become anisotropic in some regions. If the material is transparent, then this stress can be revealed by observations through polarization filters. This is shown in Figure 14.5.

Figure 14.5. A bar of plexiglass under stress, viewed through crossed polarizers, shows fringes.

The principle involved is the production of birefringence within the stressed sample (see also Chapter 5). When linearly polarized light hits the sample, optical anisotropy leads to the production of two rays of light within the sample. When these rays emerge from the sample and reach the analyzer (another piece of polarizing film), light of some wavelengths will pass through, and the hue of the observed light will depend on the stress. This is shown schematically in Figure 14.6.

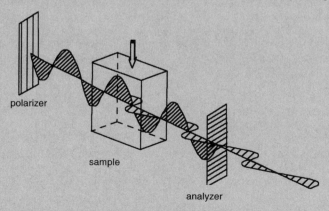

Figure 14.6. Polarized light passing through an anisotropic material (e.g., a material under stress) leads to two exiting polarizations that can be separated by an analyzer. Each color (wavelength) will behave differently and the observed colour will depend on the stress.

"HAPPY" AND "UNHAPPY" BALLS

The importance of structure to mechanical properties can be illustrated by bouncing "happy" (neoprene) and "unhappy" (polynorbornene) balls,[2] as shown in Figure 14.7. The structures of the polymers are shown in Figure 14.8.

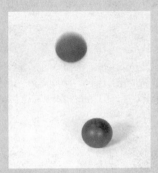

Figure 14.7. Although they look very similar, the neoprene ball (on the left) bounces while the polynorbornene ball (on the right) does not.

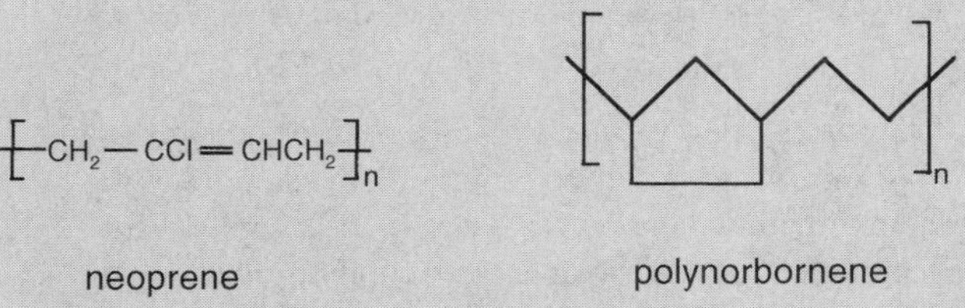

$$\left[CH_2 - CCl = CHCH_2 \right]_n$$

neoprene   polynorbornene

Figure 14.8. Molecular structures of the repeat units of neoprene and polynorbornene. Both materials are polymers.

The differences in their elastic behavior stem from their compositions and their chain architectures: neoprene is a three-dimensional network polymer with linear chains and a small portion of cross-links that prevent chain slippage; polynorbornene contains linked cyclic units and a relatively large quantity of nonvolatile liquid such as naphthenic oil. The rigidity of the ring makes polynorbornene less elastic than neoprene (which can coil and uncoil in response to different stresses). Furthermore, the liquid in polynorbornene can absorb the energy imparted the "unhappy ball" when it is dropped. Thus, two materials that look and feel very similar can have remarkably different mechanical properties.

---

[2] Available from Arbor Scientific, Ann Arbor, MI.

MOLECULAR ORIGINS OF THE STRENGTH OF SPIDER SILK

Spider silk has amazing mechanical properties, and only now are scientists beginning to understand this at a molecular level. For example, recent ^{2}H nmr studies of the dragline silk produced by deuterium-fed spiders showed that the crystalline fraction of the silk consists of two types of alanine-rich regions, one highly oriented and the other less densely packed and poorly oriented. The weakly oriented sheets may account for the high compressive strength of spider silk, and the coupling of the poorly oriented crystallites to the highly oriented crystalline domains and the amorphous regions could be the source of spider silk's toughness.

ARTIFICIAL BONES

Our bones are among the most important mechanical parts in our bodies. Although many broken bones will eventually heal themselves, there are applications, such as weakened load-carrying bones, where replacements are required. Hydroxyapatite, the main part of the rigid structural material in natural bones, has now been combined in the laboratory with organic polymers to produce strong artificial bones that promote natural bone growth by incorporating organic matter. The complex organoapatite has stiffness comparable with apatite; it also has more than twice the tensile strength.

HUMIDITY PLAYS A ROLE IN BASEBALL

Humidity can change the elastic properties of polymers, especially those in which hydrogen bonds play an important role. For example, due to its polyamide (wool) core, a baseball can lose as much as 12 g of water on heating at 100°C for 12 hours, and this leads to a harder ball, which is less deformable due to the loss of water, and this causes the collisions to be more elastic. This is why it is easier to hit a home run on a cool, dry evening than on a hot, humid night.

14.3 BEYOND THE ELASTIC LIMIT

In the region of reversible stress–strain relations, i.e., the elastic region, a strained material will return to its original dimensions once the stress is removed. Furthermore, below the elastic limit, the strain observed is independent of the rate at which the stress is applied.

At larger stress, a **brittle** material fractures (i.e., the reversible stress–strain region ends abruptly with **failure**) whereas a **ductile** material is permanently deformed (i.e., the deformation is no longer reversible). The distinction between brittle and ductile materials is illustrated in the stress–strain relations of Figure 14.9.

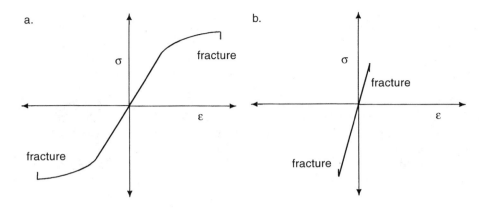

Figure 14.9. Stress–strain relations, including compression ($\varepsilon < 0$) and tension ($\varepsilon > 0$), for (a) a ductile material and (b) a brittle material. The brittle material shown is stronger in compression than in tension.

When a large stress (beyond the elastic limit) is applied to a ductile material, on removal and subsequent reapplication of the stress, the strain path does not retrace its original route and hysteresis results (Figure 14.10), i.e., there is residual strain within the material even when the stress is removed. The area within the hysteresis loop is proportional to the energy given to the system. For example, in the case of a fly landing in a spider web, the web extends due to the force of the fly landing and then, when the stress is over, the kinetic energy of the fly has been transferred to the silk of the web, and it is manifest as a slight increase in both the temperature and the extension of the silk. The high values of Young's modulus and tensile strength of the spider silk ensure that the web does not break when the fly contacts the web.

High stress can lead to inelastic deformation; in addition, at high stress the rate of application of the stress becomes important. For example, a bullet shot into ice will penetrate to about the same depth as one shot into water, since, at this high stress, the H_2O molecules do not have time to rearrange themselves in either ice or water. In a less extreme example, polymers beyond their elastic limit and subjected to stress can **creep,** i.e., they continue to deform with time.

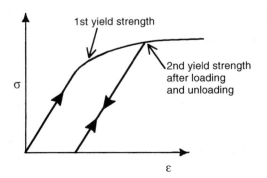

Figure 14.10. The application of stress beyond the elastic limit of a ductile material can lead to permanent deformation, as shown by the hysteresis on removal of a high stress. The increase in yield strength after loading and unloading is due to work hardening.

The temperature of a material can greatly affect the way in which it behaves at very high stress. For a polymer below its glass transition temperature, brittle fracture will occur, i.e., there will be fracture with little deformation. Amorphous materials tend to be brittle at relatively low temperature (often room temperature is low enough) because the energy required to deform the material by moving atoms or molecules past one another is much greater than the energy required to propagate cracks (see Section 14.5). For a polymer above its glass transition temperature, the fracture mode will be ductile (i.e., with considerable plastic deformation) and dependent on the strain rate, with slow strain rates allowing time for rearrangement of the polymer chains leading to greater deformation.

Far beyond the elastic limit, pure metals or **alloys** (mixtures of several metals) can often be manipulated without fracture. For example, they can be rolled, extruded into wires, etc. The high ductility of metals can be attributed to the atomic packing, i.e., the rather nondirectional arrangement of the atoms, which provides mobility of **dislocations** (packing irregularities; also see Section 14.4). For example, if a piece of metal is bent, the atoms rearrange their bonds with other atoms by sending clouds of dislocations flying around, but with little overall disruption to the packing. On the other hand, if a sheet of ice is bent, considering that the bonds in ice are more directional in their nature, much more reorganization of the bonding is required than for the piece of metal. Therefore, dislocations, which could allow bending, are less mobile in a solid such as ice because of the directional nature of its bonds. It might seem therefore that ice will simply continue to deform elastically, but in fact it fails in an entirely different way by propagation of a crack. This happens at rather low stresses. As a result, the nonductile (brittle) ice will dissipate the energy by failure at low stress, whereas metal will bend.

COMMENT

BULLET-PROOF VESTS

Kevlar, a form of poly(p-phenylene terephthalamide) (Figure 14.11) was invented by DuPont chemist Stephanie Kwolek; she was inducted into the National Inventor's Hall of Fame in 1995 for this important invention. Kevlar has an ultra-high Young's modulus (1.8×10^5 MPa), as well as high tensile strength and high thermal stability, and low extension at the break point. The strength of Kevlar is related to its structure: fibrils are interconnected by small tie bundles, in a pleated sheet with a 500 nm repeat distance. The polymer is highly crystalline, and it would take a very large amount of energy to disturb the conformation of the chains. Conversely, the chains can absorb significant amounts of energy (buckling from *trans* to *cis* conformations) without much damage: this allows Kevlar to stop bullets in bulletproof vests. Kevlar is an example of a material in which ease of microscopic deformation leads to macroscopic strength.

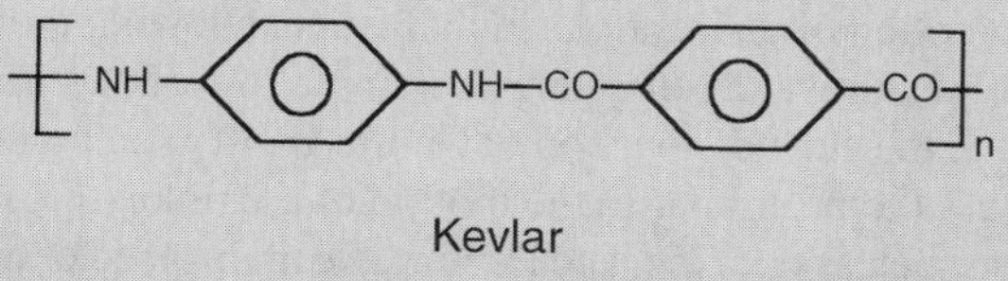

Kevlar

Figure 14.11. Molecular structure of Kevlar, showing the repeat unit of this important polymer.

COMMENT

CONTROLLED STRESS

Prince Rupert drops, named after a seventeenth-century prince who saw them at a Bavarian glassworks, are formed when drops of molten glass are rapidly cooled by dropping them into cold water. The exterior cools more quickly than the interior, so the outer surface of the drop is in compression. This stress, which makes the drops very strong, can be seen by the view of the glass through crossed polarizers (as in the stressed plexiglass in Figure 14.5). If the tail of the droplet is cut, the stress is released explosively and the glass shatters into many very small pieces. Since the glass breaks into so many pieces, these drops are much safer to handle than glass that breaks into sharp shards.

Glass that has been **tempered** (toughened by having its outer surface in compression), either by quick cooling or by chemical treatment, has many applications, including eyeglasses, windshields, and cookware.

14.4 DEFECTS AND DISLOCATIONS

No material is perfect, and it is often the type and location of imperfections that determine the mechanical properties of a material. Examples of **point defects** [atoms missing or in irregular places in the lattice; examples include **lattice vacancies, substitutional impurities, interstitial impurities,** and **proper** (or **self**) **intersitials**] are shown in Figure 14.12. Examples of **linear defects** (groups of atoms in irregular positions, including **screw dislocations** and **edge dislocations**) are shown in Figures 14.13 and 14.14. An indication of the concentration of imperfections is that if point defects and linear imperfections in 1 kg of Al were placed side by side, the defects would cover the equator more than a thousand times!

The **strength** of a material could be calculated from the force required to break individual chemical bonds (see Figure 14.3), per unit area. For copper this yields a theoretical strength of 2.5×10^9 Pa, but copper is known to deform and break at much lower stresses. In fact, many pure metals start deforming at less than 1% of their theoretical strength. The reason for the difference is the ease of motion of dislocations in the lattice, allowing the material to deform, for example, through the breaking and reforming of a single row of atomic bonds along the dislocation line, as shown in the edge dislocation of Figure 14.14. The motion of a dislocation in a crystal can be considered to be similar to the motion of a wrinkle in a rug when the wrinkle is pushed. In some cases, such as tin, zinc, and indium, the high speed of dislocation-assisted motion leads to a "cry" that can be heard when rods of these materials are bent. (See Section 14.5 for further discussion of the role of defects in strength of a material.)

Imperfections also can play a role in the hardness of a material. For example, a major prac-

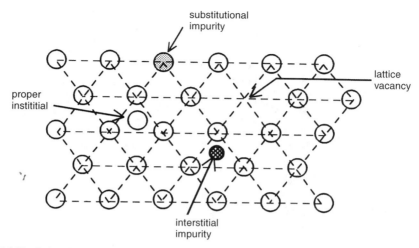

Figure 14.12. Point defects in an otherwise periodic monatomic crystal. A **lattice vacancy** is the absence of an atom at a lattice site; a **substitutional impurity** indicates the presence of an impurity atom at a lattice site; an **interstitial impurity** indicates the presence of an impurity atom at a location other than a lattice site; a bulk atom at a position other than at a lattice site is called a **proper interstitial.**

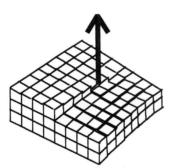

Figure 14.13. A **screw dislocation** in a crystal. This defect arises when atoms spiral around an axis, shown here with an arrow. Note that at the front of the crystal there is a step, but there is no step at the back edge.

tical aspect of metallurgy is development of harder materials, and one method involves addition of impurities that will reduce mobility of dislocations (see Figure 14.15). This can be enhanced if there are small aggregates of impurities (e.g., gold alloy with 10% copper is 10 times harder than pure gold). Thermal history can play an important role because impurities can be induced to clump together by heating, reducing the hardness of the material. Dislocations are immobilized at **grain boundaries** (the interfaces of homogeneous regions of the material), so decreasing the grain size can help harden a material. In a **work-hardened** material, a mechanical process has introduced more dislocations, reducing the overall dislocation mobility, making the material harder.

To summarize, a ductile material such as a metal bends well below its theoretical strength because of dislocations, e.g., those that allow planes to slip past one another easily. A metal can be strengthened by decreasing the number of dislocations (e.g., by heat treatment that anneals out dislocations), or pinning dislocations by work hardening (deformation causes dislocation tangles), precipitation hardening, or grain refining (by mechanical and/or heat treatment).

slip plane

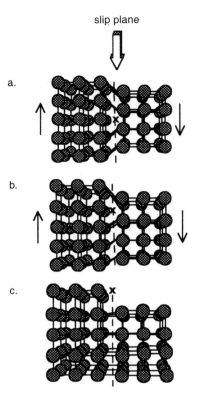

a.

b.

c.

Figure 14.14. The motion of an **edge dislocation** (a layer of extra atoms within the crystal structure), indicated by x, from (a) start to (b) intermediate time to (c) finish. In the process of moving, bonds are broken and re-formed along the **slip plane** (the plane along which the slippage takes place, shown here as a dashed line). Arrows indicate that one side of the crystal is moving with respect to the other. This process is the basic way in which ductile materials rearrange bonds between atoms.

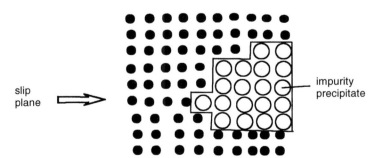

slip plane

impurity precipitate

Figure 14.15. A dislocation associated with a slip plane can proceed no further when it encounters a grain boundary at an impurity precipitate, just as a wrinkle in a rug cannot move further along the rug when it encounters a piece of heavy furniture. The presence of the precipitate hardens the material. **Annealing** the material, i.e. heating it to a high enough temperature to mobilize impurity precipitates allowing them to become larger but fewer in number, will allow dislocations to move more easily and therefore annealing decreases the hardness of the material.

However, if dislocations are too immobilized the metal no longer can bend, and it will snap in a brittle fashion under load.

COMMENT

HEAT TREATMENT OF ALUMINIUM

The binary Al–Cu phase diagram (Figure 14.16) illustrates the effect of thermal treatment on mechanical properties. From the phase diagram, cooling 96 mass% Al/4 mass% Cu (this is so-called **Duralumin,** or 2000 series aluminium alloy) will give, at room temperature, a phase of approximate composition $CuAl_2$ dispersed in an aluminium-rich phase. If this mixture is cooled slowly from the melt, precipitates of $CuAl_2$ are large and far apart; dislocations can rather easily avoid the precipitates in this material, and the material is soft. If cooled quickly, the dispersion of $CuAl_2$ in the Al-rich phase is much finer (smaller grains), and the high concentration of (small) precipitates ensures more interaction of dislocations, resulting in a harder material.

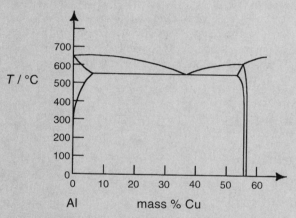

Figure 14.16. A portion of the Al–Cu phase diagram. An intermetallic compound of approximate composition $CuAl_2$ is formed at 55 mass% Cu.

14.5 CRACK PROPAGATION

One important factor concerning mechanical strength of a material is the response of its structure to stress. Many materials will crack when they are stressed, and in this section we examine this in terms of both structure and energetics.

The propagation of a crack in a material both releases energy (the strain energy) and costs energy [due to increased surface area (recall that surfaces are higher in energy than layers within the bulk materials) and energy to reorganize the material in the region of the crack]. The competition between the energy cost and the energy gain can lead, in certain circumstances, to crack propagation.

First, consider the energy that is released when a stressed material is cracked. (The analysis

Figure 14.17. Schematic view of a crack of depth d in a material. Stress, σ, is released primarily in the shaded areas, which have a total area of $\approx d^2$.

that follows applies to brittle materials.) In the schematic two-dimensional crack shown in Figure 14.17, a crack of depth d can release stress on both sides of the crack, such that a total area of about d^2 has its stress released. Therefore, the internal energy gained by the surroundings in this cracking process, ΔU_{stress}, is

$$\Delta U_{stress} = -ad^2 \tag{14.17}$$

where a is a proportionality constant ($\approx \sigma^2/E$) and the negative sign indicates that the energy of the system is lowered by ΔU_{stress}. Lowering energy is always a favorable process. Since a is proportional to σ^2, the more stress put on a body, the easier it is for a crack to occur.

However, there are also the energy costs of crack growth. One is kinetic energy of the mobilized atoms, but we ignore this here and concentrate on the energy costs of increasing the surface area, ΔU_{surf}. For the two-dimensional crack shown in Figure 14.17, this is

$$\Delta U_{surf} = 2Wd \tag{14.18}$$

where W is the **work of fracture** and the sign of ΔU_{surf} now shows that this costs energy. The work of fracture includes a contribution for the energy to create a new surface and a contribution to reorganize the material in the region of the crack. These two components mean that account must be taken not only of the surface energy of the material, but also how far into the material the damage extends. W can be expressed in terms of the **surface energy** of a material, G_s, where G_s is typically 1 J m^{-2} (see Chapter 10). For a ductile material (e.g., a metal; i.e., a material that can undergo permanent strain before it fractures), W is 10^4 to 10^6 times G_s, since the crack motion is accompanied by defects[3] well below the surface of the material.[4] (Note that the equations here are for a two-dimensional simplification of a three-dimensional effect.) On the other hand, W for a brittle material such as glass is only about $10\,G_s$, since the effects of stress do not reach very far below its surface. Such a low value of W means that the material could be susceptible to brittle failure. This can be very important in practice: evidence indicates that the Titanic's plating was brittle due to impurities in the steel, lowering W relative to better steel, and leading to catastrophic failure (brittle fracture) on hitting an iceberg.

---

[3] The defects are clouds of dislocations emerging from the crack tip.

[4] We are familiar with the fact that defects decrease strength in that metal can be broken easily by repeated bending. When a piece of metal is bent, the dislocation concentration at the bend site increases considerably. They tangle around one another and this eventually leads to the material becoming brittle, so W is decreased and the material breaks. A piece of metal that has been bent is quite stiff (have you ever tried to bend a piece of copper tubing that has already been bent? it is very difficult!) but it can be returned to its original flexible state by heating it to anneal out the defects.

At the balance point between the release of stress energy and the cost of surface energy, the crack has a particular length, l_G, called the **Griffith length,**[5] such that with stress σ and Young's modulus E,

$$l_G = \frac{2WE}{\pi\sigma^2}.$$

(14.19)

(Equation 14.19 holds for a three-dimensional crack; for an edge crack, $l_G = WE/\sigma^2$.) At crack lengths less than the Griffith length, the cost in surface energy and internal rearrangement energy exceeds the energy gain from release of stress. (However, if the stress on the material is increased, more stress energy is released, and the crack will grow further.) For cracks longer than the Griffith length, stress energy releases more energy than the energetic costs of increased surface area and internal rearrangement, and the crack grows spontaneously, provided that the stress remains.

Typical values of l_G can vary considerably from one material to another. For example, a 1-m crack in the hull of a ship can be "safe" (less than the Griffith length under normal sea-going stresses; it would cost more energy to grow further, so it does not propagate spontaneously). On the other hand, a microscopic crack in a glass material can grow spontaneously when stressed. Anyone who has used a file to start a crack in a piece of glass tubing will be familiar with this; the crack will propagate across the tube with relatively little stress. This also is seen in automobile windshields: a small crack can exist for some time, but when the stress is sufficient (e.g., from hitting a pothole or from having the increased air pressure in a sun-warmed car) the crack can grow rapidly and spontaneously, causing the windshield to shatter.

The state of the surface of a crack also is important in determining the rate at which it will grow. For example, it is common to see a glassblower wet the scratch on a glass rod before breaking it. The reason for this is that the water wets the surface in the crack, and this lowers the surface energy of the glass, thus requiring less stress to achieve crack propagation. Indeed, moisture content can be one of the important variables in crack propagation rates.

COMMENT

ADDED STRENGTH IN FIBERS

In 1920, Griffith suggested that the strengths of materials are limited by microscopic cracks on their surfaces. Griffith based his conclusion on studies of glass rods and fibers, and his finding of remarkably high strength when the diameter was less than about 10^{-6} m. He concluded that these fibers were relatively free of surface cracks and therefore much stronger than glass rods. We now understand that the increased strength arises from the decreased probability of finding a strength-limiting flaw in a fiber compared with a bulk material.

An additional factor adding to the strength of crystalline fibers is the reduction in the number of dislocations and/or their mobilities in fibers, compared with bulk materials, particularly since dislocations reduce the strength of the material.

(continued)

---

[5] Alan A. Griffith (1893–1963) was a British researcher who studied strengths of materials while employed at the Royal Aircraft Establishment, Farnborough, England.

COMMENT

ADDED STRENGTH IN FIBERS (continued)

Today fibers are commonly used for strength; for example, sporting equipment such as fishing rods, bicycles, and tennis rackets often contains carbon fibers for added strength. Graphite/metal composite fibers with values of Young's modulus over 600 GPa have been produced. Studies of carbon nanotubes (tubular analogues of Fullerene) have shown that their regular structure could lead to fibers with strengths far in excess of that of steel.

COMMENT

IMPURITIES IMPORTANT IN FAILURE

Impurities tend to be concentrated at grain boundaries, where they can be 10^5 times more abundant than in the bulk of the sample. This can make the material harder because dislocation motion is more difficult, but the material will be more susceptible to brittle failure because the grain boundaries cannot deform with the impurities there and cracks will tend to begin at the grain boundaries. For example, phosphorous segregates at grain boundaries in steel, and, to maintain mechanical properties, steel must have less than 0.05% phosphorous.

COMMENT

MECHANICAL PROPERTIES OF GLASS CERAMICS

Small crystallites in a glass have been found to inhibit crack propagation. This is utilized in the production of glass ceramics known as Corning Ware. Very fine particles of titanium oxide act as highly dispersed nuclei (10^{12} nuclei/mm^3 of molten glass) on which SiO_2 crystallites can form. In production, the object is first made by blowing or molding to achieve the appropriate shape. Then controlled heat treatment is used to **devitrify** (i.e., to crystallize) up to 90% of the material. Because the small grains scatter visible light, the material loses its transparency and becomes white and opaque following devitrification. It is thought that the small crystallites create many grain boundaries, and these immobilize dislocations (making the material harder) and also prevent crack propagation (making the material more resistant to failure). Furthermore, glass ceramics have low thermal expansion so that thermally induced stress is reduced. These two factors—high tensile strength and low thermal expansion—make it possible to place a cold Corning Ware dish in a hot oven without catastrophic failure. Glass ceramics were discovered in 1957 after the accidental overheating of a furnace.

SUPERHARD POLYMERS

Polymers are generally rather soft materials, but they can be hardened by cross-linking. Ordinarily this does not make them comparable to very hard materials such as steel, but researchers have found that irradiation with high-energy ions can harden a polymer from a tensile strength of about 0.1 GPa (unhardened) to 22 GPa (hardened), which is better than steel (2 GPa). The irradiation process breaks chemical bonds that can re-form with cross-linkage. The new bonds also can cause further delocalization of the π-bonds and therefore increase both the strength and the electrical conductivity of the polymers after irradiation.

14.6 ADHESION

Archaeologists have shown that adhesives were used thousands of years ago: pitch and natural resins held spear heads to shafts. Glue, one of the oldest adhesives known, is derived from animal skins and bones.

Adhesion of two materials results when the atoms or molecules are in such intimate contact that weak intermolecular forces (such as van der Waals interactions, which fall off as $1/r^6$ where r is the intermolecular distance) can collectively strengthen the contact. Even when two pieces of ceramic are broken and can be held back in place, the distances between all but a few contact points are too large to allow adhesion to take place on its own. An adhesive can flow into all the crevices and make intimate contact between both parts; when the adhesive hardens (due to evaporation of the solvent, or polymerization processes, for example), the interactions allow adhesion.

To adhere, an adhesive must first **wet** the substrate, i.e., it must spread across its surface. For this reason, the surface tension of the liquid adhesive is an important consideration. A familiar experience is that liquids bead up on Teflon® surfaces, and nothing will adhere to the surface.

A second criterion for a good adhesive is the strength of the interaction, both between adhesive and substrate (**adhesive strength**) and within the adhesive (**cohesive strength**). For example, increased cross-linkage in polymers enhances cohesive strength.

Flexibility can be an important factor in a successful adhesive; e.g., pure epoxy resins are brittle and subject to cohesive failure. However, the addition of a toughening agent, such as a liquid rubber, leads to microphase separation during the cure stage, with rubber-rich domains dispersed in the epoxy matrix. This leads to a tougher adhesive as fracture energy can be dissipated in the rubber particles. Of course, there must also be good adhesion between the rubber particles and the epoxy.

Although strong adhesives have such diverse applications as bookbindings from the last century to holding on the wings of modern airplanes, there are also applications in which only mild adhesion is required. The most notable example of this is the development of Post-it® notes. Although the **low-tack** (i.e., weak) adhesive was developed by 3M scientist Dr. Spence Silver, it was another 3M scientist, Art Fry, who, in seeking a way to temporarily mark his pages in a music book, found its first use. In this case, the adhesion is rather weak, and the bond can be easily broken.

COMMENT

UNUSUAL VISCOELASTIC BEHAVIORS

Some polymers behave elastically at low stress (or slow application of stress), yet behave like viscous liquids at high stress (or fast application of stress). This allows these materials to be liquid-like while under stress, yet solid-like when left alone. An example of this is "no-spill" paint, which is solid-like in the paint can (exhibiting elastic behavior at low stress), yet can be applied with a brush (viscous behavior under application of stress). The origin of this unusual behavior is the arrangement of the chains of the polymer (Figure 14.18): at low (or slow) stress, the polymer molecules cluster together and the material is "solid-like," while under shear stress the molecules align and the network can be broken, allowing for flow as a viscous "liquid."

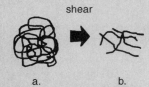

shear

a.   b.

Figure 14.18. The arrangement of polymer chains changes under shear stress. (a) Without stress, there is a rigid network. (b) Under shear stress, the polymer chains can be aligned and this allows easier flow, e.g., in no-spill paint.

Related principles explain why a mixture of corn starch and water thickens on rapid stirring (the molecules do not have time to align) but flows when left alone. Similarly, Silly Putty™ bounces (i.e., acts elastically at high stress) yet can be slowly extruded into threads (given sufficient time, molecules align). In these two cases, due to the molecular structures, the structure responds elastically to rapid shearing forces, but inelastically with slower application of stress.

COMMENT

THE IMPORTANCE OF OTHER PHASES AND MICROSTRUCTURE

When a material is composed of two or more other materials, some of its properties are additive and can be determined by suitably weighted values of the properties of the phases present. Color, density, heat capacity, and often thermal and electrical conductivity, can all be considered to be additive, at least to a first approximation. However, mechanical properties are highly interactive, i.e., different from the weighted average of the constituent phases, and this can be used to great advantage in designing materials with particular mechanical properties.

(*continued*)

THE IMPORTANCE OF OTHER PHASES AND MICROSTRUCTURE (continued)

For example, materials can be strengthened by the addition of fillers, such as carbon with rubber, sand with clay, sand with tar or asphalt, or wood flour to plastic. The additives increase the resistance of the material to deformation or flow. Similarly, the additions of ferrite and carbide to steel increase the hardness and the tensile strength.

The coarseness of the **microstructure** also plays an important role in mechanical properties. Very fine sand, when added to asphalt, produces a more viscous mixture than an equal amount of gravel. Similarly, steel with a very fine microstructure of ferrite and carbide will be much harder and stronger than steel with the same carbon content but a much coarser microstructure. The microstructure of a material can be controlled by heat treatment.

14.7 ELECTROMECHANICAL PROPERTIES: THE PIEZOELECTRIC EFFECT

One very interesting effect that relates mechanical and electrical properties is the **piezoelectric effect,** first discovered by Pierre and Jacques-Paul Curie in 1877. This effect is responsible for such seemingly diverse phenomena as Wint-O-Green lifesavers™ lighting up when crunched and LA Gear™ running shoes illuminating when the foot in them moves.

The requirement of the piezoelectric effect is an ionic crystal, such as quartz, that has a **noncentrosymmetric** structure (in quartz this is due to the tetrahedral arrangement of four oxygens around each silicon atom). The difference between an ionic centrosymmetric structure (positive ions symmetrically surrounded by negative ions; no net dipole or electric polarization of the crystal) and an ionic noncentrosymmetric crystal (positive ions not symmetrically surrounded by negative ions; nonzero net dipole and electric polarization of the crystal) is shown schematically in Figure 14.19. As already seen in Chapter 12, a polar noncentrosymmetric ionic crystal can be ferroelectric, meaning it has a nonzero electric dipole. (Recall that the term "ferroelectric" was coined to be analogous to the term "ferromagnetic," which refers to a collective net magnetic dipole moment in the structure, originally referring to iron-based permanent magnets.)

If stress is applied to a centrosymmetric ionic crystal, this stress does not change the (zero) electric polarization of the crystal (see Figure 14.19a). However, if the structure is noncentrosymmetric, the stress would change the electric polarization (Figure 14.19b). In a noncentrosymmetric (piezoelectric) ionic crystal, stress has the effect of changing the electric field across the crystal. If this happens to be part of an electrical circuit, this can induce electrical changes. For example, the gravitational force of a 0.5-kg mass applied across a 0.5-mm-thick quartz plate 10 cm × 2 m induces opposing charges of 2×10^{-9} C on the faces. The electric field from a piezoelectric device can amount to more than 10^4 V and the resulting sparks can be used for ignition in piezoelectric propane barbecue and cigarette lighters. The circuit for the LA Gear™ shoes that make use of the piezoelectric effect is shown schematically in Figure 14.20.

The converse also is true: if an electric field is applied across a piezoelectric crystal, the

a.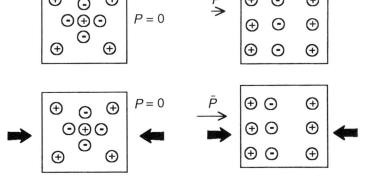

Figure 14.19. A schematic representation of (a) a centrosymmetric ionic crystal with and without an applied stress, showing no change in electric polarization, P, i.e., no piezoelectric effect and (b) a noncentrosymmetric ionic crystal with and without an applied stress, showing a change in electric polarization, P, on application of stress, i.e., the piezoelectric effect.

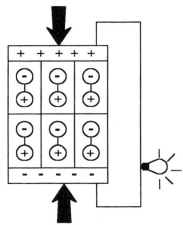

Figure 14.20. Schematic representation of a piezoelectric crystal in use in an LA Gear™ running shoe. Stress induces an electric field, which causes an electrical potential and a momentary burst of current and illumination of the light in the circuit.

crystal will have an induced stress. This can be used to control very fine manipulations. For example, very fine motions (fractions of an angstrom) can be controlled using a piezoelectric crystal in the circuit: a small change in the electric field across the crystal can stress the crystal, changing the crystal's dimensions. This can be coupled to a mechanical part (e.g., the tip of the scanning tunneling microscope) to control its motion. The head of an inkjet printer contains hundreds of piezoelectric inkwells, and each can be electrically induced to contract and squirt its ink.

If the applied electric field is alternating, the thickness of the crystal oscillates with the same frequency, which can be of considerable amplitude if it is the frequency of a normal lattice mode. For quartz, a 1.5-V silver oxide battery can be used to give 10^6 oscillations per second. The natural frequency in one particular direction in quartz is particularly independent of temperature such that frequency stabilities of 1 part in 3×10^8, corresponding to an accuracy of 1 s per 10 years in a quartz timepiece, can be achieved. Before about 1980, wristwatches (based on mechanical springs) required adjustment at least once a week. Now quartz-crystal watches require adjustment less often than once a year.

The conversion of mechanical energy in a piezoelectric material into electrical energy (and vice versa) is the basis for many **transducers.** One example is the piezoelectric crystals in our ears, which turn sound waves (mechanical motion) into electrical impulses that are carried along auditory nerves to our brain.

INTELLIGENT MATERIALS

Materials that respond to their environment are known as **intelligent** (or **smart**) **materials.** For example, buildings could respond to the strains induced by an earthquake, bandages could contract and apply pressure to a wound in response to bleeding, and other devices could change when under chemical stress or in magnetic fields. Using composite materials, sensors, piezoelectric devices, and shape-memory alloys, such devices, which rely on phase transformations for their various "states," are already reality.

SOUNDING OUT IMPROVED MATERIALS

Researchers at the National Institute of Science and Technology in Gaithersburg, MD have developed an inexpensive method to investigate mechanical properties of new composite materials or film coatings. A test sample is submerged in water and an acoustic wave transducer sends a pulsed sound wave through it. The speed of the reflected wave reveals the Young's modulus, and the direction of the reflected wave provides information concerning crystal planes or defects in the material. The instrument makes use of a curved transducer (for a lensing effect) made of an inexpensive piezoelectric plastic film.

MODELING FRACTURE

There is a very successful theory for fracture that describes when a crack will begin to move in response to complicated applied stresses. It was created by George Irwin, who, as a researcher at the Naval Research Laboratories in the 1940s, was asked to discover why a large fraction of the first U.S. missiles exploded during launch. The theory, called **fracture mechanics,** needs as input only the energy necessary to move a crack a small distance. However, calculating this energy requirement from a microscopic level is largely an unsolved problem.

There is one case in which the problem is starting to be understood: crack motion in a brittle crystal at low temperatures. Cracks can move in two ways under these circumstances. At moderate velocities, up to some fraction of the speed of sound, the crack **cleaves** the crystal (i.e., breaks it along a natural line of division), leaving only phonons in its wake. However, when there is too much energy in the crack tip, the motion of the crack becomes unstable and the crack begins to generate increasing amounts of subsurface damage. Figure 14.21 shows these two types of crack propagation.

(continued)

MODELING FRACTURE (continued)

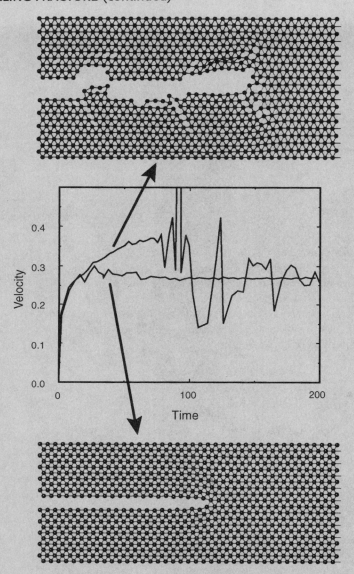

Figure 14.21. Two simulations of crack propagation in a brittle crystal at low temperatures. Smooth, steady motion (lower panel) leads to cleavage, whereas erratic, unstable motion (upper panel) leads to subsurface damage. Velocities shown are fractions of the speed of sound, and time is in arbitrary units. Diagram provided by M. Marder.

Results similar to those of this two-dimensional simulation have been observed in three-dimensional simulations and in experiments.

Memory Metals: A Tutorial

It is possible that a given material can exist in different polymorphs, with very different mechanical properties. One example is carbon, in its graphite and diamond phases.

a. Name another example of a material with different properties in different phases.

Another material that can change its mechanical properties with a change in phase is an alloy of nickel and titanium, called **Nitinol** (after *Ni*ckel *Ti*tanium *N*aval *O*rdnance *L*aboratory, where it was discovered in 1965). This material has two phases, a high-temperature **austenite** phase that is very difficult to bend and a low-temperature **martensite** phase that is very flexible. The high-temperature phase can be achieved by immersing a piece of Nitinol in boiling water. On cooling to room temperature, the austenite phase can remain, but it is metastable with respect to the martensite phase. This metastability leads to a hysteresis loop in the physical properties.

b. Sketch Young's modulus, *E,* for Nitinol as a function of temperature (heating and cooling paths), from —50 to 100°C.

The structure of Nitinol in the austenite phase is more symmetric than in the martensite phase, as shown in Figure 14.22.

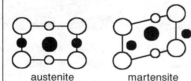

austenite   martensite

Figure 14.22. Nitinol in its austenite and martensite phases. (○) Ni in higher layer; (○) , Ni in lower layer; (●), Ti in higher layer;(●), Ti in lower layer.

There are several variations in the way that martensite cells can be oriented; two ways are shown schematically in Figure 14.23, in contrast with the single arrangement of the austenite form.

a.   b.

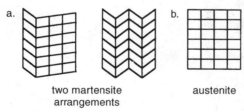

two martensite
arrangements   austenite

Figure 14.23. Packing arrangements in (a) martensite and (b) austenite structures of Nitinol.

c. Based on Figure 14.23, can you explain why the martensite phase is more flexible than the austenite phase?

d. Nitinol can be bent easily into a new shape when it is in the martensite phase. What do you expect will happen to the shape of a Nitinol wire when it is heated into the austenite phase?

e. Can you suggest applications for this interesting property of Nitinol (see Figure 14.24)?

Figure 14.24. A sculpture made of Nitinol, "Le Totem du Futur," by French sculptor Jean-Marc Philippe. The photos are of the sculpture at (a) 5°C; (b) 20°C; (c) 35°C. Photos by J.-M. Philippe.

f. It would be useful to be able to shape Nitinol wire in the high-temperature form, so that when it is heated it returns to this particular shape (i.e., not just a straight wire). This can be accomplished by annealing. Annealing takes place at high temperature, where there is sufficient thermal energy to allow easier mobility of the atoms, although the material does not melt. For example, intricate laboratory glassware is annealed to remove stress after the glass is blown and before it is used. Why?

g. Explain what happens to a Nitinol wire when it is heated and shaped to a "V" in a flame (austenite phase). Use a diagram to show a schematic view of the microscopic structure. Also show the structure in the martensite phase.

h. Although many applications of Nitinol can be imagined, these would be limited if the transition temperature were fixed at 70°C, as it is for the alloy $Ni_{0.46}Ti_{0.54}$. Can you suggest a way to modify the transition temperature?

i. Ni–Ti is not the only **memory metal** (martensitic) system; others include Cu–Zn–Al and Cu–Al–Ni. Ni–Ti has an advantage in that it is more corrosion resistant. However, Ni–Ti is more difficult to machine. Discuss how corrosion resistance and machining ability could be important in applications.

FURTHER READING

General References

Materials for Sports. Special Issue of *MRS Bulletin,* March 1998.

February 1996 issue of *Materials Research Society Bulletin* devoted to Interatomic Potentials for Atomistic Simulations.

M.F. Ashby and D.H.R. Jones (1987). *Engineering Materials 2: An Introduction to Microstructures, Processing and Design.* Pergamon Press.

D.R. Askeland (1994). *The Science and Engineering of Materials,* 3rd Ed. PWS.

P. Ball (1994). *Designing the Molecular World.* Princeton University Press.

P. Ball (1997). Made to Measure: New Materials for the 21st Century. Princeton University Press.

D.J. Barber and R. Loudon (1989). *An Introduction to the Properties of Condensed Matter.* Cambridge University Press.

J.D. Birchall and A. Kelly (1983). New Inorganic Materials. *Scientific American,* May 1983, 104.

W.D. Callister, Jr. (1994). *Materials Science and Engineering,* 4th Ed. John Wiley & Son, Ltd.

B.S. Chandrasekhar (1998). *Why Things Are the Way They Are.* Cambridge University Press.

R. Cotterill (1985). *The Cambridge Guide to the Material World.* Cambridge University Press.

J.H.W. de Wit, A. Demaid, and M. Onillon, Eds. (1992). *Case Studies in Manufacturing with Advanced Materials,* Vol. 1. North-Holland.

J.P. Den Hartog (1952). *Advanced Strength of Materials.* Dover.

A.B. Ellis, M.J. Geselbracht, B.J. Johnson, G.C. Lisensky, and W.R. Robinson (1993). *Teaching General Chemistry: A Materials Science Companion.* American Chemical Society.

J.E.Gordon (1978). *Structures or Why Things Don't Fall Down.* Penguin Books.

J.E. Gordon (1976). *The New Science of Strong Materials,* 2nd Ed. Princeton University Press.

A. Guinier and R. Jullien (1989). *The Solid State.* Oxford University Press.

R.A. Higgins (1994). Properities of Engineering Materials, 2nd Ed., Arnold.

B.H. Kear (1986). Advanced Materials. *Scientific American,* October 1986, 158.

J. Krim (1996). Friction at the Atomic Scale. *Scientific American,* October 1996, 74.

C.N.R. Rao and K.J. Rao (1992). Ferroics. Chapter 8 in *Solid State Chemistry: Compounds.* A.K. Cheetham and P. Day, Eds. Clarendon Press.

H.M. Rosenberg (1988). *The Solid State,* 3rd Ed. Oxford University Press.

G.L. Schneberger (1983). Metal, Plastic and Inorganic Bonding: Practice and Trends. *Journal of Materials Education,* **5,** 363.

J.F. Shackelford (1988). *Introduction to Materials Science for Engineers,* 2nd Ed. Macmillan.

W.F. Smith (1996). *Principles of Materials Science and Engineering,* 3rd Ed. McGraw-Hill.

M.A. Steinberg (1986). Materials for Aerospace. *Scientific American,* October 1986, 66.

P.A. Thrower (1992). *Materials in Today's World.* McGraw-Hill.

A.J. Walton (1983). *Three Phases of Matter,* 2nd Ed. Oxford University Press.

Adhesives

A.R.C. Baljon and M.O. Robbins (1996). Energy Dissipation During Rupture of Adhesive Bonds. *Science,* **271,** 482.

H.R. Brown (1996). Adhesion of Polymers. *Materials Research Society Bulletin,* **21**(1), 24.

N.A. de Bruyne (1962). The Action of Adhesives. *Scientific American,* April 1962, 114.

D. Eagland (1990). What Makes Stuff Stick? *CHEMTECH,* April 1990, 248.

G.F. Krueger (1983). Design Methodology for Adhesives Based on Safety and Durability. *Journal of Materials Education,* **5,** 411.

W.J. Moore (1967). *Seven Solid States.* W.A. Benjamin.

James T. Rice (1983). The Bonding Process. *Journal of Materials Education,* **5,** 95.

R.V. Subramanian (1983). The Adhesive System. *Journal of Materials Education,* **5,** 31.

Ceramics

H.K. Bowen (1986). Advanced Ceramics. *Scientific American,* October 1986, 168.

R.E. Newnham (1984). Structure-Property Relations in Electronic Ceramics. *Journal of Materials Education,* **6,** 807.

K.M. Prewo, J.J. Brennan, and G.K. Layden (1986). Fibre Reinforced Glasses and Glass-Ceramics for High Performance Applications. *American Ceramic Society Bulletin,* **65,** 305.

Composites

T.-W. Chou, R.L. McCullough, and R.B. Pipes (1986). Composites. *Scientific American,* October 1986, 192.

J. Economy (1988). High-Strength Composites. Chapter 11 in *Biotechnology and Materials Science,* M.L. Good, Ed. American Chemical Society.

Fracture

M. Bain and S. Solomon (1991). Cracking the Secret of Eggshells. *New Scientist,* March 30 1991, 27.

J.J. Gilman (1960). Fracture in Solids. *Scientific American,* February 1960, 95.

D.G. Holloway (1968). The Fracture of Glass. *Physics Education,* **3,** 317.

M. Marder and J. Fineberg (1996). How Things Break. *Physics Today,* September 1996, 24.

T.A. Michalske and B.C. Bunker (1987). The Fracturing of Glass. *Scientific American,* December 1987, 122.

J.W. Provan (1988). An Introduction to Fracture Mechanics. *Journal of Materials Education,* **10,** 325.

C.N. Reid (1988). An Introduction to Crack Resistance (R-Curves). *Journal of Materials Education,* **10,** 483.

E. Wilson (1997). New Ceramic Bends Instead of Breaking. *Chemical & Engineering News,* September 8, 1997, 12.

Hardness and Toughness

R. Baum (1993). Carbon Nitride Film Synthesized, Solid May Be Harder Than Diamond. *Chemical & Engineering News,* July 19, 1993, 34.

F.P. Bundy (1974). Superhard Materials. *Scientific American,* August 1974, 62.

R. Dagani (1995). Superhard-Surfaced Polymers Made by High-Energy Ion Irradiation. *Chemical & Engineering News,* January 9, 1995, 24.

H. Schubert and G. Petzow (1988). What Makes Ceramics Tougher? *Journal of Materials Education,* **10,** 601.

D.M. Teter (1998). Computational Alchemy: The Search for New Superhard Materials. *MRS Bulletin,* January 1998, 22.

Intelligent Materials, Including Memory Metals

R. Abeyaratne, C. Chu, and R.D. James (1996). Kinetics of Materials with Wiggly Energies: Theory and Application of Twinning Microstructures in a Cu-Al-Ni Shape Memory Alloy. *Philosophical Magazine A,* **73,** 457.

R. Dagani (1995). Polymeric "Smart" Materials Respond to Changes in Their Environment. *Chemical & Engineering News,* September 18 1995, 30.

R. Dagani (1997). Intelligent Gels. *Chemical & Engineering News,* June 9, 1997, 26.

K.R.C. Gisser, M.J. Geselbracht, A. Cappellari, L. Hunsberger, A.B. Ellis, J. Perepezko, and G.C. Lisensky (1994). Nickel-Titanium Memory Metal. A "Smart" Material Exhibiting a Solid-State Phase Change and Superelasticity. *Journal of Chemical Education,* **71,** 334.

T.C. Halsey and J.E. Martin (1993). Electrorheological Fluids. *Scientific American,* October 1993, 58.

A.J. Hudspeth and V.S. Markin (1994). The Ear's Gears: Mechanoelectrical Transduction by Hair Cells. *Physics Today,* February 1994, 22.

A.J. Hudpeth (1983). The Hair Cells of the Inner Ear. *Scientific American,* January 1983, 54.

T.W. Lewis and G.G. Wallace (1997). Communicative Polymers: The Basis for Development of Intelligent Materials. *Journal of Chemical Education,* **74,** 703.

R.E. Newnham (1997). Molecular Mechanisms in Smart Materials. *MRS Bulletin,* May 1997, 20.

J. Ouellette (1996). How Smart Are Smart Materials? *The Industrial Physicist,* December 1996, 10.

C.A. Rogers (1995). Intelligent Materials. *Scientific American,* September 1995, 154.

L. McDonald Schetky (1979). Shape-Memory Alloys. *Scientific American,* November 1979, 98.

D.W. Urry (1995). Elastic Biopolymer Machines. *Scientific American,* January 1995, 64.

T.A. Witten (1990). Structured Fluids. *Physics Today,* July 1990.

P. Zurer (1996). Self-Powered 'Smart' Window Could Control Own Transparency. *Chemical & Engineering News,* October 21, 1996, 10.

Metals

A.H. Cottrell (1967). The Nature of Metals. *Scientific American,* September 1967, 90.

R.D. Doherty (1984). Stability of the Grain Structure in Metals. *Journal of Materials Education,* **6,** 841.

Polymers

E. Baer (1986). Advanced Polymers. *Scientific American,* October 1986, 192.

F.L. Buchholz (1994). A Swell Idea. *Chemistry in Britain,* August 1994, 652.

G.B. Kauffman, S.W. Mason, and R.B. Seymour (1990). Happy and Unhappy Balls: Neoprene and Polynorbornene. *Journal of Chemical Education,* **67,** 199.

C. O'Driscoll (1996). Spinning a Stronger Yarn. *Chemistry in Britain,* December 1996, 27.

W. Peng and B. Reidl (1995). Thermosetting Resins. *Journal of Chemical Education,* **72,** 587.

R.B. Seymour and G.B. Kauffman (1990). Piezoelectric Polymers. *Journal of Chemical Education,* **67,** 763.

Spider Webs

L. Lin (1995). Studying Spider Webs: A New Approach to Structures. *Siviele Ingnieurswese,* May 1995, 14.

A.H. Simmons, C.A. Michal, and L.W. Jelinski (1996). Molecular Orientation and Two-Component Nature of the Crystalline Fraction of Spider Dragline Silk. *Science,* **271,** 84.

P. Vollrath (1992). Spider Webs and Silks. *Scientific American,* March 1992, 70.

Viscoelastic Behavior

R.J. Hunter (1993). *Introduction to Modern Colloid Science.* Oxford University Press.

T.W. Huseby (1983). Viscoelastic Behavior of Polymer Solids. *Journal of Materials Education,* **5,** 491.

■ ───

PROBLEMS

1. Many polymers change their mechanical properties as a function of temperature. With the same temperature (x) axis, sketch (a) the DSC trace of a polymer passing through its glass transition temperature, T_g, and (b) its Young's modulus.

2. Two model bridges have been built identical in every way except that one bridge is one-half scale in each dimension with respect to the other bridge. Which bridge would support a larger load? Explain.

3. Based on your answer to Question 2, which is better able to contain a gas at very high pressure: a small diameter tube or a large diameter tube?

4. Two materials, a and b, differ in their values of Young's modulus, with $E_a \gg E_b$. The two materials are polymers of the same molecular structure, and the only difference between them is that one material is crystalline and the other is amorphous.
   a. Which material is crystalline?
   b. What are their relative values of (i) Debye temperatures, (ii) velocity of sound, (iii) heat capacities at room temperature, and (iv) thermal conductivities at room temperature?

5. a. Explain how the mechanical properties of a material can influence its applicability as a toner for use in the photocopying process.
   b. What role does temperature play in consideration of the properties of toner?

6. The Leaning Tower of Pisa was being restored. During the work, the ground was frozen with liquid nitrogen ($T = 77$ K) to "prevent dangerous vibrations during the next phase of the salvage project." What is the effect of temperature on the mechanical properties (Young's modulus, sound velocity, Debye temperature) of the ground?

7. Most materials expand when heated. However, stressed rubber contracts when heated. Explain this phenomenon. *Hint:* Consider whether entropy increases or decreases on contraction of rubber.

8. A tub is to be made of metal, and in its use it will be under some (calculable) maximum stress. For this application it does not matter if there is a crack in the tub, but it is important that the material not fail, i.e., any cracks must not propagate. Explain how knowing the Griffith length for several potential tub materials will aid in the selection of an appropriate metal for this application.

9. The resin of fiber-reinforced composite materials (e.g., FibreGlass®) can be damaged by water. For example, the glass transition temperature, T_g, can be reduced by swelling of the resin caused by uptake of water. Explain how lowering T_g can adversely influence the mechanical properties of this material.

10. One of the most common construction materials is **concrete,** usually made by mixing Portland cement with sand, crushed rock, and water. The resulting material is very complex, resulting primarily from the reaction of Portland cement with water, as the adhesion results from hydrates. The setting process of concrete is exothermic. In some situations, such as outdoor use in winter, the heat source is useful. In other applications, such as large concrete structures, the exothermicity of the reaction can lead to stress. What is the source of the stress? Discuss factors that could be considered to lower the stress during concrete setting.

11. A recent advance in materials for turbine blades is the production of a "**superalloy**" that forms as single crystals, i.e., without grain boundaries. What properties will be affected by the absence of grain boundaries?

12. Glass can be strengthened chemically by ion exchange to give a surface of lower thermal expansion coefficient than the underlying (bulk) glass. How does this strengthen the glass?

13. Glass can be strengthened thermally by prestressing it so that all components are under compression at high temperatures. This is followed by cooling.
    a. Would this technique work best for low-thermal expansion glass or for high-thermal expansion glass?
    b. Explain how the shape and size of the glass item can play an important role in the success of this tempering method.

14. Window glass that has been **tempered** (i.e., heat treated for strengthening) will show patterns when viewed through a polarizer. This can be observed for side and rear car windows, for example. [Windshields are laminated (layered) structures and do not show this effect.] Explain the origin of the polarization patterns.

15. **Piezomagnetism** is the induction of a magnetic moment by stress. For example, some materials become magnetic after they are struck with another object. By analogy to the term piezoelectricity, explain the origins of piezomagnetism. Include a discussion of any necessary conditions.

16. **Piezoresistivity** is the variation in electrical resistivity of a material, produced by applied mechanical stress. It is said that this effect is observed to some extent in all crystals, but it is maximal in semiconductors. Why is it largest in semiconductors?

17. Explain how stress on a crystal can change its degree of birefringence. (This is a **piezooptic effect.**)

18. The addition of a second metal to a pure metal usually leads to an alloy with higher yield strength, tensile strength, and hardness. However, the addition of the impurity can change other properties. Would you recommend adding a second component to copper wires used for transmission of electrical power? (The alloy could have greater strength than pure copper.) Explain.

19. When a rubber band is immersed in liquid nitrogen, it loses its elastic properties and can be shattered easily with a hammer. Why?

20. a. Diamond is a very hard material, yet if a diamond is hit with a hammer, it will shatter. Explain why the diamond, not the hammer, breaks.
     b. If a diamond is squeezed between two metal surfaces, the diamond will become imbedded in the metal. Why do the diamond and metal not shatter as in (a)?
     c. In designing diamond drills, it is important to prevent fracture of the diamonds. Explain why polycrystalline diamond is often used for these applications. Consider the force per unit area on the diamond surfaces.

21. Copper can be drawn into wires that are strengthened by work hardening due to increased concentration and entanglement of dislocations.
     a. How would drawing a copper wire affect its electrical resistivity?
     b. How would the electrical resistivity of a drawn copper wire change with annealing?

22. Polyethylene bags are more difficult to stretch (i.e., plastically deform) along the polymer backbone direction that in perpendicular directions (between polymers). Explain.

23. Polymers are inherently weak materials. They can be strengthened by various measures, e.g., cross-linking (an example is the S–S bonds formed in vulcanization of rubber), chain stiffening (e.g., addition of large groups such as C_6H_5 that introduce steric hindrance), and/or chain crystallization (by suitable preparation and/or heat treatment). Discuss why each of these methods leads to stronger polymers.

24. Give the relative values of Young's modulus for a clean hair and a hair coated with (dried) hair gel.

25. Draw stress–strain diagrams for two materials, one very tough and one not so tough.

26. Explain how the temperature dependence of ductility could make a ship unsafe in cold waters yet seaworthy in warm waters.

1

FUNDAMENTAL PHYSICAL
CONSTANTS

Quantity	Symbol	Value with units
Speed of light in vacuum	c	299792458 m s^{-1}
Permeability of vacuum	μ_0	$4\pi \times 10^{-7}$ N A^{-2}
Permittivity of vacuum	ε_0	$8.854187817 \times 10^{-12}$ F m^{-1}
Newtonian constant of gravitation	G	$6.67259(85) \times 10^{-11}$ m^3 kg^{-1} s^{-2}
Planck's constant	h	$6.6260755(40) \times 10^{-34}$ J s
Avogadro constant	N_A	$6.0221367(36) \times 10^{23}$ mol^{-1}
Molar gas constant	R	$8.3141510(70)$ J K^{-1} mol^{-1}
Boltzmann constant, R/N_A	k	$1.38065812 \times 10^{-23}$ J K^{-1}

2

UNIT
CONVERSIONS

	Energy / J	$\tilde{\nu}$ / m^{-1}	ν / s^{-1}	T / K	E / eV
1 J =	1	$1/(hc)$ 5.03411×10^{24}	$1/h$ 1.509189×10^{33}	$1/k$ 7.242925×10^{22}	$1/e$ 6.241506×10^{18}
1 m^{-1} =	(hc) 1.9645×10^{-25}	1	c 299792458	$(hc)/k$ 0.01438769	$(hc)/e$ 1.2398242×10^{-6}
1 s^{-1} =	h 6.62608×10^{-34}	$1/c$ $3.35640952 \times 10^{-9}$	1	h/k 4.799216×10^{-11}	h/e $4.1356692 \times 10^{-15}$
1 K =	k 1.380658×10^{-23}	$k/(hc)$ 69.50388	k/h 2.083674×10^{10}	1	k/e 8.617386×10^{-5}
1 eV^a =	e 1.602177×10^{-19}	$e/(hc)$ 806554	e/h 2.417988×10^{14}	e/k 11604.45	1

^{a} 1 eV corresponds to 806544 m^{-1}.

A P P E N D I X **3**

THE 14 THREE-DIMENSIONAL
LATTICE TYPES

CUBIC
$a = b = c$
$\alpha = \beta = \gamma = 90°$

simple

body-centered

face-centered

TETRAGONAL
$a = b \neq c$
$\alpha = \beta = \gamma = 90°$

simple

body-centered

ORTHORHOMBIC
$a \neq b \neq c$
$\alpha = \beta = \gamma = 90°$

simple

body-centered

end-centered

face-centered

RHOMBOHEDRAL
$a = b = c$
$\alpha = \beta = \gamma \neq 90° < 120°$

HEXAGONAL
$a = b \neq c$
$\alpha = \beta = 90°$
$\gamma = 120°$

MONOCLINIC
$a \neq b \neq c$
$\alpha = \gamma = 90° \neq \beta$

simple

end-centered

TRICLINIC
$a \neq b \neq c$
$\alpha \neq \beta \neq \gamma$

4

THE GREEK
ALPHABET

Name	Lower case	Upper case
alpha	α	A
beta	β	B
gamma	γ	Γ
delta	δ	Δ
epsilon	ε	E
zeta	ζ	Z
eta	η	H
theta	θ	Θ
iota	ι	I
kappa	κ	K
lambda	λ	Λ
mu	μ	M
nu	ν	N
xi	ξ	Ξ
omicron	ο	O
pi	π	Π
rho	ρ	P
sigma	σ	Σ
tau	τ	T
upsilon	υ	Y
phi	φ	Φ
chi	χ	X
psi	ψ	Ψ
omega	ω	Ω

SOURCES OF LECTURE
DEMONSTRATION MATERIALS

Aldrich Chemical
P.O. Box 2060
Milwaukee, WI 53201
USA
Tel: (800) 558-9160

Arbor Scientific
P.O. Box 2750
Ann Arbor, MI 48106-2750
USA
Tel: (800) 367-6695

Edmund Scientific
101 E. Gloucester Pike
Barrington, NJ 08007-1380
USA
Tel: (609) 573-6250

Educational Innovations
151 River Road
Cos Cob, CT 06807
USA
Tel: (203) 629-6049

Flinn Scientific Ltd.
P.O. Box 219
Batavia, IL 60510-0219
USA
Tel: (708) 879-6962

Institute for Chemical Education
Department of Chemistry
University of Wisconsin–Madison
1101 University Ave.
Madison, WI 53706
USA
Tel: (608) 262-3033

NADA Scientific Ltd.
P.O. Box 1336
Champlain, NY 12919-1336
USA
Tel: (800) 233-5381

INDEX